*TM 3-34.44 (FM 5-428/18 June 1998)/MCRP 3-17.7D

Technical Manual No. 3-34.44/Marine Corps
Reference Publication 3-17.7D

Headquarters
Department of the Army
Washington, D.C., 23 July 2012

Concrete and Masonry

Contents

		Page
PREFACE		xiii

PART ONE   CONCRETE

Chapter 1	GENERAL	1-1
	Section I - Basic Consideration	1-1
	Concrete Composition	1-1
	Concrete As A Building Material	1-2
	Section II - Desirable Concrete Properties	1-3
	Plastic Concrete	1-3
	Hardened Concrete	1-4
Chapter 2	CONCRETE COMPONENTS	2-1
	Section I - Cements	2-1
	Portland Cements	2-1
	Section II - Water	2-3
	Purpose	2-3
	Impurities	2-3
	SECTION III - Aggregates	2-5
	Characteristics	2-5
	Impurities	2-12
	Handling and Storing	2-12
	Section IV - Admixtures	2-13
	Definition and Purpose	2-13
	Air-Entrained Concrete	2-13
Chapter 3	PROPORTIONING CONCRETE MIXTURES	3-1
	Section I - Method Considerations	3-1
	Selecting Mix Proportions	3-1
	Basic Guidelines	3-1

DISTRIBUTION RESTRICTION: Approved for public release; distribution is unlimited.

*This publication supersedes FM 5-428, 18 June 1998.

Contents

	Section II - Trial-Batch Method	3-6
	Basic Guidelines	3-6
	Example Using the Trial-Batch Method	3-6
	Section III - Absolute-Volume Method	3-10
	Basic Guidelines	3-10
	Example Using the Absolute-Volume Method	3-11
	Variation in Mixtures	3-13
	Adjustments for Moisture on Aggregates	3-14
	Summary	3-18
Chapter 4	**FORM DESIGN AND CONSTRUCTION**	4-1
	Section I - Principles	4-1
	Importance of Form Design	4-1
	Form Characteristics	4-1
	Form Materials	4-1
	Forming	4-2
	Section II - Design	4-5
	Design Considerations	4-5
	Basis of Form Design	4-5
	Panel-Wall Form Design	4-5
	Bracing for Wall Forms	4-16
	Overhead Slab Form Design	4-19
	Concrete Slab on Grade Thickness Design	4-26
	Column Form Design	4-31
	Section III - Construction	4-33
	Foundation Forms	4-33
	Column and Footing Forms	4-36
	Panel-Wall Forms	4-36
	Stair Forms	4-39
	Steel Pavement Forms	4-41
	Oiling and Wetting Forms	4-41
	Safety Precautions	4-42
	Form Failure	4-42
	Section IV - Joints and Anchors	4-43
	Joints	4-43
	Anchor Bolts	4-48
Chapter 5	**CONSTRUCTION PROCEDURES**	5-1
	Section I - Reconnaissance	5-1
	Determining Possible Difficulties	5-1
	Specific Considerations	5-1
	Section II - Site Preparation	5-2
	Building Approach Roads	5-2
	Clearing and Draining The Site	5-2
	Ensuring Adequate Drainage	5-2
	Locating the Building Site	5-2
	Stockpiling Construction Materials	5-2

Locating Batching Plants 5-3
Constructing Safety Facilities 5-3
Section III - Excavation **5-3**
Excavation and Shoring Considerations 5-3
Section IV - Form Work **5-5**
Management Aspects 5-5
Time Element 5-5
Section V - Mixing **5-5**
Principles 5-5
Measuring Mix Materials... 5-5
Hand Mixing.. 5-6
Machine Mixing and Delivery.. 5-7
Section VI - Handling and Transporting **5-13**
Principles 5-13
Delivery Methods... 5-14
Section VII - Placement **5-19**
Importance Of Proper Procedures 5-19
Preliminary Preparation 5-19
Placing Concrete 5-20
Consolidating Concrete 5-21
Placing Concrete Underwater... 5-24
Section VIII - Finishing **5-25**
Finishing Operations... 5-25
Finishing Pavement 5-28
Cleaning the Surface 5-29
Section IX - Curing **5-30**
Importance of Curing to Hydration 5-30
Curing Methods 5-31
Section X - Temperature Effects **5-34**
Hot-Weather Concreting... 5-34
Cold Weather Concreting 5-37
Section XI - Form Removal **5-41**
Form Stripping 5-41
Form Removal Procedures... 5-42
Section XII - Repairing **5-42**
New Concrete 5-42
Old Concrete 5-45

Chapter 6 REINFORCED CONCRETE... 6-1
Section I - Development And Design 6-1
Principles and Definitions 6-1
Reinforced-Concrete Design 6-3
Section II - Structural Members 6-4
Reinforcement 6-4
Slab and Wall Reinforcement... 6-7

Section III - Reinforcing Steel 6-7
  Grades, Designations, and Methods... 6-7
  Placement 6-13
Section IV - Precast Concrete 6-16
  Definition and Characteristics 6-16
  Design 6-18
  Prefabrication 6-18
  Transportation 6-19
  Erection 6-20

PART TWO   MASONRY

Chapter 7   BASIC EQUIPMENT AND COMPONENTS 7-1
Section I - Mason's Tools And Equipment 7-1
  Definition 7-1
  Tools 7-1
  Equipment 7-3
Section II - Mortar 7-4
  Desirable Properties... 7-4
  Mixing Mortar 7-6
Section III - Scaffolding 7-7
  Construction and Safety 7-7
  Types of Scaffolding... 7-7
  Materials Tower 7-11
  Elevator 7-12

Chapter 8   CONCRETE MASONRY 8-1
Section I - Characteristics of Concrete Block 8-1
  Nature and Physical Properties 8-1
  Concrete Block Masonry Unit 8-3
Section II - Construction Procedures 8-5
  Modular Coordination and Planning... 8-5
  Walls and Wall Footings 8-7
  Subsurface Drainage 8-10
  Basement Walls 8-10
  Floor And Roof Support 8-11
  Weathertight Concrete Masonry Walls 8-11
  Intersecting Walls.. 8-24
  Lintels... 8-25
  Sills... 8-25
  Patching and Cleaning Block Walls 8-25
Section III - Rubble 8-29
  Rubble Stone Masonry 8-29
  Random Rubble Masonry Materials 8-30
  Laying Rubble Stone Masonry... 8-30

Contents

Chapter 9   **BRICK AND TILE MASONRY** 9-1
          Section I - Characteristics of Brick 9-1
             Physical Properties and Classification 9-1
             Strength of Brick Masonry 9-3
             Weather Resistance 9-4
             Fire Resistance 9-4
             Abrasion Resistance 9-5
             Insulating Qualities of Brick Masonry 9-5
          Section II - Bricklaying Methods 9-6
             Fundamentals 9-6
             Brick Masonry Terms 9-6
             Types of Bonds 9-7
             Flashing 9-9
             Making And Pointing Mortar Joints 9-9
             Mortar Joint 9-9
             Picking Up And Spreading Mortar 9-9
             Making Bed and Head Joint 9-11
             inserting a brick in a wall 9-15
             Making Cross Joints and Joints Closure 9-16
             Cutting Brick 9-18
             Finishing Joints 9-20
          Section III - Brick Construction 9-21
             Bricklayer's Duties 9-21
             bricktender's duties 9-21
             Laying Footings 9-21
             Laying An 8-Inch Common-Bond Brick Wall 9-23
             Laying A 12-Inch Common-Bond Brick Wall 9-31
             Protecting Work Inside Walls 9-31
             Using A Trig 9-32
             Constructing Window And Door Openings 9-33
             Lintels 9-35
             Corbeling 9-37
             Arches 9-38
             Watertight Walls 9-40
             Fire-Resistant Brick 9-42
             Mortar 9-42
             Types of Walls 9-42
             Manholes 9-44
             Supporting Beams On A Brick Wall 9-46
             Maintaining and Repairing Brick Walls 9-46
             Cleaning New Brick and Removing Stains 9-47
             Cleaning Old Brick 9-47
             Flashing 9-49
             Freeze Protection During Construction 9-50
             Material Quantities Required 9-52

Contents

Section IV - Reinforced Brick Masonry 9-52
    Applications and Materials 9-52
    Construction Methods 9-53
    Beams And Lintel Construction 9-54
    Foundation Footings 9-55
    Columns and Walls 9-56
Section V - Structural Clay-Tile Masonry 9-58
    Structural Clay Tile 9-58
    Physical Characteristics of Structural Clay Tile 9-63
    Applications 9-63
    Mortar Joints 9-64
    Laying An 8-Inch Brick Wall With A 4-Inch Hollow-Tile Backing 9-65
    Laying An 8-Inch Structural Clay-Tile Wall 9-65

Appendix A   **CONVERSION TABLE** A-1
Appendix B   **METHOD OF MAKING SLUMP TEST** B-1
Appendix C   **FIELD TEST FOR MOISTURE DEFORMATION ON SAND** C-1
                 GLOSSARY Glossary-1
                 REFERENCES References-1
                 INDEX Index-1

Figures

Figure 2-1. Limits specified in ASTM C33 for FA and one size of CA 2-7
Figure 2-2. Water requirements for concrete of a given consistency as a function of CA size 2-10
Figure 2-3. Moisture conditions of aggregates 2-11
Figure 2-4. Variation in fine-aggregate bulking with moisture and aggregate grading .. 2-11
Figure 2-5. Correct and incorrect aggregate handling and storage 2-13
Figure 3-1. Work sheet for concrete trial-mix data 3-8
Figure 3-2. Relationship between percentage of FA and cement content for a given W/C ratio and slump 3-10
Figure 3-3. Bulking factor curves 3-15
Figure 3-4. Retaining wall 3-17
Figure 4-1. Wood form for a concrete panel wall 4-3
Figure 4-2. Form for a concrete column and footing 4-4
Figure 4-3. MCP graph 4-7
Figure 4-4. Elements of diagonal bracing 4-17
Figure 4-5. Typical overhead slab forms 4-20
Figure 4-6. Maximum wheel loads for industrial floors 4-29
Figure 4-7. Typical foundation 4-34
Figure 4-8. Small footing form 4-34
Figure 4-9. Footing and pier form 4-35

Figure 4-10. Typical footing form 4-35
Figure 4-11. Methods of bracing bearing wall footing form 4-36
Figure 4-12. Method of connecting panel wall unit form together 4-37
Figure 4-13. Detail at corner of panel wall form 4-37
Figure 4-14. Form for panel walls and columns 4-38
Figure 4-15. Wire ties for wall forms 4-38
Figure 4-16. Tie-rod and spreader for wall form 4-39
Figure 4-17. Removing wood spreader 4-40
Figure 4-18. Stairway form 4-40
Figure 4-19. Typical isolation and control joints 4-44
Figure 4-20. Joints at columns and walls 4-45
Figure 4-21. Expansion/contraction joint for a bridge 4-46
Figure 4-22. Control joint 4-46
Figure 4-23. Keyed, wall construction joint (perspective view) 4-47
Figure 4-24. Keyed, wall construction joint (plan view) 4-48
Figure 4-25. Construction joint between wall and footing showing keyway 4-48
Figure 4-26. Types of construction joints 4-48
Figure 4-27. Anchor bolt with pipe sleeve 4-49
Figure 4-28. Hooked anchor bolt 4-49
Figure 4-29. Suspended anchor bolt 4-50
Figure 4-30. Anchor bolts held in place by a template 4-50
Figure 5-1. Measuring aggregate by weight 5-6
Figure 5-2. Hand mixing of concrete 5-7
Figure 5-3. M919 concrete mobile mixer 5-8
Figure 5-4. Typical on-site arrangement of mixing equipment and materials 5-9
Figure 5-5. Charging a 16 cubic-foot mixer 5-11
Figure 5-6. Concrete handling techniques to prevent segregation 5-14
Figure 5-7. Handling concrete by buggy 5-15
Figure 5-8. Handling concrete by power buggy 5-15
Figure 5-9. Runways for wheelbarrow or buggy use 5-16
Figure 5-10. Hopper and chute for handling concrete 5-17
Figure 5-11. Placing concrete using a bucket and crane 5-18
Figure 5-12 Piston pump and discharge pipeline 5-19
Figure 5-13. Concrete placing technique 5-22
Figure 5-14. Using a vibrator to consolidate concrete 5-23
Figure 5-15. Consolidation by spading and a spading tool 5-23
Figure 5-16. Placing concrete underwater with a tremie 5-24
Figure 5-17. Screeding operation 5-26
Figure 5-18. Wood floats and floating operations 5-27
Figure 5-19. Steel finishing tools and troweling operation 5-28
Figure 5-20. Moist curing effect on compressive strength of concrete 5-31

Contents

Figure 5-21. Curing a wall with wet burlap sacks 5-33
Figure 5-22. Waterproof paper used for curing.. 5-33
Figure 5-23. Concrete mix water requirements as temperature increases...................... 5-35
Figure 5-24. Effect of high temperature on concrete compressive strength at various ages 5-36
Figure 5-25. Mixing water temperatures required to produce concrete of required temperature 5-36
Figure 5-26. Effect of low temperature on concrete compressive strength at various ages 5-38
Figure 5-27. Relationship between early compressive strengths of portland cement types and low curing temperatures 5-39
Figure 5-28. Effect of temperature of materials on temperature of fresh concrete........ 5-40
Figure 5-29. Repairing large and deep holes in new concrete 5-44
Figure 5-30. Patching flat surface in new concrete... 5-45
Figure 6-1. Cross section of a concrete beam showing location of reinforcement in tension 6-2
Figure 6-2. Deformed steel-reinforced bars 6-2
Figure 6-3. Concrete beams subjected to vertical load without and with steel reinforcement.. 6-3
Figure 6-4. Typical shapes of reinforcing steel for beams 6-5
Figure 6-5. Supports for horizontal reinforcing steel... 6-5
Figure 6-6. Reinforcing concrete columns 6-7
Figure 6-7. Reinforcing bar grade-marking systems.. 6-9
Figure 6-8. Method of splicing reinforcing bars... 6-11
Figure 6-9. Bar bending table 6-14
Figure 6-10. Reinforcing steel arrangement for a floor slab 6-15
Figure 6-11. Beam reinforcing steel.. 6-15
Figure 6-12. Wall and footing reinforcement... 6-16
Figure 6-13. Schematic layout of on-site or temporary prefabrication yard for military operations 6-19
Figure 7-1. Basic mason's tools... 7-2
Figure 7-2. Square, mason's level, and straightedge 7-3
Figure 7-3. Mortar box and mortar board 7-4
Figure 7-4. Trestle scaffold 7-8
Figure 7-5. Foot scaffold 7-8
Figure 7-6. Putlog scaffold 7-9
Figure 7-7. Outrigger scaffold 7-10
Figure 7-8. Prefabricated steel scaffold 7-11
Figure 7-9. Materials tower and elevator 7-12
Figure 8-1. Typical unit sizes and shapes of concrete masonry units 8-2
Figure 8-2. Elements of modular design... 8-6
Figure 8-3. Planning concrete masonry wall openings 8-7
Figure 8-4. Dimensions of masonry wall footings 8-10

Figure 8-5. Laying first course of blocks for a wall 8-12
Figure 8-6. Leveling and plumbing first course of blocks for a wall................................... 8-13
Figure 8-7. Vertical joints.. 8-14
Figure 8-8. Checking each course at the corner 8-15
Figure 8-9. Using a story or course pole 8-16
Figure 8-10. Checking horizontal block spacing... 8-16
Figure 8-11. Filling in the wall between corners 8-17
Figure 8-12. Installing a closure block 8-18
Figure 8-13. Cutting off excess mortar from the joints 8-18
Figure 8-14. Tooled mortar joints for weathertight exterior walls 8-19
Figure 8-15. Tooling mortar joints 8-20
Figure 8-16. Installing anchor bolts for wood plates... 8-21
Figure 8-17. Making a control joint 8-22
Figure 8-18. Control joints made using roofing felt or control joint blocks 8-23
Figure 8-19. Tying intersecting bearing walls... 8-24
Figure 8-20. Tying intersecting nonbearing walls... 8-26
Figure 8-21. Installing precast concrete lintels without and with steel angles................... 8-27
Figure 8-22. Installing precast concrete sills 8-28
Figure 8-23. Cleaning mortar droppings from a concrete block wall................................. 8-28
Figure 8-24. Rubble stone masonry.. 8-29
Figure 8-25. Bonding stones extend through a rubble stone masonry wall...................... 8-29
Figure 9-1. Common cut brick shapes 9-2
Figure 9-2. Names of brick surfaces 9-2
Figure 9-3. Masonry units and mortar joints... 9-7
Figure 9-4. Types of masonry bond 9-8
Figure 9-5. Metal ties... 9-9
Figure 9-6. Correct way to hold trowel 9-10
Figure 9-7. Picking up and spreading mortar 9-11
Figure 9-8. A poorly bonded brick 9-12
Figure 9-9. Making a bed joint in a stretcher course 9-13
Figure 9-10. Proper way to hold a brick when buttering the end....................................... 9-14
Figure 9-11. Making a head joint in a stretcher course 9-15
Figure 9-12. Inserting a brick in a wall.. 9-16
Figure 9-13. Making a cross joint in a header course 9-17
Figure 9-14. Making a closure joint in a header course 9-18
Figure 9-15. Making a closure joint in a stretcher course 9-19
Figure 9-16. Cutting brick with a bolster... 9-19
Figure 9-17. Cutting a brick with a hammer 9-20
Figure 9-18. Joint finishes 9-20
Figure 9-19. Laying a wall footing 9-22
Figure 9-20. Column footing 9-23

Contents

Figure 9-21. Determining number of bricks in one course and head joint widths 9-23
Figure 9-22. Laying first course of corner lead for 8-inch common-bond brick wall 9-27
Figure 9-23. Laying second course of corner lead for 8-inch common-bond
brick wall 9-28
Figure 9-24. Plumbing a corner 9-29
Figure 9-25. Using a line to lay face tier of brick between corner lead 9-30
Figure 9-26 Laying backup brick for the corner lead of an 8 inch common bond
brick wall 9-31
Figure 9-27. Laying a 12-inch common-bond brick wall 9-32
Figure 9-28. Use a trig to support the line when building a long wall 9-33
Figure 9-29. Constructing a window opening 9-34
Figure 9-30. Constructing a door opening 9-35
Figure 9-31. Installing a double-angle steel lintel in an 8 inch wall 9-36
Figure 9-32. Installing lintels in a 12 inch wall 9-37
Figure 9-33. Constructing a corbelled brick wall 9-38
Figure 9-34. Common shapes 9-39
Figure 9-35. Using a template to construct an arch 9-39
Figure 9-36. Parging the back of the face tier for watertightness 9-40
Figure 9-37. Draining a wall around its foundation 9-41
Figure 9-38. Construction details of a cavity wall 9-43
Figure 9-39. Construction details of a rowlock wall 9-44
Figure 9-40. Construction details of a sewer manhole 9-45
Figure 9-41. Supporting a wood beam on a brick wall 9-46
Figure 9-42. Installing flashing at window opening 9-50
Figure 9-43. Flashing installation at intersection of roof and walls 9-51
Figure 9-44. Reinforced brick-masonry beam construction 9-54
Figure 9-45. Reinforced brick-masonry lintel construction 9-55
Figure 9-46. Reinforced brick masonry-wall footing construction 9-56
Figure 9-47. Reinforced brick masonry-column footing construction 9-57
Figure 9-48. Reinforced brick masonry-column construction 9-58
Figure 9-49. Corner lead for reinforced brick masonry wall 9-58
Figure 9-50. Types of structural clay tile 9-60
Figure 9-51. Laying end-construction clay tile 9-64
Figure 9-52. Laying side-construction clay 9-65
Figure 9-53. Laying a corner lead with hollow-tile backing 9-66
Figure 9-54. Constructing the corner lead of an 8 inch structural clay tile wall 9-67
Figure B-1. Measuring slump B-1
Figure C-1. Damp sand C-1
Figure C-2. Wet sand C-2
Figure C-3. Very wet sand C-2

Tables

Table 2-1. Aggregate characteristics and standard test 2-5
Table 2-2. Typical fineness modulus calculation 2-8
Table 2-3. Fitness modulus ranges for FAs 2-8
Table 2-4. Grading requirement for coarse aggregate 2-9
Table 2-5. Impurities in aggregates 2-12
Table 2-6. Total air content for frost-resistant concrete 2-15
Table 2-7. Maximum chloride ion content for corrosion protection 2-16
Table 3-1. Maximum W/C ratios for various exposure conditions 3-2
Table 3-2. Maximum permissible W/C ratios for concrete 3-3
Table 3-3. Approximate mixing-water and air-content requirements for different slumps and maximum sizes of aggregate 3-4
Table 3-4. Slumps for various types of construction (with vibration) 3-6
Table 3-5. Results of laboratory trial mixes 3-9
Table 3-6. Volume of CA per cubic yard of concrete 3-11
Table 3-7. Concrete Used in flatwork 3-14
Table 4-1. Maximum stud (joist) spacing for board sheathing 4-8
Table 4-2. Maximum stud (joist) spacing for plywood sheathing, in inches 4-9
Table 4-3. Maximum spacing, in inches, for wales, ties, stringers, and 4"x 4" or larger shores where member to be supported is a single member 4-10
Table 4-4. Maximum spacing, in inches, for ties and 4" x 4" or larger shores where member to be supported is a double member 4-12
Table 4-5. Average breaking load of tie material, in pounds 4-13
Table 4-6. J factor 4-17
Table 4-7. Allowable load (in pounds) on wood shores, based on shore strength 4-22
Table 4-8. Allowable load on specified shores, in pounds, based on bearing stresses where the maximum shore area is in contact with the supported member 4-23
Table 4-9. Concrete floor classifications 4-26
Table 4-10. Recommended slumps and compressive strengths 4-27
Table 4-11. Minimum cement contents and percentages of entrained air 4-31
Table 4-12. Column yoke spacing using 2 by 4s and 1 inch sheathing 4-32
Table 4-13. Spacing of control joints 4-47
Table 5-1. Machine excavation 5-4
Table 5-2. Hand excavation 5-4
Table 5-3. Physical characteristics of a typical 16-cubic feet mixer 5-10
Table 5-4. M919 Concrete-mobile-mixer unit 5-12
Table 5-5. Curing methods 5-31
Table 5-6. Recommended concrete temperatures for cold-weather construction 5-37
Table 5-7. Recommended duration of protection for concrete placed in cold weather (air-entrained concrete) 5-38
Table 5-8. Recommended form stripping time 5-43

Contents

Table 6-1. Standard steel reinforcing bar 6-8
Table 6-2. Minimum concrete cover requirements for steel reinforcement 6-10
Table 6-3. Recommended end hooks (all grades) 6-10
Table 6-4. Minimum splice overlap 6-12
Table 6-5. Recommended precasting team personnel requirements 6-20
Table 7-1. Recommended mortar mix proportions by unit volume 7-6
Table 8-1. Nominal length of modular-concrete masonry walls in stretchers 8-7
Table 8-2. Nominal height of modular-concrete masonry walls in courses 8-9
Table 8-3. Unit weight and quantities for modular concrete masonry walls 8-9
Table 9-1. Fire resistance of brick load-bearing walls laid with Portland-cement-lime mortar 9-5
Table 9-2. Height of course using 2 1/4-inch brick, 3/8 inch joint 9-24
Table 9-3. Height of course using 2 1/4-inch brick, 1/2-inch joint 9-25
Table 9-4. Height of course using 2 1/4-inch brick, 5/8-inch joint 9-26
Table 9-5. Lintel sizes for 8-inch and 12-inch walls 9-36
Table 9-6. Quantities of materials required for brick walls 9-53
Table 9-7. Quantities of bars required for lintels 9-55
Table 9-8. Quantities of materials required for side construction of hollow clay-tile walls 9-62
Table 9-9. Quantities of materials required for end construction of hollow clay-tile walls 9-62

Preface

TM 3-34.44 is primarily a training guide and reference text for engineer personnel using concrete and masonry materials in field construction. The manual has two parts: Concrete (Part One) and Masonry (Part Two).

Part One covers the physical characteristics, properties, and ingredients of concrete; mixtures, design and construction of forms; and with reinforced concrete and field construction procedures.

Part Two addresses the mason's tools and equipment as well as the physical characteristics and properties of concrete blocks, bricks, and structural clay tiles. It further explains construction procedures and methods for these masonry units.

Appendix A contains an English to metric measurement conversion chart.

The proponent of this publication is HQ TRADOC. Send comments and recommendations on Department of Army (DA) Form 2028 (Recommended Changes to Publications and Blank Forms) directly to Commandant, USAES, ATTN: ATSE-TD-D, Fort Leonard Wood, MO 65473-6650.

Unless this publication states otherwise, nouns and pronouns do not refer exclusively to men.

This page intentionally left blank.

PART ONE
Concrete

Concrete is the type of construction material most widely used by both military and civilian personnel. Part one covers the physical characteristics, properties, and ingredients of concrete; the form design and construction principles; construction procedures with emphasis on concrete placement, finishing, and curing techniques; and the proportions of ingredients.

Chapter 1
General

Concrete is produced by mixing a paste of cement and water with various inert materials. The most commonly used inert materials are sand and gravel or crushed stone. A chemical process begins as soon as the cement and water are combined. Different amounts of certain materials will effect the desirable properties of workability, nonsegregation, and uniformity of the mix. There are numerous advantages and some limitations for using concrete as a building material.

SECTION I - BASIC CONSIDERATION

CONCRETE COMPOSITION

1-1. *Concrete* is a mixture of aggregate, and often controlled amounts of entrained air, held together by a hardened paste made from cement and water. Although there are other kinds of cement, the word cement in common usage refers to portland cement. A chemical reaction between the portland cement and water—not drying of the mixture—causes concrete to harden to a stone-like condition. This reaction is called hydration. Hydration gives off heat, known as the *heat of hydration*. Because hydration—not air drying—hardens concrete, freshly placed concrete submerged underwater will harden. When correctly proportioned, concrete is at first a plastic mass molded into nearly any size or shape. Upon hydration of the cement by the water, concrete becomes stone-like in strength, durability, and hardness.

PORTLAND CEMENT

1-2. Portland cement is the most commonly used modern hydraulic cement. In this case, the word hydraulic is the cement's characteristic of holding aggregate together by using water or other low-viscosity fluids. Portland cement is a carefully proportioned and specially processed chemical combination of lime, silica, iron oxide, and alumina. It was named after the Isle of Portland in the English Channel.

WATER

1-3. Unless tests or experience indicates that a particular water source is satisfactory, water should be free from acids, alkalis, oils, and organic impurities. The basic ratio of cement to water determines the concrete's strength; generally, the less water in the mix, the stronger, more durable and watertight the concrete. The concrete should be workable but not too stiff to use. Too much water dilutes the cement paste (binder), resulting in weak and porous concrete. Concrete quality varies widely, depending on the characteristics of its ingredients and the proportions of the mix.

Chapter 1

AGGREGATES

1-4. Inert filler materials (usually sand and stone or gravel) make up between 60 and 80 percent of the volume of normal concrete. In air-entrained concrete, the air content ranges up to about 8 percent of the volume. Aggregate is often washed when impurities are found that can retard cement hydration or deteriorate the concrete's quality. All aggregate is screened to ensure proper size gradation, because concrete differs from other cement-water-aggregate mixtures in the size of its aggregate. For example, when cement is mixed with water and aggregates passing the number 4 sieve (16 openings per square inch), it is called mortar, stucco, or cement plaster. When cement is mixed with coarse aggregate (CA) of more than 1/4 inch plus fine aggregate (FA) and water, the product is concrete. The aggregate's physical and chemical properties also affect concrete properties; aggregate size, shape, and grade influence the amount of water required. For example, limestone aggregate requires more water than similar size marble aggregate. Aggregate surface texture influences the bond between the aggregate and the cement paste. In properly mixed concrete, the paste surrounds each aggregate particle and fills all spaces between the particles. The elastic properties of the aggregate influence the elastic properties of the concrete and the paste's resistance to shrinkage. Reactions between the cement paste and the aggregate can either improve or harm the bond between the two and consequently, the concrete's quality.

ADMIXTURES

1-5. These substances are added to the concrete mixture to accelerate or retard the initial set, improve workability, reduce water requirements, increase strength, or otherwise alter concrete properties. They usually cause a chemical reaction within the concrete. Admixtures are normally classified into accelerators, retardants, air-entraining agents, workability agents, dampproofing and permeability-reducing agents, pozzolans, color pigments, and miscellaneous materials. Many admixtures fall into more than one classification.

CONCRETE AS A BUILDING MATERIAL

1-6. Concrete has a great variety of applications; it not only meets structural demands but lends itself readily to architectural treatment. In buildings, concrete is used for footings, foundations, columns, beams, girders, wall slabs, and roof units—in short, all important building elements. Other important concrete applications are in road pavements, airport runways, bridges, dams, irrigation canals, water-diversion structures, sewage-treatment plants, and water-distribution pipelines. A great deal of concrete is used in manufacturing masonry units such as concrete blocks and concrete bricks.

ADVANTAGES

1-7. Concrete and cement are among the most important construction materials. Concrete is fireproof, watertight, economical, and easy to make. It offers surface continuity (absence of joints) and solidity; it will bond with other materials. Concrete is usually locally available worldwide.

LIMITATIONS

1-8. Certain limitations of concrete cause cracking and other structural weaknesses that detract from the appearance, serviceability, and useful life of concrete structures. Listed below are some principal limitations and disadvantages of concrete.

Low Tensile Strength

1-9. Concrete members subject to tensile stress must be reinforced with steel bars or mesh to prevent cracking and failure.

Thermal Movements

1-10. During setting and hardening, the heat of hydration raises the concrete temperature and then it gradually cools. These temperature changes can cause severe thermal strains and early cracking. In addition, hardened concrete expands and contracts with changes in temperature (at roughly the same rate as steel); therefore, expansion and contraction joints must be provided in many types of concrete structures to prevent failures.

Drying Shrinkage And Moisture Movements

1-11. Concrete shrinks as it dries and as it hardens. It expands and contracts with wetting and drying. These movements require that control joints be provided at intervals to avoid unsightly cracks. To prevent drying shrinkage in newly placed concrete, its surface is kept moist continuously during the curing process. Moisture is applied when the concrete is hard enough so as not to damage the concrete's surface.

Creep

1-12. Concrete deforms gradually (creeps) under load. This deformation does not recover completely when the load is removed.

Permeability

1-13. Even the best quality concrete is not entirely impervious to moisture. Concrete normally contains soluble compounds that are leached out in varying amounts by water, unless properly constructed joints allow water to enter the mass. Impermeability is particularly important in reinforced concrete—the concrete must prevent water from reaching the steel reinforcement.

INGREDIENTS

1-14. The unit of measure for concrete is cubic foot (cu ft). Thus, a standard sack of portland cement weighs 94 pounds and equals 1 loose cubic foot. FAs and CAs are measured by loose volume, whereas water is measured by the gallon. Concrete is usually referred to by cubic yards (cu yd).

SECTION II - DESIRABLE CONCRETE PROPERTIES

PLASTIC CONCRETE

1-15. Plastic Concrete is a concrete in a relatively fluid state that is readily molded by hand, like a lump of modeling clay. A plastic mix keeps all the grains of sand and the pieces of gravel or stone encased and held in place (homogeneous). The degree of plasticity influences the quality and character of the finished product. Significant changes in the mix proportions affect plasticity. Desirable properties of plastic concrete are listed below.

WORKABILITY

1-16. Workability is the relative ease or difficulty of placing and consolidating concrete in the form. It is largely determined by the proportions of FAs and CAs added to a given quantity of paste. One characteristic of workability is consistency, which is measured by the slump test (see appendix B). A specific amount of slump is necessary to obtain the workability required by the intended conditions and method of placement. A very stiff mix has a low slump and is desirable for many uses, although difficult to place in heavily reinforced sections. A more fluid mix is necessary when placing concrete around reinforcing steel.

NONSEGREGATION

1-17. Plastic concrete must be homogeneous and carefully handled to keep segregation to a minimum. For example, plastic concrete should not drop (free fall) more than 3 to 5 feet nor be transported over long distances without proper agitation.

UNIFORMITY

1-18. The uniformity of plastic concrete affects both its economy and strength. It is determined by how accurate the ingredients are proportioned and mixed according to specifications. Each separate batch of concrete must be proportioned and mixed exactly the same to ensure that the total structural mass has uniform structural properties.

HARDENED CONCRETE

1-19. Hardened Concrete is the end product of any concrete design. The essential properties that it must have are strength, durability, and watertightness.

STRENGTH

1-20. The concrete's ability to resist a load in compression, flexure, or shear is a measure of its strength. Concrete strength is largely determined by the ratio of water to cement in the mixture (pounds of water to pounds of cement). A sack of cement requires about 2 1/2 gallons of water for hydration. Additional water allows for workability, but too much water (a high water:cement [W/C] ratio) reduces the concrete's strength. The amount of water in economical concrete mixes ranges from 4 gallons minimum to 7 gallons maximum per sack.

DURABILITY

1-21. Climate and weather exposure affect durability. Thus, the concrete's ability to resist the effects of wind, frost, snow, ice, abrasion, and the chemical reaction of soils or salts are a measure of its durability. As the W/C ratio increases, durability decreases correspondingly. Durability should be a strong consideration for concrete structures expected to last longer than five years. Air-entrained concrete (see paragraph 2-55) has improved freeze-thaw durability.

WATERTIGHTNESS

1-22. Tests show that the watertightness of a cement paste depends on the W/C ratio and the extent of the chemical-reaction process between the cement and water. The Corps of Engineers specifications for watertightness limit the maximum amount of water in concrete mixtures to 5.5 gallons per sack of cement (W/C = 0.48) for concrete exposed to fresh water and 5 gallons per sack (W/C = 0.44) for concrete exposed to saltwater. The watertightness of air-entrained concrete is superior to that of nonair-entrained concrete.

Chapter 2
Concrete Components

Cement is the primary binder used in concrete. There are various manufacturers for cement, but portland cement is the standard for the building trade. Water for concrete should be clean and free from damaging amounts of oil, acid, alkali, organic matter, or other harmful substances. In general, any drinking water free from pronounced odors or taste is satisfactory. Aggregates, sand and gravel, must adhere to specific standards (clean, hard, strong, and durable), as they make up the majority of the concrete mix. Aggregate size, distribution, and grading will also affect the concrete's workability. Admixtures used in limited quantities, modify the workability, strength, and durability of the concrete.

SECTION I - CEMENTS

PORTLAND CEMENTS

2-1. Portland cement contains lime and clay minerals (such as limestone, oyster shells, coquina shells, marl, clay, and shale), silica sand, iron ore, and aluminum.

MANUFACTURE

2-2. The raw materials are finely ground, carefully proportioned, and then heated (calcined) to the fusion temperature (2,600 to 3,000 degrees Fahrenheit [F]) to form hard pellets called clinkers, which are ground to a fine powder. Because the powder is extremely fine, nearly all of it will pass through a number 200 sieve (200 meshes to the linear inch or 40,000 openings per square inch). Regardless of the manufacturer, portland cement is the standard for the trade.

AMERICAN SOCIETY FOR TESTING AND MATERIALS (ASTM) TYPES

2-3. ASTM specifications cover five types of portland cements in ASTM C 150.

Type I

2-4. Type I cement is a general-purpose cement for concrete that does not require any of the special properties of the other types. It is intended for concrete that is not subjected to a sulfate attack or when the heat of hydration will not cause minimum temperature rise. It is used in pavement and sidewalk construction. It reinforces concrete buildings and bridges, railways, tanks, reservoirs, sewers, culverts, water pipes, masonry units, and soil-cement mixtures. Type I cement will reach its design strength in 28 days and is more available than other types.

Type II

2-5. Type II cement is modified to resist a moderate sulfate attack. It usually generates less heat of hydration and at a slower rate than Type I cement. Typical applications are drainage structures where the sulfate concentrations in either the soil or ground waters are higher than normal but not severe; large structures produces only a slight temperature rise in the concrete that moderately affects the heat of hydration. However, temperature rise can be a problem when concrete is placed in warm weather. Type II cement will reach its design strength in 45 days.

Type III

2-6. Type III cement is a high-early-strength cement that produces design strengths at an early age, usually seven days or less. It has a higher heat of hydration and is ground finer than Type I cement. Type III cement permits fast form removal and in cold weather construction, reduces the period of protection against low temperatures. Although richer mixtures of Type I cement can obtain high early strength, Type III cement produces it more satisfactorily and more economically. Use this material cautiously in concrete structures having a minimum dimension of 2 1/2 feet or more; the high heat of hydration can cause shrinkage cracking.

Type IV

2-7. Type IV cement is unusual in that it has a low heat of hydration. It is intended for applications requiring a minimal rate and a minimal amount of the heat of hydration. Its strength also develops at a slower rate than the other types. Type IV cement is used primarily in very large concrete structures, such as gravity dams, where the temperature rise from the heat of hydration could damage the structure. It will reach its design strength in 90 days.

Type V

2-8. Sulfate resistant cement (Type V) is used mainly where the concrete is subject to severe sulfate action, such as when the soil or ground water contacting the concrete has a high sulfate content. Type V cement will reach its design strength in 60 days.

OTHER ASTM CEMENTS

2-9. ASTM specifications cover the following specific types of portland cements:

Air-Entraining Portland Cement

2-10. Cement types IA, IIA, and IIIA correspond in composition to cement types I, II, and III, respectively, with the addition of small quantities of air-entraining materials interground with the clinker during manufacturing. Especially useful for sidewalks, air-entraining portland cements produce concrete having improved resistance to freeze-thaw action and to scaling caused by snow and ice removal chemicals. Such concrete contains extremely small (as many as 3 billion per cubic yard), well-distributed, and completely separate air bubbles.

White Portland Cement

2-11. White portland cement is a white-finished concrete product used mainly for architectural purposes.

Portland Blast-Furnace Slag Cement

2-12. Cement types IS and IS-A are used in ordinary concrete construction. During manufacturing, granulated blast-furnace slag is either interground with the portland-cement clinker or blended with the ground portland cement. Cement type IS-A contains an air-entraining additive.

Portland Pozzolana Cement

2-13. During the manufacturing of cement types P, IP, PA, and IP-A, pozzolana (containing silica and alumina) is blended with the ground portland-cement clinker. Types PA and IP-A also contain an air-entraining additive. Finely divided siliceous aluminous material reacts chemically with slake and lime at an ordinary temperature and in the presence of moisture, forms a strong slow-hardening cement.

Masonry Cement

2-14. Masonry cements (also called mortar cement) are a typical mixture of portland cement, hydrated lime, and other materials that improve workability, plasticity, and water retention.

Concrete Components

SPECIAL PORTLAND CEMENTS

2-15. Other special types of portland cement not covered by ASTM specifications include—

- Oil-well portland cement. It hardens at the high temperatures that prevail in very deep oil wells.
- Waterproofed portland cement. It contains water-repellent materials that are ground with the portland- cement clinker.
- Plastic cement. It contains plastic agents that are added to the portland-cement clinker; it is commonly used to make plaster and stucco.

PACKAGING AND SHIPPING

2-16. Cement is shipped by railroad, truck, or barge in standard sacks weighing 94 pounds, or in bulk. Cement quantities for large projects may be stated in tons.

STORAGE

2-17. Portland cement retains its quality indefinitely if kept dry. Store- sacked cement in an airtight warehouse or shed. If no shed is available, place the sacks on raised, wooden platforms. Place them close together to reduce air circulation and away from exterior walls. Sacks stored outside for long periods should be covered with tarpaulins or other waterproof coverings so that rain cannot reach either the cement or the platforms. Rain-soaked platforms can damage the bottom layers of the sacks. Cement should be free-flowing and free from lumps when used. Stored, sacked cement sometimes develops warehouse pack. This is a slightly hardened condition caused by packing sacks extremely tight or excessively high. Such cement still retains its quality and is usually restored to a free-flowing condition by rolling the sacks on the floor. However, cement sometimes does develop lumps that are difficult to break up; the cement should be tested to determine its quality. Hard lumps indicate partial hydration that reduces both the strength and durability of the finished concrete. Do not use partially hydrated cement in structures where strength is a critical factor. Store bulk cement in weatherproof bins.

SECTION II - WATER

PURPOSE

2-18. Water has two functions in the concrete mix—it affects hydration and it improves workability.

IMPURITIES

2-19. Mixing water should be clean and free from organic materials, alkalis, acids, and oil. Potable water is usually suitable for mixing with cement. However, water containing many sulfates may be drinkable, but it makes a weak paste that leads to concrete deterioration or failure. You can use water of unknown quality if concrete cylinders made with it have 7- and 28-day strengths equaling at least 90 percent of the control cylinders made with potable water. Test batches can also determine if the cement's setting time is badly affected by water impurities. Too many impurities in mixing water can affect not only setting time, but can cause surface efflorescence and corrosion of the steel reinforcement. In some cases, you can increase the concrete's cement content to offset the impurities. The effects of certain common water impurities on the quality of plain concrete are described below.

ALKALI CARBONATE AND BICARBONATE

2-20. Carbonates, bicarbonates of sodium, and potassium can either accelerate or retard the set of different cements. In large concentrations, these salts can materially reduce concrete strength. A test must be performed to determine the effect on setting time if the sum of these dissolved salts exceeds 1,000 parts per million (ppm).

Sodium Chloride and Sodium Sulfate

2-21. A highly dissolved solid content in a natural water usually results from a high content of sodium chloride or sodium sulfate. Both can be tolerated in rather large quantities in the concrete mix. Concentrations of 20,000 ppm of sodium chloride are generally tolerable in concrete that will dry in service and have low potential for corrosive reactions. Water used in concrete that will have aluminum embodiments should not contain large amounts (more than 550 ppm) of chloride ion.

Other Common Salts

2-22. Carbonates of calcium and magnesium are seldom concentrated enough to affect the concrete's strength. Bicarbonates of calcium and magnesium in concentrations up to 400 ppm will not affect strength, nor will magnesium sulfate and magnesium chloride in concentrations up to 40,000 ppm. Calcium chloride accelerates both hardening and strength gain.

Iron Salts

2-23. Although natural groundwater usually contains only small amounts of iron, acid mine waters can contain large amounts. Iron salts in concentrations up to 40,000 ppm are acceptable.

Miscellaneous Inorganic Salts

2-24. Salts of manganese, tin, zinc, copper, and lead can greatly reduce concrete strength and cause large variations in setting time. Sodium iodate, sodium phosphate, sodium arsenate, and sodium borate can greatly retard both the set and the strength development. Concentrations of these salts up to 500 ppm are acceptable, whereas concentrations of sodium sulfide as small as 100 ppm can be harmful.

Seawater

2-25. Up to 35,000 ppm of salt in seawater are normally suitable for un-reinforced concrete. You can offset the degree of strength reduction that occurs by reducing the W/C ratio, but seawater can corrode the steel in reinforced concrete. The risk of corrosion is reduced if the steel has sufficient cover (4 or more inches), the concrete is watertight, and the concrete contains enough entrained air.

Acid Waters

2-26. The acceptance of acid water is based on the concentration of acids (ppm) in the water, rather than on the hydrogen-ion activity (pH) of the water. Hydrochloric, sulfuric, and other common inorganic acids in concentrations up to 10,000 ppm generally have no harmful effect on concrete strength.

Alkaline Waters

2-27. Concentrations of sodium hydroxide above 0.5 percent by weight of cement can reduce concrete strength. Potassium hydroxide in concentrations up to 1.2 percent by weight of cement has little effect on the strength of the concrete that is developed by some cements but can substantially reduce the strength of others. When making concrete with water having a pH greater than 11, test the concrete for strength.

Industrial Waste Waters

2-28. Less than 4,000 ppm of total solids in industrial waste waters generally reduces compressive strength no more than 10 percent. Test any water that contains unusual solids.

Sanitary Waste Water

2-29. Sewage diluted in a good disposal system generally has no significant effect on the concrete's strength.

SUGAR

2-30. Small amounts of sugar (0.03 to 0.15 percent by weight of cement) usually retard setting time, whereas larger amounts (about 0.20 percent by weight of cement) usually accelerate the set. However, sugar in quantities of about 0.25 percent by weight of cement can substantially reduce strength. Perform strength tests if the sugar concentration in the water exceeds 500 ppm.

SILT OR SUSPENDED PARTICLES

2-31. Mixing water can contain up to 2,000 ppm of suspended clay or fine rock particles without adverse effect. Higher concentrations may adversely affect other concrete properties and should be avoided.

OILS

2-32. Pure mineral (petroleum) oil probably affects strength development less than other oils. However, mineral oil in concentrations greater than 2 percent by weight of cement can reduce the concrete's strength by more than 20 percent.

ALGAE

2-33. Water containing algae is unsuitable for making concrete.

SECTION III - AGGREGATES

CHARACTERISTICS

2-34. Aggregates make up 60 to 80 percent of the concrete's volume. Their characteristics considerably influence the mix proportions and economy of the concrete. For example, very rough-textured or flat and elongated particles re-quire more water to produce workable concrete than do rounded or cubed particles. Angular particles require more cement paste to coat them, making the concrete more expensive. Aggregates should be clean, hard, strong, durable, and free from chemicals or coatings of clay or other fine materials that affect the bond of the cement paste. The most common contaminating materials are dirt, silt, clay, mica, salts, humus (decayed plant matter), or other organic matter that appears as a coating or as loose, fine material. You can remove many contaminants simply by washing the aggregate. However, test CAs containing easily crumbled or laminated particles. The most commonly used aggregates are sand, gravel, crushed stone, and blast-furnace slag. They produce normal-weight concrete (concrete that weighs 135 to 160 pounds per cubic foot). Normal-weight aggregates should meet the specifications for concrete aggregates that restrict contaminating substances and provide standards for gradation, abrasion resistance, and soundness. Aggregate characteristics, their significance, and standard tests for evaluating are given in table 2-1.

Table 2-1. Aggregate characteristics and standard test

Characteristics	Significance or Importance	ASTM Test or Practice Designation	Specification Requirement
Resistance to abrasion	Use in the index of aggregate quality; warehouse floors, loading platforms, and pavements.	C131	Maximum percent loss*

Table 2-1. Aggregate characteristics and standard test

Characteristics	Significance or Importance	ASTM Test or Practice Designation	Specification Requirement
Resistance to freezing and thawing	Use in the structures subjected to weathering	C666	Maximum number of cycles
Chemical stability	Use in all types of structures for strength and durability.	C227 - motor bar C289 - chemical C589 - aggregate prism C295 - petrographic	Maximum expansion of mortar bar* Aggregates must be reactive with cement alkalis*
Particles shape and surface texture	Importance for workability of fresh concrete.		Maximum percent flat and elongated pieces
Grading	Importance for workability of fresh concrete economy.	C136	Maximum and minimum percent passing standard sieves
Bulk unit weight	Use in mix—design calculations classification	C29	Maximum and minimum unit weight (special concrete)
Specific gravity	Use in mix—design calculations	C127- CA C128 - FA	
Absorption and surface moisture	Use in the control of concrete quality	C70, C127, C128	

* Aggregates not conforming to specification requirements can be used if either service records or performance test indicate that they produce concrete having the desired properties.

ABRASION RESISTANCE

2-35. Abrasion resistance is essential when the aggregate is subject to abrasion, such as in a heavy-duty concrete floor.

FREEZE AND THAW RESISTANCE

2-36. The resistance to freezing and thawing relates to aggregate porosity, absorption, and pore structure. If particles absorb too much water, there will not be enough pore space for the water to expand during freezing. The undesirable result is that this water will expand anyway, cracking the concrete. You can predict aggregate performance during freezing and thawing in two ways: pasted performance and freeze-thaw tests of concrete specimens. If aggregates from the same source have been satisfactory under those conditions in the past, use the aggregate. You can determine the performance of unknown aggregates by subjecting concrete specimens to both freeze-thaw and strength tests.

Chemical Stability

2-37. In aggregates, this characteristic means that they do not react unfavorably with cement, and that external sources do not affect them chemically. Good field service records are generally the best predictor of nonreactive aggregates. If no service record exists and you suspect an aggregate is chemically unstable, laboratory tests are necessary.

Particle Shape and Surface Texture

2-38. The particle shape and surface texture affect the properties of plastic concrete more than those of hardened concrete. Very sharp, rough aggregate particles or flat, elongated particles require more fine materials (hence more cement) to produce workable concrete than do rounded or cubed particles. Avoid stones that break up into long sliver pieces or limit them to a maximum of 15 percent in either FAs or CAs.

Aggregate Size Distribution and Grading

2-39. The aggregate's size distribution and grading affect the concrete's workability, economy, porosity, and shrinkage. Prior experience has shown that exclusive use of very fine sands produces uneconomical mixes, whereas exclusive use of very coarse sands produces harsh, unworkable mixes. The proportioning of the different particle sizes is called grading an aggregate. Grading is controlled by the aggregate producer. The aggregates particle-size distribution is determined by separation with a series of standard sieves. The standard sieves are numbers 4, 8, 16, 30, 50, and 100 for FAs and 6, 3, 1 1/2, 3/4, and 3/8 inch and number 4 for CAs (other sieves can be used for CAs). The number of a FA sieve corresponds to the number of meshes (square openings) to the linear inch that the sieve contains; the higher the number, the finer the FA sieve. Any material retained in the number 4 sieve is considered CA, and any material that passes the number 200 sieve is too fine for concrete. The finest CA sieve is the same number 4 used as the coarsest FA sieve. With this exception, a CA sieve is designated by the size of one of its mesh openings. The size of the mesh openings in consecutive sieves is related by a constant ratio. Size-distribution graphs show the percentage of material passing each sieve (see figure 2-1). Figure 2-1 also gives the grade limits for FA and for one designated size of CA. Normal CA consist of gravel or crushed stone, whereas normal FA is sand.

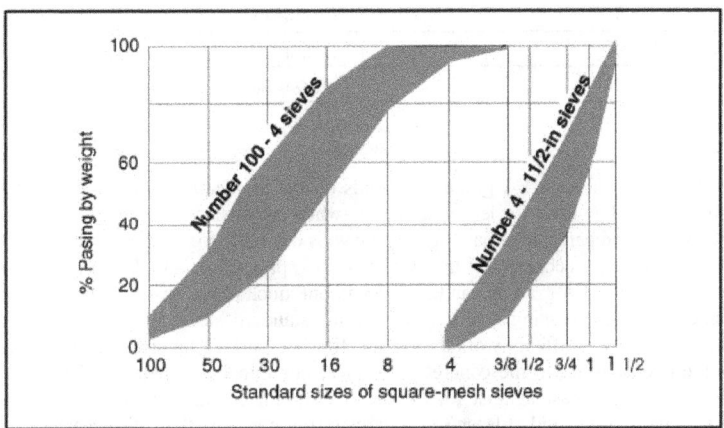

Figure 2-1. Limits specified in ASTM C33 for FA and one size of CA

Fineness Modulus

2-40. The fineness modulus indicates the fineness of a FA but is not the same as its grade. Many FA gradings can have the same fineness modulus. To obtain the fineness modulus of a FA (see table 2-2, page 2-8), quarter a sample of at least 500 grams of sand and sieve it through the numbers 4, 8, 16, 30, 50, and 100 sieves. Record the individual weights of the materials retained on each sieve and the cumulative retained weights. Add the cumulative percents and divide by 100. The result is the fineness modulus of the sample. A sand with a fineness modulus falling between 2.3 and 3.1 is suitable for concrete (see table 2-3). FA having either a very high or a very low fineness modulus is not as good a concrete aggregate as medium sand. Coarse sand is not as workable, and fine sands are uneconomical. Take care to obtain representative samples. The fineness modulus of the aggregate taken from one source should not vary more than 0.20 from all test samples taken at that source.

Table 2-2. Typical fineness modulus calculation

Screen Size	Weight Retained (Grams)		Cumulative Percentage Retained
	Individual	Cumulative	
Number 4	40	40	4.0
Number 8	130	170	17.0
Number 16	130	300	30.0
Number 30	250	550	55.0
Number 50	270	820	82.0
Number 100	100	920	92.0
Pan	80	—	—
Total Weight	1,000		280.0

Note: Fitness Modulus = 280/100 = 2.80.

Table 2-3. Fitness modulus ranges for FAs

Fineness Modulus	Designation
2.3 to 2.59	Fine sand
2.6 to 2.89	Medium Sand
2.9 to 3.10	Coarse sand

FA Grading

2-41. The selection of the best FA grading depends on the application, the richness of the mix, and the maximum size of the CA used. In leaner mixes or when small-size CA is used, a FA grading near the maximum recommended percentage passing each sieve is desirable for workability. In richer mixes, coarser FA grading is desirable for economy. If the W/C ratio is kept constant and the ratio of FA to CA is chosen correctly, a wide range of FA grading can be used without much affect on strength. Grading is expressed as the percentage by weight passing through the various standard sieves. The amount of FA passing the number 50 and 100 sieves affects workability, the finished-surface texture, and water gain or bleeding. Bleeding is the rise of water to the concrete's surface. For thin walls, hard-finished concrete floors, and smooth concrete surfaces cast against forms, the FA should contain not less than 15 percent material passing the number 50 sieve and at least 3 or 4 percent (but not more than 10 percent) material passing the number 100 sieve. These minimum amounts of FA give the concrete better workability, making it more cohesive, and they produce less water gain or bleeding than lower percentages of FA. In no case should the percentage passing a number 200 sieve exceed 5 percent, and only 3 percent if the structure is exposed to abrasive wear. Aggregate grading falling within those limits are generally satisfactory for most concrete.

CA Grading

2-42. The grading of CA of a given maximum size can vary over a wide range without much effect on cement and water requirements if the proportion of FA produces concrete having good workability. Table

Concrete Components

2-4 gives the grading requirements for CA. If CA grading varies too much, the mix proportions will need to vary to produce workable concrete. If the variance continues, it is more economical to request that the producer adjust his operation to meet the grading requirements. CA should be graded up to the largest practicable size for the job conditions. According to the American Concrete Institute (ACI) 318-83, the nominal maximum size of the CA should not be larger than 1/5 the narrowest dimension between the sides of forms, 1/3 the depth of slabs, and 3/4 the minimum clear spacing between individual reinforcing bars or wires, bundles of bars, or prestressing tendons or ducts. These limitations may be waived if, in the judgment of the engineer, workability and consolidation methods are such that concrete can be placed without honeycomb or voids. (These are undesirable areas; however, a smooth finish is desired even though CA is visible. Honeycomb or voids are usually first observed when the formwork is removed.) The larger the maximum size of the CA, the less paste (water and cement) that is required to produce a given quality. Field experience shows that the amount of water required per unit volume of concrete for a given consistency and given aggregates is nearly constant, regardless of the cement content or relative proportions of W/C. Furthermore the amount of water required decreases with increases in the maximum size of the aggregate. The water required per cubic yard for concrete with a slump of 3 to 4 inches is shown figure 2-2, page 2-10. The figure demonstrates that for a given W/C ratio, the amount of cement required decreases as the maximum size of CA increases. In some instances, especially in higher-strength ranges, concrete containing smaller maximum-size aggregate has a higher compressive strength than concrete with larger maximum-size aggregate at the same W/C ratio.

Table 2-4. Grading requirement for coarse aggregate

Size Number	Nominal Size (Sieve) With Square Openings	4 Inches	3 1/3 Inches	3 Inches	2 1/2 Inches	2 Inches	1 1/2 Inches	1 Inch
1	3 1/2 to 1 1/2 inch	100	90 to 100		25 to 60		0 to 15	
2	2 1/2 to 1 1/2			100	90 to 100	35 to 70	0 to 15	
357	2 to 4			100		95 to 100		35 to 70
467	1 1/2 to No 4					100	95 to 100	
57	1 inch to No 4						100	95 to 100
67	3/4 inch to No 4							100
7	1/2 inch to No 4							
8	3/8 inch to No 8							
3	2 to 1 inch				100	90 to 100	35 to 70	0 to 15
4	1 1/2 to 3/4					100	90 to 100	20 to 55

Chapter 2

Table 2-4. Grading requirement for coarse aggregate

Size Number	Nominal Size (Sieve) With Square Openings	4 Inches	3 1/3 Inches	3 Inches	2 1/2 Inches	2 Inches	1 1/2 Inches	1 Inch

Notes.
1. Specifications for concrete aggregate as described in ASTM-C33.
2. Amounts are finer than each laboratory sieve or square openings. Percentages are by weight.

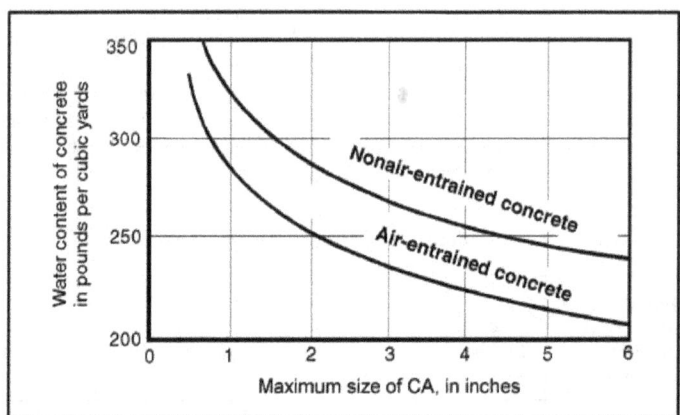

Figure 2-2. Water requirements for concrete of a given consistency as a function of CA size

Gap-Graded Aggregates

2-43. Certain particle sizes are entirely or mostly absent in gap-graded aggregates. The lack of two or more successive sizes can create segregation problems, especially in nonair-entrained concrete having slumps greater than 3 inches. However, for a stiff mix, gap-graded aggregates can produce higher strengths than normal aggregates in concrete mixes having comparable cement contents.

BULK UNIT WEIGHT

2-44. This is the weight of the aggregate that fills a 1-cubic foot container. This term is used because the volume contains both aggregate and voids of air spaces.

SPECIFIC GRAVITY

2-45. This is the ratio of aggregate weight to the weight of an equal volume of water. Normal-weight aggregates have specific gravities ranging from 2.4 to 2.9. The internal structure of an aggregate particle is made up of both solid matter and pores or voids that may or may not contain water. The specific gravities used in concrete calculations are generally for saturated, surface dry (SSD) aggregates; that is, when all pores are filled with water, but no excess moisture is present on the surface.

ABSORPTION AND SURFACE MOISTURE

2-46. Both absorption and surface moisture must be known to control the net water content of the concrete and determine correct batch weights. Clearly, dry aggregate requires more concrete mixing water. The four moisture conditions of aggregates are as follows (see figure 2-3):

- Oven dry. Surface and pores are bone-dry and fully absorbent.
- Air dry. Surface is dry but contains some interior moisture and is therefore somewhat absorbent.
- Saturated, surface-dry. Pores are saturated but surface is dry—neither absorbing water from nor contributing water to the concrete mix.
- Damp or wet. Aggregate contains an excess of moisture on the surface.

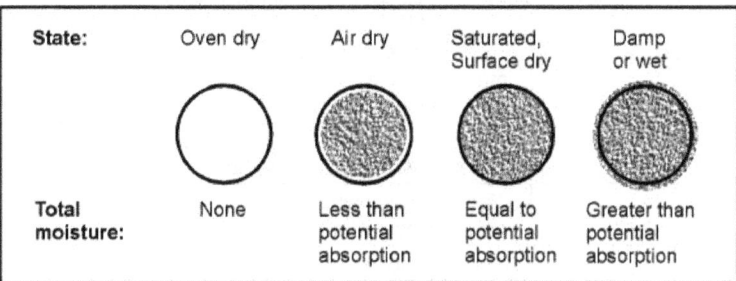

Figure 2-3. Moisture conditions of aggregates

Bulking

2-47. Bulking occurs when damp FA is handled. Bulking is the undesirable increase in volume caused by surface moisture holding the particles apart. Figure 2-4 shows the variation in the amount of bulking with moisture content and grading. Sand is normally delivered in batch quantities in a damp condition; but, due to bulking, actual sand content can vary widely in a batch volume, often not in proportion to the moisture content of the sand. Therefore, be very careful when proportioning by volume. Too much moisture on the aggregate surfaces also adds to the concrete mixing water. The amount can be considerable, especially the excess water in FA.

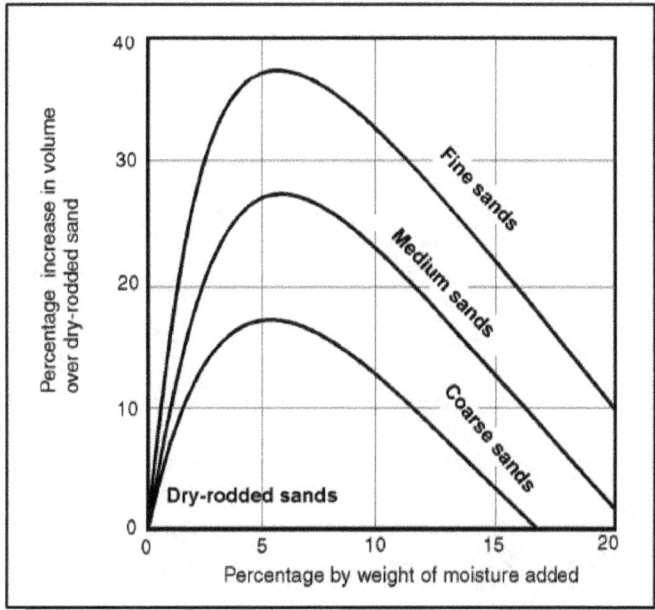

Figure 2-4. Variation in fine-aggregate bulking with moisture and aggregate grading

IMPURITIES

2-48. Aggregates can contain such impure substances as organic matter, silt, clay, coal, lignite, and certain light-weight and soft particles. Table 2-5 summarizes the effects of these substances on concrete.

Table 2-5. Impurities in aggregates

Impure Substances	Effects on Concrete	ASTM Test Designation
Organic impurities	Affects setting and hardening time and may cause deterioration	C40 C87
Materials finer than number 200 sieve	Affects bonding and increases water requirements	C117
Coal, lignite, or other lightweight materials	Affect durability and may cause stains and pop outs	C1213
Soft particles	Affects durability	C235
Friable particles	Affects workability and durability and may cause pop outs	C142

HANDLING AND STORING

2-49. Figure 2-5 shows both the correct and incorrect methods of handling and storing aggregates. You must handle and store aggregates to minimize segregation and prevent contamination by impure substances. Aggregate normally stored in stockpiles builds up in layers of uniform thickness. Stockpiles should not be built up in high cones or allowed to run down slopes because this causes segregation. FA remains at the top of the stockpile while heavier aggregate rolls toward the bottom. Do not allow aggregate to fall freely from the end of a conveyor belt. To minimize segregation, remove aggregates from stock-piles in horizontal layers. When you are using batch equipment and storing some aggregate in bins, load the bins by allowing the aggregate to fall vertically over the outlet. Chuting the materials at an angle against the side of the bin causes particle segregation.

Concrete Components

 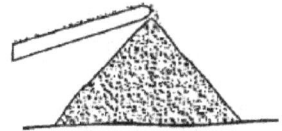

Stockpiling of coarse aggregate

Preferable
Use a crane or other equipment to stockpile material in separate batches, each no larger than a truckload, so that it remains where placed and does not run down slopes.

Objectionable
Do not use stockpiling methods that permit the aggregate added to the pile to roll down the slope or permit hauling equipment to operate over the same level repeatedly.

Marginally acceptable
Generally, do not use a bulldozer to build a pile in horizontal radial layers when working with materials falling from a conveyor belt. You may need a rock ladder in this setup.

Generally objectionable
Using a bulldozer to stack progressive layers on a slope not flatter than 3:1 is usually objectionable, unless materials strongly resist breakage.

Figure 2-5. Correct and incorrect aggregate handling and storage

SECTION IV - ADMIXTURES

DEFINITION AND PURPOSE

2-50. An admixture is any material other than cement, water, or aggregate that is added to concrete in small quantities—either immediately before or during mixing—to modify such properties as workability, strength, durability, watertightness, or wear resistance. Admixtures can also reduce segregation, the heat of hydration, and entrained air, and can either accelerate or retard setting and hardening. Similar results are obtainable by changing the concrete-mix proportions instead of using admixtures (except air-entraining ones). When possible, examine all alternatives before using an admixture to determine which is more economical or convenient.

AIR-ENTRAINED CONCRETE

2-51. A major advance in concrete technology in recent years is the introduction of tiny disconnected air bubbles into concrete called air entrainment. Air-entrained concrete results from using either an air-entraining cement or an air-entraining admixture during mixing. Adding entrained air to concrete provides important benefits in both plastic and hardened concrete, such as resistance to freezing and thawing in a saturated environment. Air entrapped in nonair-entrained concrete fills relatively large voids that are not uniformly distributed throughout the mix.

Chapter 2

PROPERTIES

2-52. The following are properties of air-entrained concrete:

- Workability. The improved workability of air-entrained concrete greatly reduces water and sand requirements, particularly in lean mixes and in mixes containing angular and poorly graded aggregates. In addition, the disconnected air bubbles reduce segregation and bleeding of plastic concrete.
- Freeze-thaw durability. The expansion of water as it freezes in concrete can create enough pressure to rupture the concrete. However, entrained air bubbles serve as reservoirs for the expanded water, thereby relieving expansion pressure and preventing concrete damage.
- Deicers resistance. Because entrained air prevents scaling caused by deicing chemicals used for snow and ice removal, air-entrained concrete is recommended for all applications where the concrete contacts deicing chemicals.
- Sulfate resistance. Entrained air improves concrete's resistance to sulfate. Concrete made with a low W/C ratio, entrained air, and cement having a low tricalcium-aluminate content is the most resistant to sulfate attack.
- Strength. The voids to cement ratio basically determines air-entrained concrete strength. For this ratio, voids are defined as the total volume of water plus air (both entrained and entrapped). When the air content remains constant, the strength varies inversely with the W/C ratio. As the air content increases, you can generally maintain a given strength by holding the voids to the cement ratio constant. To do this, reduce the amount of mixing water, increase the amount of cement, or both. Any strength reduction that accompanies air entrainment is minimized because air-entrained concrete has lower W/C ratios than nonair-entrained concrete having the same slump. However, it is sometimes difficult to attain high strength with air-entrained concrete, such as when slumps remain constant while the concrete's temperature rises when using certain aggregates.
- Abrasion resistance. Air-entrained concrete has about the same abrasion resistance as that of nonair-entrained concrete of the same compressive strength. Abrasion resistance increases as the compressive strength increases.
- Watertightness. Air-entrained concrete is more watertight than nonair-entrained concrete since entrained air prevents interconnected capillary channels from forming. Therefore, use air-entrained concrete where watertightness is a requirement.

AIR-ENTRAINING MATERIALS

2-53. Air-entraining admixtures are usually liquid derivatives of natural wood resins, animal or vegetable fats or oils, alkali salts of sulfated or sulfonated organic compounds, and water-soluble soaps intended for use in the mixing water. The manufacturer provides instructions to produce a specified air content. Some manufacturers market automatic dispensers that accurately control the quantities of air-entraining agents in a mix. Air is incorporated in concrete by using air-entraining cement, an air-entraining admixture at the mixer, or both methods. Air-entraining cements should meet the specifications in ASTM C 175. Add commercial air-entraining admixtures at the mixer. They should comply with ASTM C 260. Use adequate controls to always ensure the proper air content. Factors affecting air content are:

- Aggregate gradation and cement content. Both significantly affect the air content of both air-entrained and nonair-entrained concrete. For aggregate sizes smaller than 1 1/2 inch, the air content in-creases sharply as the aggregate size decreases due to the increase in cement volume. As cement content increases, the air content decreases but remains within the normal range of the cement content.
- FA content. This affects the percentage of entrained air in concrete. Increasing the FA content incorporates more air in a given amount of air--entraining cement or admixture.
- Slump and vibration. These affect the air content of air-entrained concrete because the greater the slump, the larger the percentage reduction in air content during vibration. At all slumps, even

a 15-second vibration causes reduced air content. However, properly applied vibration mainly eliminates large air bubbles and little of the intentionally entrained tiny, air bubbles.

- Concrete temperature. Its effects become more pronounced as slump increases. Less air is entrained as the concrete's temperature increases.
- Mixing action. This is the most important factor in producing air-entrained concrete. The amount of entrained air varies with the mixer type and condition, the amount of concrete mixed, and the mixing rate. Stationary and transit mixers may produce concrete having very different amounts of entrained air. Mixers not loaded to capacity can increase air content, whereas overloading can decrease air content. Generally, more air is entrained as the mixing speed increases.
- Admixtures and coloring agents. These can reduce the amount of entrained air, particularly fly ash having high percentages of carbon. To prevent a chemical reaction with certain air-entraining admixtures, you must add calcium-chloride solutions to the mix separately.
- Premature finishing operations. This can cause excess water to work itself to the concrete surface. If this occurs, the surface zone may not contain enough entrained air and be susceptible to scaling.

RECOMMENDED AIR CONTENT

2-54. Air contents for frost-resistant concrete must be as stated in table 2-6. Such concrete must be used when there is a danger of concrete freezing while saturated or nearly saturated with water. Air content produced by air-entraining admixtures meeting ASTM requirements will give about 9 percent air in the fraction of the concrete mixture passing the 4.75 millimeter (mm) number 4 sieve; that is, the mortar fraction.

Table 2-6. Total air content for frost-resistant concrete

Nominal Maximum Aggregate Size, in Inches*	Air Content, in Percentage	
	Severe Exposure	Moderate Exposure
3/8	7 1/2	6
½	7	5 1/2
¾	6	5
1	6	4 1/2
1 ½	5 1/2	4 1/2
2**	5	4
3**	4 1/2	3 1/2
* See ASTM C 33 for tolerance on oversize of various nominal maximum size designations.		
** The air content applies to the total mix. When testing concrete aggregates larger than 1 1/2 inch, removal is by hand or sieve. Air Content is determined on the minus 1 1/2 fraction of the mix. Tolerance on air content as delivered applies to this value. Air content of the total mixture is computed from values determined on the minus 1 1/2 fraction.		

TESTS FOR AIR CONTENT

2-55. Tests that determine air entrainment in freshly mixed concrete measure only air volume, not air-void characteristics. This indicates the adequacy of the air-void system when using air-entraining materials meeting ASTM specifications. Tests should be made regularly during construction, using plastic samples taken immediately after discharge from the mixer and from already placed and consolidated concrete. The following are the standard methods to determine the air content of plastic concrete.

- Pressure method. This method is practical for field testing all concrete except those containing highly porous and lightweight aggregates.
- Volumetric method. This method is practical for field testing concrete, particularly concrete containing lightweight and porous aggregates.

Chapter 2

- Gravimetric method. Impractical for field testing because it requires accurate knowledge of specific gravities and absolute volumes of concrete ingredients. It is satisfactory for laboratory use.

OTHER ADMIXTURES

2-56. When added in small qualities, admixtures change the uniformity of the concrete. The following are admixtures that modify concrete:

- Water-reducing admixtures. These reduce the quantity of mixing water required to produce concrete of a given consistency. The slump is increased for a given water content.
- Retarding admixtures. These are sometimes used to reduce the rate of hydration to permit placing and consolidating concrete before the initial set. They also offset the accelerating effect of hot weather on the set. These admixtures generally consist of fatty acids, sugars, and starches.
- Accelerating admixtures. These hasten the set and strength development. Calcium chloride is the most common. Add in solution form as part of the mixing water, but don't exceed 2 percent by weight of cement. Do not use calcium chloride or other admixtures containing soluble chlorides in prestressed concrete or concrete containing embedded aluminum, in permanent contact with galvanized steel, subject to alkali-aggregate reaction, or exposed to soils or water containing sulfates. Table 2-7 shows the limitations.
- Pozzolans. These materials contain considerable silica or much silica and alumina. They are combined with calcium hydroxide to form compounds having cement-like properties. Pozzolans should be tested first to determine their suitability, because the properties of pozzolans and their effects on concrete vary considerably.
- Workability agents. These improve the workability of fresh concrete. They include entrained air, certain organic materials, and finely divided materials. When used as work-ability agents, fly ash and natural pozzolans should conform to ASTM C 618.
- Dampproofing and permeability-reducing agents. These are water-repellent materials that are used to reduce the capillary flow of moisture through concrete that contacts water or damp earth. Pozzolans are also permeability-reducing agents.
- Grouting agents. These are various air-entraining admixtures, accelerators, retarders, and workability agents that alter the properties of portland cement grouts for specific applications.
- Gas-forming agents. When added to concrete or grout in very small quantities, they cause a slight expansion before hardening in certain applications. However, while hardening, the concrete or grout decreases in volume in an amount equal to or greater than that of normal concrete or grout.

Table 2-7. Maximum chloride ion content for corrosion protection

Type of Members	Maximum Water Soluble (Chloride Ion in Concrete)*
Prestressed concrete	0.05
Reinforced concrete exposed to chloride in service	0.15
Reinforced concrete that will be dry or protected from moisture in service	1.00
Other reinforced concrete construction	0.30
*Percentages are by weight of cement.	

Chapter 3
Proportioning Concrete Mixtures

The two primary methods used to proportion a mix design are the trial-batch method and the absolute-volume method. Proportions of ingredients for concrete should be selected to make the most economical use of available materials that will produce concrete of the required plausibility, durability, and strength. Basic relationships already established and the laboratory tests provide guidance for optimum combinations. Recommended and typical mixes of concrete for various types or classes of work will determine how you should proportion your concrete mix. Factors affecting this include—
- W/C ratio.
- Type and size of aggregate.
- Air or nonair-entrained concrete.
- Slump of the mix.

SECTION I - METHOD CONSIDERATIONS

SELECTING MIX PROPORTIONS

3-1. Concrete proportions for a particular application are determined by the concrete's end use and by anticipated conditions at the time of placement. Find a moderate balance between reasonable economy and the requirements for placeability, strength, durability, density, and appearance that may be in the job specifications.

BASIC GUIDELINES

3-2. Before proportioning a concrete mixture, certain information about a job is necessary, such as the size and shape of the structural members, the concrete strength required, and the exposure conditions. Other important factors discussed below are the W/C ratio, aggregate characteristics, the amount of entrained air, and the slump.

W/C Ratio

3-3. The W/C ratio is determined by the strength, durability, and watertightness requirements of the hardened concrete. Strength, durability and watertightness are usually specified by the structural-design engineer, but a tentative mix proportion can be determined from knowledge of a prior job. Always remember that a change in the W/C ratio changes the characteristics of the hardened concrete. Use table 3-1, page 3-2, to select a suitable W/C ratio for normal-weight concrete that will meet the anticipated exposure conditions. The W/C ratios in table 3-1 are based on concrete's strength under certain exposure conditions. If possible, perform tests using job materials to determine the relationship between the W/C ratio selected and the strength of the finished concrete. If laboratory-test data is not obtainable or experience records for the relationship are unavailable, use the data in table 3-2, page 3-3, as a guide. In table 3-2, locate the desired compressive strength of concrete in pounds per square inch (psi) and read across to determine the maximum W/C ratio. If possible, interpolate between the values. When both exposure conditions and strength must be considered, use the lower of the two indicated W/C ratios. If flexural strength rather than compressive strength is the basis for a design, such as a pavement, perform the necessary tests to determine the relationship between the W/C ratio and flexural strength. An approximate relationship between flexural and compressive strength is as follows:

Table 3-1. Maximum W/C ratios for various exposure conditions

Exposure Condition	Normal-Weight Concrete (Absolute W/C Ratio by Weight)
Concrete protected from exposure to freezing and thawing or the application of deicer chemicals	Select a W/C ratio on the basis of strength, workability, and finishing needs
Watertight concrete* • In freshwater • In seawater	0.50 0.45
Frost-resistant concrete* • Thin sections; any section with less than a 2-inch cover over reinforcement and any concrete exposed to deicing salts • All other structures	0.45 0.50
Exposure to sulfates* • Moderate • Severe	0.50 0.45
Concrete placed underwater	Do not use less than 650 pounds of cement per cubic yard (386 kg/m3).
Floors on grade	Select W/C ratio for strength, plus minimum cement requirements described in table 3-7, page 3-14.
* For the properties of watertight concrete, frost-resistant concrete and exposure to sulfates, use designing strength for air-entrained concrete.	

Table 3-2. Maximum permissible W/C ratios for concrete

Specified Compressive Strength, in psi*	Maximum Absolute Permissible W/C Ratios by Water	
	Nonair-Entrained Concrete	Air-Entrained Concrete
2,500	0.67	0.54
3,000	0.58	0.46
3,500	0.51	0.40
4,000	0.44	0.35
4,500	0.38	**
5,000	**	**

Note. 1,000 psi = 7 MPa.

*28-day strength. The W/C ratios will provide average strengths that are greater than the specified strengths.

**For strength above 4,500 psi (nonair-entrained concrete) and 4,000 psi (air-entrained concrete), proportions should be established by the trial-batch method.

$$f'_c = \frac{R^2}{K}$$

Where—

f'_c = compressive strength, in psi

R = flexural strength (modulus of rupture), in psi

K = a constant, usually between 8 and 10

AGGREGATE

3-4. Use FA to fill the spaces between the CA particles and to increase the workability of the mix. Aggregate that does not have a large grading gap nor an excess of any size but gives a smooth grading curve produces the best mix. Fineness modulus and FA grading are discussed in paragraph 2-45.

3-5. Use the largest practical size of CA in the mix. The maximum size of CA that produces concrete of maximum strength for a given cement content depends upon the aggregate source as well as the aggregate shape and grading; thus, in most cases, a decrease will take place in the overall cost. The larger the maximum size of the CA, the less paste (water, cement and usually entrained air) that is required for a given concrete quality. The maximum size of aggregate should not exceed one-fifth the minimum dimension of the member or three-fourths the space between reinforcing bars. For pavement or floor slabs, the maximum size of aggregate should not exceed one-third the slab thickness.

Entrained Air

3-6. To improve workability, use entrained air in all concrete that is exposed to freezing and thawing and, sometimes, to mild conditions. Always use entrained air in paving concrete, regardless of climatic conditions. Table 3-3 gives the recommended total air contents of air-entrained concrete. When mixing water remains constant, air entrainment increases slump. When the cement content and slump remain constant, less water is required. The resulting decrease in the W/C ratio helps to offset possible strength decreases and improves other paste properties, such as permeability. The strength of air-entrained concrete may equal, or nearly equal, that of nonair-entrained concrete when cement contents and slumps are the same. The upper half of table 3-3 gives the percentage of entrapped air in nonair-entrained concrete. The lower half of the table gives the recommended average, total air content, and percent for air-entrained concrete based exposure levels.

Table 3-3. Approximate mixing-water and air-content requirements for different slumps and maximum sizes of aggregate

Maximum Aggregate Size:	3/8 Inch	1/2 Inch	3/4 Inch	1 Inch	1 1/2 Inches	2** Inches	3** Inches	6** Inches
Water in Pounds Per Cubic Yard of Concrete*								
Slump, in Inches	Nonair-Entrained Concrete							
1 to 2	350	335	315	300	275	260	240	210
3 to 4	385	365	340	325	300	285	285	230
6 to 7	410	385	360	340	315	300	285	
Air-Entrained Concrete								
1 to 2	305	295	280	270	250	240	225	200
3 to 4	340	325	305	295	275	265	250	220
6 to 7	365	345	325	310	290	280	270	—
Approximate percentage amount of entrapped air in nonair-entrained concrete								
	3	2.5	2	1.5	1	0.5	0.3	0.2
Recommended percentage average and total air content for air-entrained concrete								
Anticipated Usage								
Mild exposure	4.5	4.0	3.5	3.0	2.5	2.0	1.5	1.0
Moderate exposure	6.0	5.5	5.0	4.5	4.5	4.0	3.5	3.0
Severe exposure	7.5	7.0	6.0	6.0	5.5	5.0	4.5	4.0

Table 3-3. Approximate mixing-water and air-content requirements for different slumps and maximum sizes of aggregate

Maximum Aggregate Size: ___	3/8 Inch	1/2 Inch	3/4 Inch	1 Inch	1 1/2 Inches	2** Inches	3** Inches	6** Inches

*These quantities of mixing water are for use in computing cement factors for trial batches. They are maximums for reasonably well-shaped, angular CA graded within limits of accepted specifications.

**The slump values for concrete containing aggregate larger than 1 1/2 inches are based on slump tests made after removal of particles larger than 1 1/2 inches by wet screening.

Mild Exposure

3-7. Mild Exposure includes indoor or outdoor service in a climate that does not expose the concrete to freezing or deicing agents. When you want air entrainment for any reason other than durability, such as to improve workability or cohesion or to improve strength in low-cement-factor concrete, you can use air contents that are lower than those required for durability.

Moderate Exposure

3-8. Moderate exposure means service in a climate where freezing is expected, but where the concrete is not continually exposed to moisture or free-standing water for long periods before freezing or to deicing agents or other aggressive chemicals. Structures that do not contact wet soil or receive direct applications of deicing salts are exterior beams, columns, walls, girders, and slabs.

Severe Exposure

3-9. Severe exposure means service where the concrete is exposed to deicing chemicals or other aggressive agents or where the concrete continually contacts moisture or free-standing water before freezing. Examples are pavements, bridge decks, curbs, gutters, sidewalks, canal linings, or exterior water tanks or slumps.

SLUMP

3-10. The *slump* test (see appendix B) measures the consistency of concrete in cubic yards. Do not use it to compare mixes having different proportions or mixes containing different sizes of aggregate. When testing different batches of the same mixture, changes in slump indicate changes in materials, mix proportions, or water content. Table 3-4, page 3-6, gives the recommended slump ranges.

Chapter 3

Table 3-4. Slumps for various types of construction (with vibration)

Concrete Construction	Slump, in Inches	
	Maximum*	Minimum*
Reinforced foundation walls and footings	3	1
Plain footing, caissons, and substructure walls	3	1
Beams and reinforced walls	4	1
Building columns	4	1
Pavements and slabs	3	1
Mass concrete	2	1
Note. 1 inch = 25 mm *May be increased 1 inch for consolidation by methods such as rods and spades.		

SECTION II - TRIAL-BATCH METHOD

BASIC GUIDELINES

3-11. In the trial-batch method of mix design, use actual job materials to obtain mix proportions. The size of the trial batch depends on the equipment you have and how many test specimens you make. Batches using 10 to 20 pounds of cement may be big enough, although larger batches produce more accurate data. Use machine mixing, if possible, because representation is close to the actual job conditions. Always use a machine to mix concrete containing entrained air. Be sure to use representative samples of aggregate, cement, water, and air-entraining admixture in the trial batch. The following steps are setup guidelines:

Step 1. Pre-wet the aggregate and allow it to dry to a saturated, surface-dry condition.

Step 2. Place the aggregate in covered containers to maintain this condition until ready for use. This simplifies calculations and eliminates error caused by variations in aggregate moisture content. When the concrete quality is specified in terms of the W/C ratio, the trial-batch procedure consists of combining paste with the proper amounts of FA and CA to produce the required slump and workability.

Step 3. Calculate the larger quantities per sack and/or per cubic yard.

EXAMPLE USING THE TRIAL-BATCH METHOD

3-12. The trial-batch method determines the mix proportions for a concrete retaining wall exposed to freshwater in a severe climate. The required compressive strength is 3,000 psi at 28 days. The minimum wall thickness is 8 inches, with 2 inches of concrete covering the reinforcement. Enter all trial-mix data in the appropriate blanks on the trial-mix-data work sheets (see figure 3-1, page 3-8). Table 3-1, page 3-2, indicates that a maximum W/C ratio of 0.50 by weight satisfies the exposure requirements. Using Type IA (air-entraining) portland cement and a compressive strength of 3,000 psi. Table 3-2, page 3-3, shows that a maximum W/C ratio of approximately 0.46 by weight satisfies the strength requirements. To meet both specifications, select a W/C ratio of 0.46. Since the maximum size of the CA must not exceed one-fifth the minimum wall thickness nor three-fourths of the space between the reinforcement and the surfaces, the maximum size of the CA you will choose is 1 1/2 inch. Because of the severe exposure conditions, the concrete should contain entrained air. Table 3-3, page 3-4, shows that the recommended air content is 5.5 +/- 0.5 percent, assuming that vibration will consolidate the concrete. Table 3-4 indicates a recommended slump ranging from 1 to 3 inches. The trial-batch proportions are now determined. For a batch containing 20 pounds of cement, the mixing water required is:

$$20 \times 0.46 = 9.2 \; pounds$$

Step 1. Select and weigh representative samples of FA and CA, and record their weights in column (2) of figure 3-1, page 3-8. Use all the measured quantities of cement, water, and, air-entraining admixture.

Note. It is not normal practice to buy air-entraining cement (Type IA) and then add an air-entraining admixture. However, if the only cement available was Type IA and it did not give the needed air content, an addition of an air-entraining admixture would be necessary to achieve frost resistance.

Step 2. Add FA and CA until you produce a workable mixture having the proper slump.

Step 3. Record the weights of all materials in column 4 of figure 3-1.

Step 4. Calculate the weights for both a 1-bag batch and 1 cubic yard. Record the results in columns 5 and 6.

Step 5. Calculate and record—

- The cement factor in pounds per cubic yard as indicated in figure 3-1.
- The percentage of FA by weight and by volume of the total aggregate.
- The yield of concrete in cubic feet per bag.

Step 6. Determine and note the slump, air content, workability, and unit weight of the concrete.

Step 7. Find the most economical proportions, by making more trial batches, varying the percentage of FA.

CONCRETE
TRIAL-MIX DATA

1. Project Number _____
2. Structure _____
3. Exposure Condition:
   Severe or Moderate _____ Mild _____
   In Air _____
   In Freshwater _____
   In Seawater _____
4. Type of Structure (A-I) _____
   Max W/C for Exposure _____
   Max W/C for Watertightness _____
5. Type of Cement _____
6. Fineness Modulus of Sand _____

7. Specific Gravity:
   Sand _____
   Gravel _____
8. Maximum-size Aggregate _____
9. Air Content _____ % ± _____
10. Desired Slump Range:
    Max _____ inches Min _____ inches
11. Strength Requirement _____ psi
    W/C for Strength _____ by wt
    W/C Use _____ by wt

Data for Trial Batch
(Saturated, surface-dry aggregates)

(1) Material	(2) Initial Wt, in Pounds	(3) Final Wt, in Pounds	(4) Wt Used, in Pounds	(5) Wt for 1- Sack Batch	(6) Wt per Cubic Yard	(7) Remarks
Cement						
Water						
FA						_____ Percentage of total aggregate
CA						
Air-entraining admixture			Total (T) =			

Measured Slump _____ inches   Air Content _____ %   Workability _____

| Wt container concrete, in pounds |
| Wt container, in pounds |
| Wt concrete = A, in pounds |
| Vol container = B, in cubic feet |
| Unit Wt of Concrete = w = A/B |

A/B = _____ = pounds/cubic feet

Yield = T/w = _____ = cubic feet/sack

Figure 3-1. Work sheet for concrete trial-mix data

3-13. In each batch, keep the W/C ratio, aggregate gradations, air content, and slump approximately the same. Table 3-5 summarizes the results of four such trial batches made in the laboratory. Figure 3-2, page 3-10, plots the percentage of FA used in these mixes against the cement factor. The minimum cement factor (538 pounds per cubic yard) occurs at a FA content of about 32 percent of total aggregate. Since the

W/C ratio is 0.46 by weight and the unit weight of the concrete for an air content of 5 percent is about 144 pounds per cubic foot, the final quantities for the mix proportions per cubic yard are as follows:

Cement	= 538 pounds
Water (538 x 0.46)	= 247 pounds
Total	= 785 pounds
Concrete per cubic yard (144 x 27)	= 3,890 pounds
Aggregates (3,890 - 785)	= 3,105 pounds

Note. 1 cubic yard equals 27 cubic feet

FA (0.32 x 3,105)	= 994 pounds
CA (3,105 - 994)	= 2,111 pounds

Table 3-5. Results of laboratory trial mixes

Batch Number	Slump, in Inches	Air-Content Percentage	Unit Weight, in Pounds Per Cubic feet	Cement Factor, in Pounds Per Cubic Yard	FA*	Workability
1	3	5.4	144	540	33.5	Excellent
2	2 3/4	4.9	144	556	27.4	Harsh
3	2 1/2	5.1	144	549	33.5	Excellent
4	3 1/4	4.7	145	540	30.5	Good

Note. The W/C ratio selected was 0.46 by weight.
*Percentage of total aggregate

Chapter 3

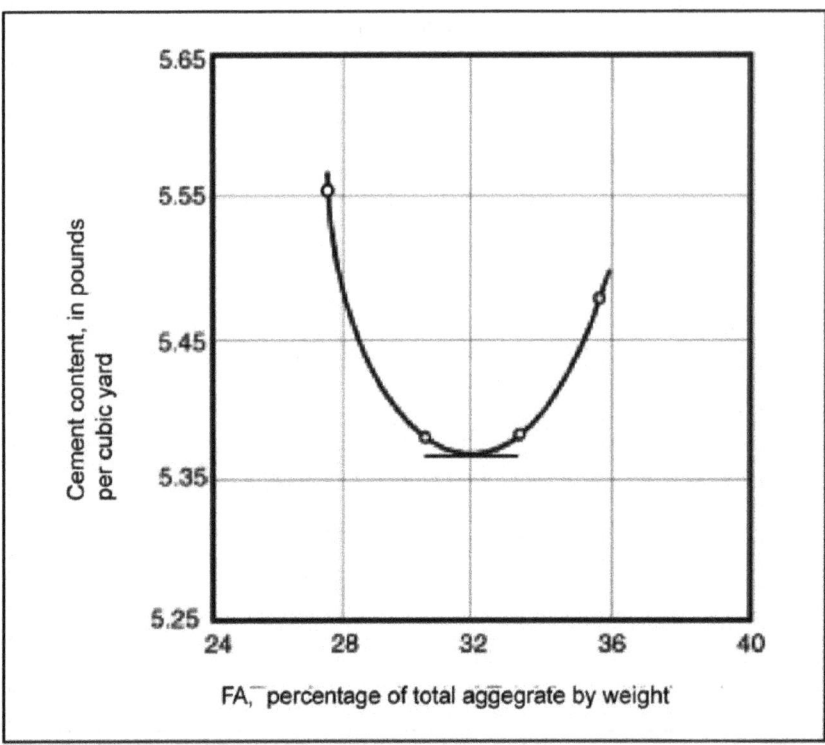

Figure 3-2. Relationship between percentage of FA and cement content for a given W/C ratio and slump

SECTION III - ABSOLUTE-VOLUME METHOD

BASIC GUIDELINES

3-14. Concrete mixtures can be proportioned using absolute volumes. This method was detailed in ACI 211.1-81. For this procedure, select the W/C ratio, slump, air content, and the maximum size of the aggregate. Then estimate the water requirement using table 3-3, page 3-4. Addition information is required before making calculations such as the specific gravities of the FA and CA, the dry-rotted unit weight (DRUW) of the CA, and the fineness modulus of the FA. If the maximum size of the aggregate and the fineness modulus of the FA are known, estimate the volume of dry-rotted CA per cubic yard. From table 3-6, calculate the quantities per cubic yard of water, cement, CA, and air, and the dry-rotted unit weight of the CA. Finally, subtract the sum of the absolute volumes of these materials in cubic feet from 27 cubic feet per cubic yard to give the specific volume of the FA.

Table 3-6. Volume of CA per cubic yard of concrete

Maximum Size of Aggregate, in Inches	Fineness Modulus of FA			
	2.40	2.60	2.80	3.00
	CA, in Cubic Feet Per Cubic Yard			
3/8	13.5	13.0	12.4	11.9
1/2	15.9	15.4	14.8	14.3
3/4	17.8	17.3	16.7	16.2
1	19.2	18.6	18.1	17.6
1 1/2	20.2	19.7	19.2	18.6
2	21.1	20.5	20.0	19.4
3	22.1	21.6	21.1	20.5

Note. Volumes are based on aggregates in a dry-rotted condition, as described in Method of Test for Unit Weight of Aggregate (ASTM C29). These volumes are selected from empirical relationships to produce concrete with a degree of workability suitable for usual reinforced construction. For less workable concrete, such as that required for concrete pavement construction, the volume may be increased about 10 percent. When placement is to be by pump, volume should be decreased about 10 percent.

EXAMPLE USING THE ABSOLUTE-VOLUME METHOD

3-15. For a retaining wall, determine the mix proportions using the following specifications and conditions:

- Required 28-day compressive strength of 3,000 psi.
- Maximum size aggregate is 3/4 inch.
- Exposure condition will be moderate freeze-thaw exposure (exposed to air).
- Fineness modulus of the FA is 2.70.
- Specific gravity of portland cement is 3.15.
- Specific gravity of the FA (SGFA) is 2.66.
- Specific gravity of the CA (SGCA) is 2.61.
- Dry-rotted unit weight (DRUW) of the CA is 104 lb/cu ft.
- DRUW of the FA is 103 lb/cu ft.
- Slump is 3 inches.
- Cement type is IA.

Step 1. Estimate the air content (refer to table 3-3) using the following information:

- Aggregate size = 3/4 inch
- Air-entrained concrete
- Moderate exposure

Chapter 3

- Air content = 5 percent

Step 2. Estimate the mixing-water content (refer to table 3-3, page 3-4) using the following information:

- Slump = 3 inches
- Aggregate size = 3/4 inch
- Air-entrained concrete
- Mixing water = 305 lb/cu yd

Step 3. Determine the W/C ratio. Check the W/C ratios for strength, durability, and watertightness using the following information:

- Strength is determined by—
    - Mix design strength = 3,000 psi
    - Specified compressive strength.
    - Design strength.
    - W/C ratio (refer to table 3-2, page 3-3) = 0.46
- Durability (refer to table 3-1, page 3-2) is determined by—
    - Frost-resistant concrete.
    - All other structures - W/C = 0.50
- Watertightness (refer to table 3-1, pg 3-2) is determined by—
    - Freshwater exposure - W/C = 0.50
    - Selection of the lowest W/C ratio - W/C = 0.46

Step 4. Calculate cement content (C) using the formula below:

$$C = \frac{water\ lb/cu\ yd}{w/c\ ratio} \qquad C = \frac{305\ lb/cu\ yd}{0.46} \qquad C = 663.0\ lb/cu\ yd$$

Step 5. Calculate the CA content (refer to table 3-6, page 3-11) by referring to the following information:

- CA size = 3/4 inch
- Fineness modulus of FA = 2.7 (given)
- Determine volume of CA by interpolation by referring to the following information:
    - - Fineness modulus.   2.6   2.7   2.8
    - For 3/4 inch:
    - - Volume of CA (cu ft/cu yd).

      17.3   17.0   16.7
- Weight of CA.
    - G = Volume of CA x unit weight of CA
    - G = 17 x 104 lb/cu ft
    - G = 1,768 lb/cu yd

Where—

G = the weight of the CA in lbs / cu yd

Step 6. Calculate the FA content by the absolute method using the following formula:

$$Absolute\ volume = \frac{weight\ of\ material}{specific\ gravity\ x\ 62.4\ lb/cu\ ft}$$

$$Absolute\ volume\ cement = \frac{C}{3.15\ x\ 62.4\ lb/cu\ ft} = \frac{663.01\ lb/cu\ yd}{3.15\ x\ 62.4\ lb/cu\ ft} = 3.37\ cu\ ft/cu\ yd$$

$$\text{Absolute volume water} = \frac{W}{1 \times 62.4 \; lb/cu\,ft} = \frac{305}{1 \times 62.4 \; lb/cu\,ft} = 4.89 \; cu\,ft/cu\,yd$$

$$\text{Absolute volume CA} = \frac{G}{SGCA \times 62.4 \; lb/cu\,ft} = \frac{1768 \; lb/cu\,yd}{2.61 \times 62.4 \; lb/cu\,ft} = 10.86 \; cu\,ft/cu\,yd$$

$Volume\ of\ air\ =\ air\ content \times 27\ cu\ ft/cu\ yd = 0.05 \times 27\ cu\ ft/cu\ yd = 1.35\ cu\ ft/cu\ yd$

- Summary
    - Cement = 3.37 cu ft/cu yd
    - Water = 4.89 cu ft/cu yd
    - CA = 10.86 cu ft/cu yd
    - Air = 1.35 cu ft/cu yd
    - Partial volume (PV) = 20.47 cu ft/cu yd

(PV is the sum of water, cement, CA and air)

Absolute volume of FA = 27 cu ft/cu yd − PV
  27 cu ft − 20.47 cu ft/cu yd = 6.53 cu ft/cu yd

- Weight of FA = Absolute vol of FA x SGFA x 62.4 lbs/cu ft =
  6.53 cu ft/cu yd x 2.66 x 62.4 lbs/cu ft = 1,083.9 lbs/cu yd

Step 7. Determine the first trial batch. Determine the proportions for the first trial batch (for 1 cubic yard) by converting the absolute volumes to dry volumes.

$$\text{Cement type IA} = \frac{663.0 \; lb/cu\,yd}{94 \; lb\,sack} = 7.05 \; sacks$$

$$Water = \frac{305 \; lb/cu\,yd}{8.33 \; lb/gal} = 36.6 \; gals$$

$$CA = \frac{1768 \; lb/cu\,yd}{104 \; lb/cu\,yd} = 17.0 \; cu\,ft$$

$$FA = \frac{1{,}083 \; lb/cu\,yd}{103 \; lb/cu\,ft} = 10.5 \; cu\,ft$$

$$Air\ content\ = 5\%$$

Step 8. If needed, mix more trial batches to obtain the desired slump and air content while keeping the W/C ratio constant.

VARIATION IN MIXTURES

3-16. The mixture proportions will vary somewhat depending on the method used. The variation is due to the empirical nature of the methods and does not necessarily imply that one method is better than another. Since the methods begin differently and use different procedures, the final proportions will vary slightly. This is to be expected and further points out the necessity of trial mixtures in determining the final mixture proportions. For variations in a mixture, such as for concrete used in slabs or other flatwork, there are minimum cement requirements, depending on the maximum size of the aggregates (see table 3-7, page 3-14).

Table 3-7. Concrete Used in flatwork

Maximum Size of Aggregate, in Inches	Cement, in Pounds Per Cubic Yard
1 1/2	470
1	520
3/4	540
1/2	590
3/8	610
Notes. 1. 1 inch = 25 mm 2. 100 lb/cu yd = 60 kg/m3	

ADJUSTMENTS FOR MOISTURE ON AGGREGATES

3-17. The initial mix design assumes that the aggregates are SSD; that is, neither the FAs nor the CAs have any free water on the surface that would be available as mixing water. This is a laboratory condition and seldom occurs in the field. The actual amount of water on the sand and gravel can only be determined from the material at the mixing site. Furthermore, the moisture content of the aggregates will change over a short period of time; therefore, their condition must be monitored and appropriate adjustments made as required. A good field test for estimating the free surface moisture (FSM) on

3-18. FA is contained in Appendix C. CAs are free-draining and rarely hold more than 2 percent (by weight) FSM even after heavy rains. FA have a tendency to bulk (expand in volume) when wetted and when the mass is disturbed. This factor becomes very important if the concrete is being batched at a mixer by volume (the initial mix design must then be adjusted). The procedure for adjusting the mixing water and sand bulking due to FSM is as follows:

Step 1. Determine the approximate FSM of the FA by the squeeze test (appendix C).

Step 2. Estimate the FSM of the CA by observation. Usually, 2 percent FSM is the maximum amount that gravel will hold without actually dripping.

Step 3. Multiply the percentages of FSM on the aggregates by their respective weights per cubic yards. This will yield the weight of the FSM on the aggregates.

Step 4. Divide the total weight of the FSM by 8.33 pounds per gallon to determine the number of gallons of water. Subtract those gallons from the water requirements in the original mix design.

- If you are batching your concrete mix by weight, you need to account for the weight contributed by the FSM by increasing the total weights of the aggregates per cubic yard by the weights of the FSM.
- If you are batching your concrete by volume, you must increase the volume of the FA by the bulking factor (BF) determined from figure 3-3. The formula for volume increase is as follows:

$$V_{wet} = V_{dry} \times (1 + BF)$$

Where—

V = volume

BF = bulking factor (CAs do not bulk, therefore no adjustment is necessary.)

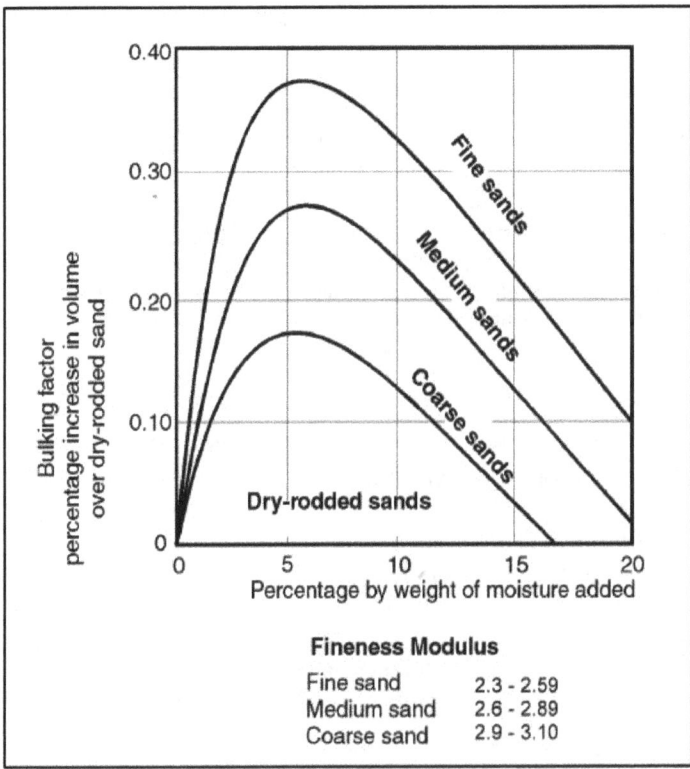

Figure 3-3. Bulking factor curves

SAMPLE PROBLEM FOR MOISTURE ADJUSTMENT ON AGGREGATES

3-19. Using the final mix proportions as determined in paragraph 3-14, adjust the design mix to account for 6 percent FSM on the FA (fineness modulus equals 2.7) and 2 percent FSM on the CA. The original mix design was—

- Cement = 7.05 sacks (Type IA)
- Water = 36.6 gal
- CA = 17.0 cu ft or 1,768 lb/cu yd
- FA = 10.5 cu ft or 1,083 lb/cu yd
- Air content = 5%

Step 1. Determine the amount of water (in gallons) on the aggregate.

CA = 1,768 lb/cu yd x 0.02 = 35.38 lb/cu yd of water

FA = 1,083 lb/cu yd x 0.06 = 64.98 lb/cu yd of water

Total weight of water = 100.36 lb/cu yd

Convert to gallons:

$$\frac{100.36\ lb/cu\ yd}{8.33\ lb/gal} = 12\ gal/cu\ yd$$

Step 2. Reduce the original amount of mixing water by the amount contributed by the aggregates as determined in step 1. Therefore, 36.6 gallons minus 12 gallons equals 24.6 gallons of water that must be added to the mix.

Step 3. Adjust the weights of the aggregates by the amount contributed by the water.

CA = 1,769 lb/cu yd + 35.38 lb/cu yd = 1,803.3 lb/cu yd

FA = 1,083 lb/cu yd + 64.98 lb/cu yd = 1,148 lb/cu yd

Step 4. Adjust the volume of the FA to reflect the BF. From figure 3-3, page 3-15, the fineness modulus equals 2.70. The FA is considered a medium sand. Select the appropriate moisture content across the bottom of the figure, read up to the appropriate sand curve and then read the correct BF on the left edge. For this example, FSM equals 6 percent; read up the figure until it intersects the medium sand line, then read left to the BF of 0.28.

The increase in FA volume is then—

$$V_{wet} = V_{dry}\ x\ (1 + BF) = 10.5\ cu\ ft\ x\ (1 + .28) = 13.44\ cu\ ft$$

Step 5. The adjusted mix design to account for the actual field conditions is now—
- Cement = 7.05 sacks (Type IA)
- Water = 24.6 gal
- CA = 17.0 cu ft or 1,803.3 lb/cu yd
- FA = 13.44 cu ft or 1,148 lb/cu yd
- Air content = 5%

3-20. It is important to check the moisture content of the aggregates and make appropriate adjustments as conditions change, such as after rains, after periods of dryness, and after the arrival of new materials. This quality-control step assures that the desired concrete is produced throughout the construction phase.

MATERIALS ESTIMATION

3-21. After designing the mix, use the following steps to estimate the total amounts of materials needed for the job:

Step 1. Determine the total volume (in cubic yards) of concrete to be placed.

Step 2. Determine the extra amount that will be added for waste. If the total volume is 200 cubic yards or less, add 10 percent. If the total volume is greater than 200 cubic yards, add 5 percent.

Step 3. Determine the total amounts of cement, FA, and CA, by multiplying the amounts of these components needed for 1 cubic yard by the adjusted total volume. Then convert the answer to tons and round up to the next whole number.

Step 4. Determine the required amount of water needed for the job. Water is required on concrete projects not only for mixing but for wetting the forms, tools, and curing the concrete.

3-22. A planning factor of 8 gallons of water for each sack of cement is usually sufficient; however, not all of this water will be used for the concrete.

SAMPLE PROBLEM FOR MATERIALS NEEDED FOR MIX DESIGN

3-23. Using the mix design, determine the total amount of materials needed to construct the retaining wall shown in figure 3-4. The 1 cubic yard mix design is recapped below.
- Cement = 7.05 sacks (Type IA)

Proportioning Concrete Mixtures

- Water = 36.6 gal
- CA = 17.0 cu ft or 1,768 lb/cu yd
- FA = 10.5 cu ft or 1,083 lb/cu yd
- Air content = 5%

Step 1. Determine the total volume of concrete required. An easy way to do this is to break the project up into simple geometric shapes. Divide the retaining wall into two sections, the wall portion and the footing. A close examination of figure 3-4 shows the wall cross-section is a trapezoid that is 14 feet, 9 inches high, 8 inches wide on one end, and 1 foot wide on the other.

Wall Volume:

$$14.75\ ft \times \frac{\frac{8\ in + 12\ in}{2\ in}}{12\ in} \times 75\ ft =$$

$$14.75\ ft \times \frac{10\ in}{12\ in} \times 75\ ft = 921.9\ cu\ ft$$

Footing volume: 1.25 ft × 8.5 ft × 75 ft = 796.9 cu ft

Total volume:

$$\frac{(921.9 + 796.9)}{27\ cu\ ft/cu\ yd} = 63.7\ cu\ yd$$

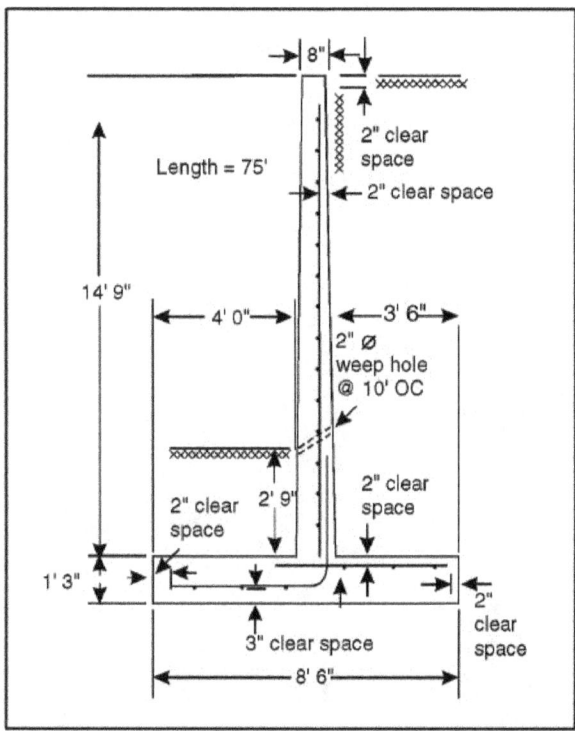

Figure 3-4. Retaining wall

Chapter 3

Step 2. Calculate the waste factor (since the volume is less than 200 cubic yards, the waste factor is 10 percent).

Total volume + waste = 63.7 cu yd x 1.10 = 70.1 cu yd

Step 3. Determine the amounts of cement and aggregates needed.

Cement 7.05 sacks/cu yd x 70.1 cu yd = 494.2 sacks = 495 sacks

Note. Round values up to the next whole number since you cannot order partial sacks.

CA 1,768 lb/cu yd x 70.1 cu yd = 123,936.8 lb or 62 tons

FA 1,083 lb/cu yd x 70.1 cu yd = 75,918.3 lb or 38 tons

Step 4. Determine the amount of water required for cleanup and mixing.

Water required = 495 sacks x 8 gallons per sack = 3,960 gallons

SUMMARY

3-24. Summary of the amounts of materials to be ordered for the project are as follows:
- Cement = 495 sacks
- Water = 3,960 gallons
- CA = 62 tons
- FA = 38 tons

Chapter 4

Form Design and Construction

An inherent property of concrete is that it can be transformed into any shape. The wet mixture is placed in forms constructed of wood, metal or other suitable material to harden or set. The form must be assembled with quality workmanship, holding to close dimensional tolerances. Formwork should be strong enough to support the concrete's weight and rigid enough to maintain its position and appearance. In addition, formwork must be tight enough to prevent the water seepage and designed to permit ready removal. Because the formwork for a concrete structure constitutes a considerable item in the cost of the completed structure, particular care should be exercised in its design. It is desirable to maintain a repetition of identical units so that the forms may be removed and reused at other locations with a minimum amount of labor.

SECTION I - PRINCIPLES

IMPORTANCE OF FORM DESIGN

4-1. Formwork holds concrete until it sets, produces the desired shapes, and develops a desired surface finish. Forms also protect concrete, aid in curing, and support any reinforcing bars or conduit embedded within the concrete. Because formwork can represent up to one-third of a concrete structure's total cost, this phase of a project is very important. The nature of the structure, availability of equipment and form materials, anticipated reuse of the forms, and familiarity with construction methods all influence the formwork design. To design forms, you must know the strength of the forming materials and the loads they support. You must also consider the concrete's final shape, dimensions, and surface finish.

FORM CHARACTERISTICS

4-2. Forms must be tight, rigid, and strong. Loose forms permit either loss of cement, resulting in honeycomb, or loss of water causing sand to streak. Brace forms enough to align them and to strengthen them enough to hold the concrete. Take special care in bracing and tying down forms used for such configurations as retaining walls that are wide at the bottom and taper toward the top. The concrete in this and other types of construction, such as the first pour for walls and columns, tends to lift the form above its proper elevation. To reuse forms, make them easy to remove and replace without damage.

FORM MATERIALS

4-3. Forms are generally made from four materials—wood, metal, earth, and fiber.

WOOD

4-4. Wood materials are the most common form materials by far, because they are economical, easy to produce and handle, and can adapt to many shapes. For added economy, you can reuse form lumber for roofing, bracing, and similar purposes. Soft wood, such as pine, fir, and spruce, makes the best and most economical form lumber because it is light, easy to work with, and available almost everywhere. Form lumber must be straight, structurally sound, strong, and only partially seasoned. Kiln-dried lumber tends to swell when soaked with water from the concrete. If the form is tightly jointed, the swelling can cause bulging and distortion. When using green lumber, either allow for shrinkage or keep the forms wet until the concrete is in place. Lumber that contacts concrete should be finished at least on one side and on both edges. The edges can be square, shiplap, or tongue and groove. The last finish makes a more watertight joint and helps prevent warping. Use plywood identified for use in concrete forms for wall and floor forms if it is made with waterproof glue. Plywood is more warp-resistant and can be reused more often than lumber. It is available in thickness of 1/4, 3/8, 9/16, 5/8, and 3/4 inch, and in widths up to 48 inches. The 5/8 and 3/4 inch thicknesses are more economical because the thinner sections require solid backing to prevent deflection. However, the 1/4 inch thickness is useful for curved surfaces. Although longer lengths are available, 8-foot plywood is the most common.

METAL

4-5. Metal materials are used for added strength or when a construction will be duplicated at more than one location. They are initially more expensive than wood forms but are more economical in the long run if reused often enough. Steel forms are common for large, unbroken surfaces such as retaining walls, tunnels, pavements, curbs, and side-walks. They are especially useful for sidewalks, curbs, pavements, and precast-concrete operations because you can use them many times.

EARTH

4-6. Earth materials are acceptable in subsurface construction if the soil is stable enough to retain the desired concrete shape. The advantages of earth forms are that they require less wood construction and resist settling better than other types of forms. Their obvious disadvantages are rough surface finish and the requirement for more excavation. Therefore, earth forms are generally restricted to footings and foundations.

FIBER

4-7. Fiber forms are prefabricated and are filled with waterproofed cardboard and other fiber materials. Successive layers of fiber are first glued together and then molded to the desired shape. Fiber forms are ideal for round concrete columns and other applications where preformed shapes are feasible because they require no form fabrication at the job site and thus save considerable time.

FORMING

4-8. Forms are assembled and used to make walls. Figure 4-1 shows the basic parts of a wood form for panel walls.

Form Design and Construction

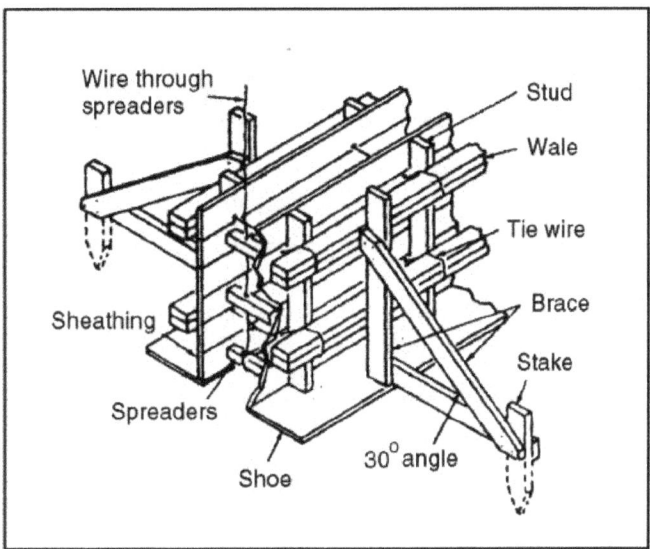

Figure 4-1. Wood form for a concrete panel wall

SHEATHING

4-9. Sheathing forms the vertical surfaces of a concrete wall but runs horizontally itself. The sheathing edges and the side that contacts the concrete should be finished as smooth as possible, especially for an exposed final concrete surface. The sheathing must also be watertight. Sheathing made from tongue-and-groove lumber gives the smoothest and most watertight concrete surface. Plywood or fiber-based hardboard can also be used.

STUDS

4-10. The weight of the plastic concrete will cause the sheathing to bulge if it's not reinforced by studs. Therefore, vertical studs add rigidity to the wall form. The studs are made from single 2 by 4s or 2 by 6s.

WALES (WALERS)

4-11. Wales reinforce the studs when they extend upward more than 4 or 5 feet. They should be made from doubled 2 by 4s or 2 by 6s, and are lapped at the form corners to add rigidity. Double wales not only reinforce the studs but also tie prefabricated panels together and keep them aligned.

BRACES

4-12. Although braces are neither part of the form design nor considered as providing any additional strength, they help stabilize the form. The most common brace is a combination of a diagonal member and a horizontal member nailed to a stake at one end and to a stud or wale at the other end. The diagonal member makes a 20- to 60-degree angle with the horizontal member. To add more bracing, place vertical members (strongbacks) behind the wales or in the angle formed by intersecting wales.

SPREADERS

4-13. Spreaders are small pieces of wood placed between the sheathing panels to maintain the proper wall thickness between them. They are cut to the same length as the wall thickness. Spreaders can be removed easily before the concrete hardens since friction, not fasteners, hold them in place. Attach a wire securely

Chapter 4

through the spreaders, as shown in figure 4-1, to pull them out when the fresh concrete exerts enough pressure against the sheathing to permit removal.

TIE WIRES

4-14. Tie wires secure the formwork against the lateral pressure of the plastic concrete. They always have double strands.

TIE-RODS (SNAP TIES)

4-15. Tie-rods sometimes replaces tie wires in the same function because working with rods are easier.

COLUMN FORMS

4-16. Figure 4-2 shows the basic parts of a wood form for a concrete column.

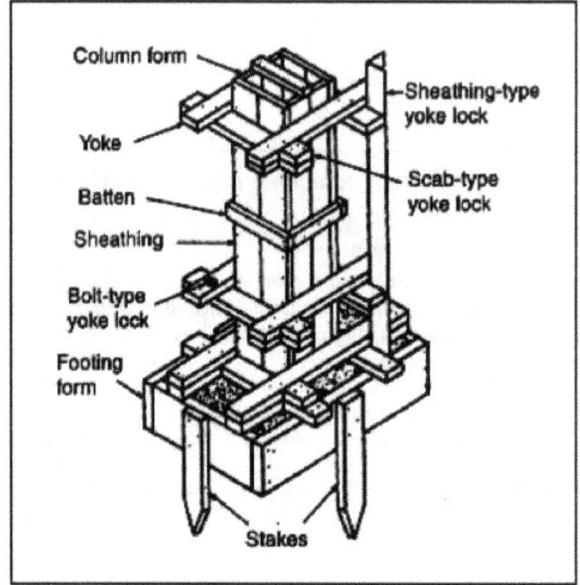

Figure 4-2. Form for a concrete column and footing

SHEATHING

4-17. Sheathing runs vertically in column forms to reduce the number of saw cuts. Nail the corner joints firmly to ensure watertightness.

YOKES

4-18. A *yoke* is a horizontal reinforcement in the form of a rectangle that wraps around a column to prevent the plastic concrete from distorting the form. It serves the same purpose as a stud in a wall form. You can lock yokes in place using the sheathing-, scab-, or bolt-type yoke lock. The small horizontal dimensions of a column do not require vertical reinforcement.

BATTENS

4-19. Battens are narrow strips of boards that are placed directly over the joints to fasten several pieces of vertical sheathing together.

FOOTING FORMS

4-20. Footing forms are shown in figure 4-2 and are discussed in paragraph 4-56.

SECTION II - DESIGN

DESIGN CONSIDERATIONS

4-21. Forms are the molds that hold the plastic concrete and support it until it hardens. They must keep deflections within acceptable limits, meet dimensional tolerances, prevent paste leakage, and produce a final product that meet appearance needs. Therefore, strength, rigidity, and watertightness are the most important form design considerations. In addition, forms must support all weights and stresses to which they are subject, including the dead load (DL) of the forms; the weight of people, equipment, and materials that transfers to the forms; and any impact due to vibration. Although these factors vary with each project, do not neglect any of them when designing a form. Ease of erection and removal are other important factors in economical form design. Sometimes you may want to consider using platforms and ramp structures that are independent of the formwork because they prevent form displacement due to loading, as well as impact shock from workers and equipment.

BASIS OF FORM DESIGN

4-22. Concrete is in a plastic state when placed in the designed form; therefore, it exerts hydrostatic pressure on the form. The basis of form design is to offset the maximum pressure developed by the concrete during placing. The pressure depends on the rate of placing and the ambient temperature. The rate of placing affects pressure because it determines how much hydrostatic head builds up in the form. The hydrostatic head continues to increase until the concrete takes its initial set, usually in about 90 minutes. However, because the initial set takes more time at low ambient temperatures, consider the ambient temperature at the time of placement. Knowing these two factors (rate of placement and ambient temperature) plus the specified type of form material, calculate a tentative design.

PANEL-WALL FORM DESIGN

4-23. It is best to design forms following a step-by-step procedure. Use the following steps to design a wood form for a concrete wall.

Step 1. Determine the materials needed for sheathing, studs, wales, braces, and tie wires or tie-rods.

Step 2. Determine the mixer output by dividing the mixer yield by the batch time. Batch time includes loading all ingredients, mixing, and unloading. If you will use more than one mixer, multiply the mixer output by the number of mixers.

$$Mixer\ output\ (cu\ ft/hr) = \frac{mixer\ yield\ (cu\ ft)}{batch\ time\ (min)} \times \frac{60 min}{hr}$$

Step 3. Determine the area enclosed by the form.

$$Plan\ area\ (sq\ ft) = Length\ (L)\ x\ Width\ (W)$$

Step 4. Determine the rate (vertical feet per hour) of placing (R) the concrete in the form by dividing the mixer output by the plan area.

Chapter 4

$$Rate\ of\ placing\ (R)(ft/hr) = mixer\ output \frac{cu\ ft/hr}{plan\ area\ (sq\ ft)}$$

Note. If using wood for an economical design, try to keep R ≤ 5 feet per hour.

Step 5. Estimate the placing temperature of the concrete.

Step 6. Determine the maximum concrete pressure (MCP) using the rate of placing (see figure 4-3 or the formula above the figure). Draw a vertical line from the rate of placing until it intersects the correct concrete-temperature line. Read left horizontally from the point of intersection to the left margin of the graph and determine the MCP in 100 pounds per square foot.

Step 7. Find the maximum stud spacing (MSS) in inches. Use table 4-1, page 4-8, for board sheathing or table 4-2, page 4-9, for plywood sheathing. Refer to the column headed maximum concrete pressure (MCP) and find the value you have for the MCP. If the value falls between two values in the column, round up to the nearest given value. Move right to the column identified by the sheathing thickness. (Always use the strong way for plywood when possible). This number is the MSS in inches.

Step 8. Determine the uniform load on a stud (ULS) by multiplying the MCP by the stud spacing divided by 12 inches per foot.

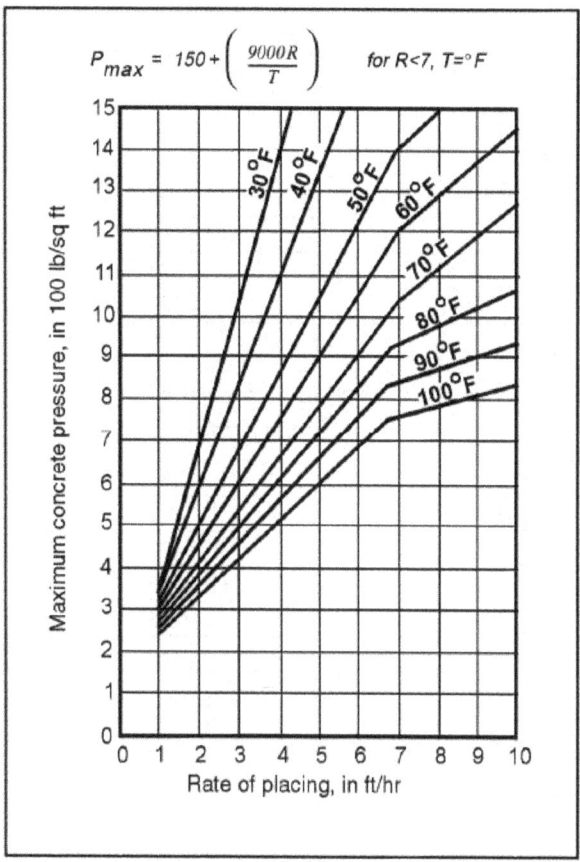

Figure 4-3. MCP graph

$$ULS\ (lb/lin\ ft) = \frac{MCP \times MSS}{12}$$

Where—

MCP = *maximum concrete pressure, in pounds per square foot*

MSS = *maximum stud spacing*

ULS = *uniform load stud, in pounds per linear foot*

Step 9. Determine the maximum wale spacing (MWS) using table 4-3, page 4-10. Refer to the column headed uniform load and find the value for the ULS. If the value falls between two values in the column, round it up to the nearest given value. Move right to the column identified by the size of the stud you are using. This number is the MWS in inches.

Table 4-1. Maximum stud (joist) spacing for board sheathing

Maximum Concrete Pressure, in Pounds per Square Foot	Nominal Thickness of S4S Boards, in Inches			
	1	1 1/4	1 1/2	2
75	30	37	44	50
100	28	34	41	47
125	26	33	39	44
150	25	31	37	42
175	24	30	35	41
200	23	29	34	39
300	21	26	31	35
400	18	24	29	33
500	16	22	27	31
600	15	20	25	30
700	14	18	23	28
800	13	17	22	26
900	12	16	20	24
1,000	12	15	19	23
1,100	11	15	18	22
1,200	11	14	18	21
1,400	10	13	16	20
1,600	9	12	15	18
1,800	9	12	14	17
2,000	8	11	14	16
2,200	8	10	13	16
2,400	7	10	12	15
2,600	7	10	12	14
2,800	7	9	12	14
3,000	7	9	11	13

Table 4-2. Maximum stud (joist) spacing for plywood sheathing, in inches

Maximum Concrete Pressure, in Pounds per Square Foot	Strong Way - 5-Ply Sanded Face, Grain Perpendicular to the Stud				Weak Way - 5-Ply Sanded Face, Grain Parallel to the Stud			
	1/2	5/8	3/4	1 (7 ply)	1/2	5/8	3/4	1 (7 ply)
75	20	24	26	31	13	18	23	30
100	18	22	24	29	12	17	22	28
125	17	20	23	28	11	15	20	27
150	16	19	22	27	11	15	19	25
175	15	18	21	26	10	14	18	24
200	15	17	20	25	10	13	17	24
300	13	15	17	22	8	12	15	21
400	12	14	16	20	8	11	14	19
500	11	13	15	19	7	10	13	18
600	10	12	14	17	6	9	12	17
700	10	11	13	16	6	9	11	16
800	9	10	12	15	5	8	11	15
900	9	10	11	14	4	8	9	15
1,000	8	9	10	13	4	7	9	14
1,100	7	9	10	12	4	6	8	12
1,200	7	8	10	11	—	6	7	11
1,300	6	8	9	11	—	5	7	11
1,400	6	7	9	10	—	5	6	10
1,500	5	7	9	9	—	5	6	9
1,600	5	6	8	9	—	4	5	9
1,700	5	6	8	8	—	4	5	8
1,800	4	6	8	8	—	4	5	8
1,900	4	5	8	7	—	4	4	7
2,000	4	5	7	7	—	—	4	7
2,200	4	5	6	6			4	6
2,400	—	4	5	6			4	6
2,600	—	4	5	5			—	5
2,800	—	4	4	5			—	5
3,000	—	-	4	5			—	5

Table 4-3. Maximum spacing, in inches, for wales, ties, stringers, and 4"x 4" or larger shores where member to be supported is a single member

Uniform Load, in Pounds per Linear Foot	Supported Members Size (S4S)*				
	2 x 4	2 x 6	3 x 6	4 x 4	4 x 6
100	60	95	120	92	131
125	54	85	440	80	124
150	49	77	100	75	118
175	45	72	93	70	110
200	42	67	87	65	102
225	40	62	82	61	97
250	38	60	77	58	92
275	36	57	74	55	87
300	35	55	71	53	84
350	32	50	65	49	77
400	30	47	61	46	72
450	28	44	58	43	68
500	27	41	55	41	65
600	24	38	50	37	59
700	22	36	46	35	55
800	21	33	43	32	51
900	20	31	41	30	48
1,000	19	30	38	29	46
1,200	17	27	35	27	42
1,400	16	25	33	25	39
1,600	15	23	31	23	36
1,800	14	22	29	22	34
2,000	13	21	27	21	32
2,200	13	20	26	20	31
2,400	12	19	25	19	30
2,600	12	19	24	18	28
2,800	11	18	23	17	27
3,000	11	17	22	17	26
3,400	10	16	21	16	25
3,800	10	15	20	15	23
4,500	9	14	18	13	21
*S4S indicates surfaced four sides.					

Form Design and Construction

Step 10. Determine the uniform load on a wale (ULW) by multiplying the MCP by the MWS and dividing by 12 inches per foot.

$$ULW = \frac{MCP \times MWS}{12 \ in/ft}$$

Where—

MCP = maximum concrete pressure, in pounds per square foot

MWS = maximum wale spacing, in inches, divided by 12 inches per foot

ULW = uniform load on wale, in pounds per linear foot

Step 11. Determine the tie spacing based on the ULW using table 4-3 or table 4-4, page 4-12 (depending on type of wale, either single or double). Refer to the column headed Uniform Load and find the value for the ULW. If the value falls between two values in the column, round it up to the nearest given value. Move right to the column identified by the size of wale you are using. This number is the maximum tie spacing, in inches, based on wale size.

Step 12. Determine the tie spacing based on the tie strength by dividing the tie breaking strength by the ULW. If the breaking strength of the tie is unknown, use table 4-5, page 4-13, to find the breaking loads for a double-strand wire and tie-rods (found in the Army supply system).

$$Tie \ wire/tie\text{-}rod \ spacing = \frac{tie \ wire \ or \ tie-rod \ strength \ (lb) \times 12 \ in \ ft}{ULW}$$

Where—

Tie wire or tie-rod = strength = average breaking load (see table 4-5, page 4-12)

ULW = uniform load on wales, in pounds per square foot

Note. If the result does not equal a whole number, round the value down to the next whole number.

Step 13. Select the smaller of the tie spacing as determined in steps 11 and 12.

Step 14. Install tie wires at the intersection of studs and wales. Reduce the stud spacing (step 7) or the tie spacing (step 13) to conform with this requirement. tie-rods may be placed along the wales at the spacing (determined in step 13) without adjusting the studs. Place the first tie at one-half the maximum tie spacing from the end of the wale.

Step 15. Determine the number of studs on one side of a form by dividing the form length by the MSS. Add one to this number and round up to the next integer. The first and last studs must be placed at the ends of the form, even though the spacing between the last two studs may be less than the maximum allowable spacing.

$$Number \ of \ studs = \frac{length \ of \ form \times 12 \ in \ ft}{stud \ spacing} + 1$$

Note. The length of the form is in feet. The stud spacing is in inches.

Table 4-4. Maximum spacing, in inches, for ties and 4" x 4" or larger shores where member to be supported is a double member

Uniform Load, in Pounds per Linear Foot	Supporting Member Size (S4S)				
	2 x 4	2 x 6	3 x 6	4 x 4	4 x 6
100	85	126	143	222	156
125	76	119	135	105	147
150	70	110	129	100	141
175	64	102	124	96	135
200	60	95	120	92	131
225	57	89	116	87	127
250	54	85	109	82	124
275	51	84	104	78	121
300	49	77	100	75	118
350	46	72	93	70	110
400	43	67	87	65	102
450	40	63	82	61	97
500	38	60	77	58	92
600	35	55	71	53	81
700	32	51	65	49	77
800	30	47	61	46	72
900	28	44	58	43	68
1,000	27	43	55	41	65
1,200	25	39	50	38	59
1,400	23	36	46	35	55
1,600	21	34	43	33	51
1,800	20	32	41	31	48
2,000	19	30	39	29	46
2,200	18	29	37	28	44
2,400	17	27	36	27	42
2,600	17	26	34	26	40
2,800	16	25	33	25	39
3,000	15	24	32	24	38
3,400	14	23	30	22	35
3,800	14	21	28	21	33
4,500	12	20	25	19	30

Form Design and Construction

Table 4-5. Average breaking load of tie material, in pounds

Steel Wire	
Size of Wire Gage Number	Minimum Breaking Load Double Strand, in Pounds
8	1,700
9	1,420
10	1,170
11	930
Barbwire	
Size of Wire Gage Number	Minimum Breaking Load
12 ½	950
13*	660
13 ½	950
14	650
15 ½	850
Tie-rod	
Description	Minimum Breaking Load, in Pounds
Snap ties	3,000
Pencil roads	3,000
*Single-strand barbwire	

Step 16. Determine the number of wales for one side of a form by dividing the form height by the MWS, and round up to the next integer. Place the first wale one-half of the maximum space up from the bottom and place the remaining wales at the MWS.

Step 17. Determine the time required to place the concrete by dividing the height of the form by the rate of placing.

SAMPLE FORM DESIGN PROBLEM NUMBER ONE

4-24. Design a form for a concrete wall 40 feet long, 2 feet thick, and 10 feet high. An M919 concrete mobile mixer is available and the crew can produce and place a cubic yard of concrete every 10 minutes. The concrete placing temperature is estimated at 70° F. The form materials are 2 by 4s, 1-inch board sheathing, and number 9 black annealed wire. The solution steps are as follows:

Step 1. Lay out available materials.

Studs: 2 x 4 (single)

Wales: 2 x 4 (double)

Sheathing: 1-inch board

Ties: Number 9 wire

Step 2. Calculate the production rate.

$$\frac{27 \; cuft}{10 \; min} x \frac{60 \; min}{hr} = 162 \; cu \; ft/hr$$

Step 3. Calculate the plan area of the form.

$$40 \; ft \; x \; 2 \; ft = 80 \; sq \; ft$$

Chapter 4

Step 4. Calculate the rate of placing.
$$\frac{162 \ cu \ ft/hr}{80 \ sq \ ft} = 2.025 \ ft/hr$$

Step 5. Determine the concrete-placing temperature.

Temperature = 70° F

Step 6. Determine the MCP (refer to figure 4-3, page 4-6).
$$MCP = 400 \ pounds \ per \ square \ foot$$

Step 7. Determine the MSS (refer to table 4-1, page 4-7).
$$MSS = 18 \ inches$$

Step 8. Calculate the ULS.
$$400 \ lb/sq \ ft \times \frac{18 in}{12 \ in/ft} = 600 \ lb/lin \ ft$$

Step 9. Determine the MWS (refer to table 4-3, page 4-9).
$$MWS = 24 \ inches$$

Step 10. Calculate the ULW.
$$ULS = 400 \ lb/sq \ ft \times \frac{24 \ in}{12 \ in/ft} = 800 \ lb/lin \ ft$$

Step 11. Determine the tie-wire spacing based on wale size (refer to table 4-4, page 4-11).
$$Tie-wire \ spacing = 30 \ inches.$$

Step 12. Determine the tie-wire spacing based on wire strength (referring to table 4-5, page 4-12, breaking strength of number 9 wire = 1,420 pounds).
$$\frac{1,420 \ lb \times 12 \ in/ft}{800 \ lb/lin \ ft} = 21.3 \ in \ (use \ 21 \ inches)$$

Step 13. Determine the maximum tie spacing.
$$Tie \ spacing = 21 \ inches.$$

Step 14. Reduce the tie spacing to 18 inches since the maximum tie spacing is greater than MSS; tie at the intersection of each stud and double wale.

Step 15. Calculate number of studs per side.
$$\frac{40 \ ft \times 12 \ in/ft}{18 \ in} + 1 = 27.7 \ (use \ 28 \ studs)$$

Step 16. Calculate the number of double wales per side.
$$10 \ ft \times \frac{12 \ in/ft}{24 \ in} = 5 \ wales$$

Step 17. Estimate the time required to place concrete.
$$\frac{10 \ ft}{2.05 \ ft/hr} = 4.93 \ (use \ 5 \ hours)$$

SAMPLE FORM DESIGN PROBLEM NUMBER TWO

4-25. Design the form for a concrete wall 40 feet long, 2 feet thick, and 10 feet high. An M919 concrete mobile mixer is available and the crew can produce and place a cubic yard of concrete every 7 minutes.

The concrete placing temperature is estimated to be 70° F. The materials for constructing the form are 2 by 4s, 3/4-inch thick plywood, and 3,000-pound (breaking strength) snap ties. The following steps will be used:

Step 1. Lay out available needed materials.

Studs: 2 x 4 (single)

Wales: 2 x 4 (doubled)

Sheathing: 3/4-inch plywood (strong)

Ties: Snap ties (3,000 pounds)

Step 2. Calculate the production rate.

$$\frac{27\ cu\ ft}{7\ min} x = \frac{60 min}{hr} = 231.4\ cu\ ft/hr$$

Step 3. Calculate the plan area of form.

$$40\ feet\ x\ 2\ feet = 80\ square\ feet$$

Step 4. Calculate rate the of placing.

$$\frac{231.4\ cu\ ft/hr}{80} = 2.9\ ft/hr$$

Step 5. Determine the concrete placing temperature.

$$Temperature = 70°\ F$$

Step 6. Determine the MCP (refer to Figure 4-3, page 4-6).

$$MCP = 500\ pounds\ per\ square\ foot$$

Step 7. Determine the MSS (refer to table 4-2, page 4-8).

$$MSS = 15\ inches$$

Step 8. Determine the ULS.

$$ULS = 500\ lb/sq\ ft\ x\ \frac{15\ in}{12\ in/ft} = 625\ lb/ft$$

Step 9. Determine the MWS (refer to table 4-3, page 4-10, using 700 pounds per foot load).

$$MWS = 22\ inches$$

Step 10. Determine the ULW.

$$ULW = 50\ lb/sq\ ft\ x\ \frac{22}{12} = 916.6\ lb/ft$$

Step 11. Determine the tie spacing based on wale size (refer to table 4-4, page 4-11, using 1,000 pounds per load).

$$Tie\ spacing = 27\ inches$$

Step 12. Determine the tie spacing based on tie-rod strength (refer to table 4-5, page 4-12).

$$\frac{3,000\ lbs\ x\ 12 in/ft}{917\ lb/ft} = 39.25\ (use\ 39\ inches)$$

Step 13. Determine the maximum tie spacing.

$$Tie\ spacing = 27\ inches$$

Chapter 4

Step 14. Insert the first tie-rod one-half the spacing from the end and one full spacing thereafter. Because tie-rods are being used, it is not necessary to adjust the tie/stud spacing.

Step 15. Calculate the number of studs per side.

$$\frac{40\ ft \times 12\ in/ft}{15\ in} + 1 = 33\ studs$$

Step 16. Calculate the number of double wales per side.

$$\frac{10\ ft \times 12\ in/ft}{22\ in} = 5.45\ (use\ 6\ wales)$$

Step 17. Estimate the time required to place concrete.

$$\frac{10\ ft}{2.9\ ft\ hr} = 3.45\ hr\ (use\ 4\ hours)$$

BRACING FOR WALL FORMS

4-26. Braces are used against wall forms to keep them in place and in alignment and to protect them from mishaps due to external forces (wind, personnel, equipment, vibration, accidents). An equivalent force due to all of these forces (the resultant force) is assumed to be acting uniformly along the top edge of the form in a horizontal plane. For most military applications, this force is assumed to be 12.5 times the wall height, in feet. As this force can act in both directions, braces to be used should be equally strong in tension as in compression, or braces should be used on both sides of the wall forms. The design procedure is based on using a single row of braces and assumes that strong, straight, seasoned lumber will be used and that the braces are properly secured against the wall forms and the ground (both ends are secured). Once we know the height of the wall to be built and have selected a material (2 inches or greater) for the braces, we need to determine the maximum safe spacing of these braces (center-to-center) that will keep the formwork aligned.

NOMENCLATURE

4-27. The following terms are used in bracing for a form wall. See figure 4-4.

- L_B = total length, in feet, of the brace member from end connection to end connection.
- L_{max} = the maximum allowable unsupported length of the brace, in feet, due to buckling and bending.
- For all 2-inch material, L_{max} = 6 1/4 feet; for all 4-inch material, L_{max} = 14 1/2 feet.
- L = the actual unsupported length, in feet, of the brace used.
- h = the overall height, in feet, of the wall form.
- y = the point of application of the brace on the wall form, measured in feet from the base of the form.
- θ = the angle, in degrees, that the brace makes with the horizontal shoe brace. For the best effect, the angle should be between 20 and 60 degrees.
- J = a factor to be applied which includes all constant values (material properties and assumed wind force). It is measured in feet[4], see table 4-6.
- S_{max} = the maximum safe spacing of braces (feet), center-to-center, to support the walls against external forces.

$$S_{max} = \frac{Jy}{h^2 L^2} \times \cos \theta$$

- cos = cosine: the ratio of the distance from the stake to the wall (x) divided by the length of the brace (L_B.)
- sin = sine: the ratio of the height of the brace (y) divided by the length of the brace (L_B.)

Form Design and Construction

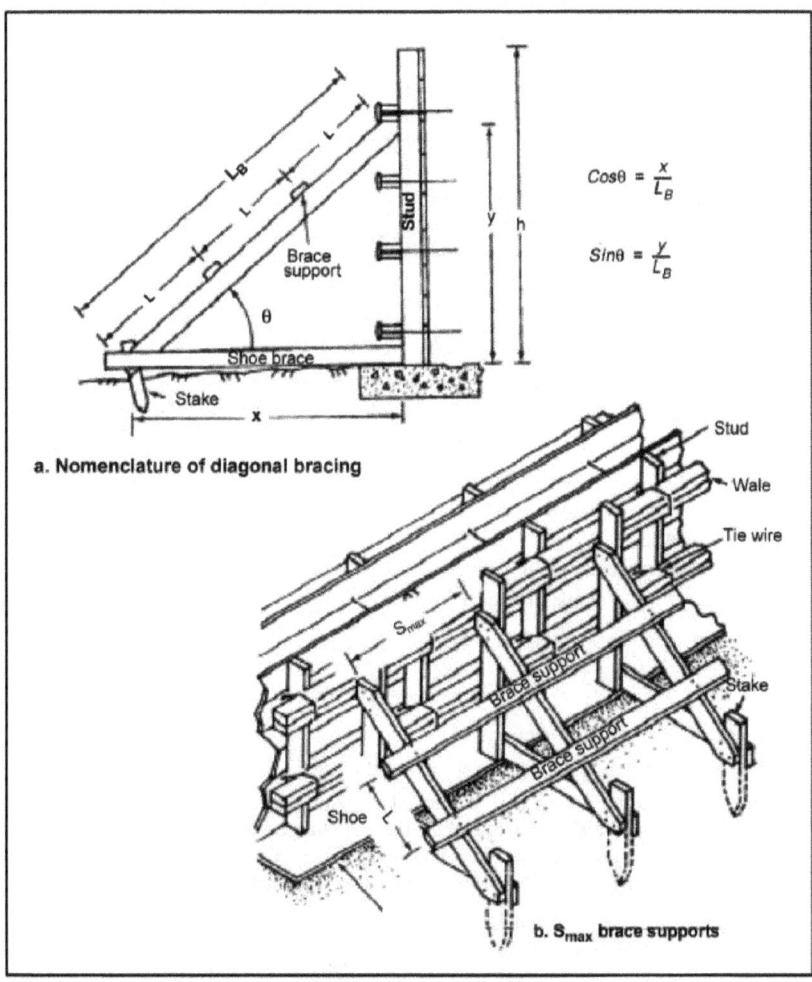

Figure 4-4. Elements of diagonal bracing

Table 4-6. J factor

Material, in Inches	J, in Feet[4]
2 x 4	2,360
2 x 6	3,710
2 x 8	4,890
2 x 10	6,240

Chapter 4

Table 4-6. J factor

Material, in Inches	J, in Feet^4
2 x 12	7,590
4 x 4	30,010

PROCEDURE

4-28. The design procedure can best be explained by the example problem that follows:

Step 1. Determine the spacing of braces for a wall 10 feet high. Use 2-by 6- inch by 10-foot material, attached 6 feet from the bottom of the form. The following selected materials are given:

- $J = 3,710$ feet^4 (from table 4-6)
- $L_{max} = 6\ 1/4$ feet (because of the 2-inch material)
- $L_B = 10$ feet
- $h = 10$ feet (from example problem)
- $y = 6$ feet (from example problem)

Step 2. Determine the angle of placement, θ

$$\sin\theta = \frac{y}{L_B} = \frac{6}{10} = .600$$

$$\theta = \sin^{-1}(.600) = 37°$$

Step 3. Determine the L which is the actual unsupported length of brace. Since the L_{max} for all 2-inch material is 6 1/4 feet and the brace in this problem is 10 feet long, we will have to use something to support the braces (usually a 1 by 4 or a 1 by 6). The best position for this support would be in the middle of the brace, thus giving $L = 5$ feet.

Step 4. Determine the S_{max} from the formula.

$$S_{max} = \frac{Jy \times \cos\theta}{h^2 L^2}$$

$$S_{max} = \frac{3,710\ ft^4 \times 6\ ft \times \cos 37°}{100\ sq\ ft \times 25\ sq\ ft} = 7.13\ feet;\ say\ 7\ feet\ (round\ down)$$

4-29. Using 2- by 6- inches by 10-feet brace applied to the top edge of the wall form at $y = 6$ feet, place these braces no further apart than 7 feet. After the braces are properly installed, connect all braces to each other at the center so deflection does not occur.

Note. This procedure determines the maximum safe spacing of braces. There is no doctrine that states the braces must be placed 7 feet apart—they can be less.

DISCUSSION

4-30. To fully understand the procedure, the following points lend insight to the formula.

$$S_{max} = \frac{Jy \times \cos \theta}{h^2 L^2}$$

- Derivation of the formula has a safety factor of 3.
- For older or green lumber, reduce S_{max} according to judgment.
- For maximum support, attach braces to the top edge of the forms (or as closely as practicable). Also, better support will be achieved when $\theta = 45°$
- Remember to use intermediate supports whenever the length of the brace (L_B) is greater than L_{max}.
- Whenever there are choices of material, the larger size will always carry greater loads.
- To prevent overloading the brace, place supports a minimum of 2 feet apart for all 2-inch material and 5 feet apart for all 4-inch material. This is necessary to prevent crushing of the brace.

OVERHEAD SLAB FORM DESIGN

4-31. There may be instances where a concrete slab will have to be placed above the ground such as bunker and culvert roofs. Carefully consider the design of the formwork because of the danger of failure caused by the weight of plastic concrete and the live load (LL) of equipment and personnel on the forms. The overhead slab form design method employs some of the same figures used in the wall-form design procedure.

NOMENCLATURE

4-32. The following terms are used in the form design (see figure 4-5, page 4-20).

- Sheathing. Shapes and holds the concrete. Plywood or solid sheet metal is best for use.
- Joists. Supports the sheathing against deflection. Performs the same function as studs in a wall form. Use 2-, 3-, or 4-inch thick lumber.
- Stringers. Supports the joists against deflection. Performs the same function as wales in a wall form. Use 2-inch thick or larger lumber. Stringers do not have to be doubled as wales are.
- Shores. Supports the stringers against deflection. Performs the same functions as ties in a wall form and also supports the concrete at the desired elevation above the ground. Lumber at least as large as the stringer should be used, but never use lumber smaller than 4 by 4s.
- Lateral bracing. Bracing may be required between adjacent shores to keep shores from bending under load. Use 1 by 6s or larger material for bracing material.
- Cross bracing. Cross bracing will always be required to support the formwork materials.
- Wedges. Wood or metal shims used to adjust shore height
- Mudsill. Board which supports shores and distributes the load. Use 2-inch thick or larger lumber.

Chapter 4

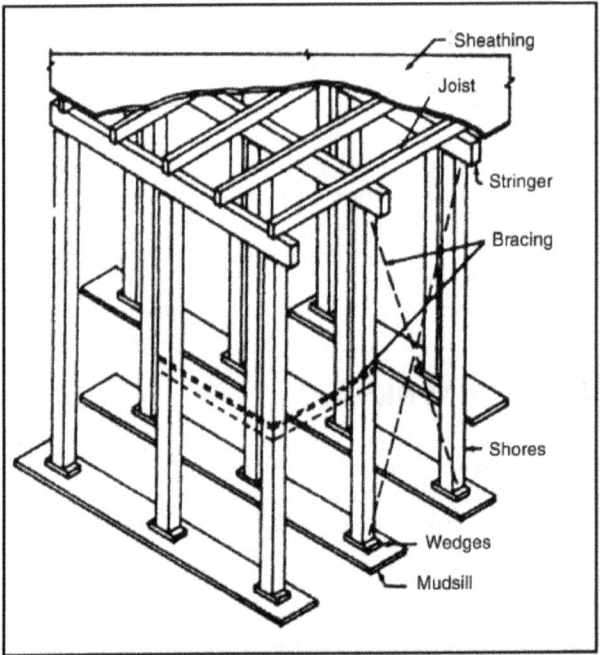

Figure 4-5. Typical overhead slab forms

Design Procedure

4-33. Use the following steps in determining form design:

Step 1. Lay out and specify the materials you will be using for the construction of the overhead roof slab. It is important that anyone using your design will know exactly the materials to use for each of the structural members.

Step 2. Determine the maximum total load (TL) the formwork will have to support.

Step 3. The LL of materials, personnel, and equipment is estimated to be 50 pounds per square foot unless the formwork will support engine-powered concrete buggies or other power equipment. In this case, a LL of 75 pounds per square foot will be used. The LL is added with the DL of the concrete to obtain the maximum TL. The concrete's DL is obtained using the concrete's unit weight of 150 pounds per cubic foot. The formulas are—

$$TL = LL + DL$$

$$LL = 50\ lb/sq\ ft\ or\ 75\ lb/sq\ ft\ with\ power\ equipment$$

$$DL\ (lb/sq\ ft) = 150\ lb/cu\ ft \times \frac{overhead\ slab\ thickness\ (in)}{12\ in/ft}$$

Step 4. Determine the maximum joist spacing. Use table 4-1, page 4-8, or table 4-2, page 4-9, and determine the joist spacing based on the sheathing material used. This is the same procedure used in determining the MSS for wall-form design. Use the maximum TL in place of the MCP.

Step 5. Calculate the uniform load on the joists (ULJ). The same procedure is used as determining uniform loads on structural members in wall-form design.

$$ULJ\ (lb/ft) = \frac{TL \times joist\ spacing\ (in)}{12\ in/ft}$$

Step 6. Determine the maximum stringer spacing. Use table 4-3, page 4-10, and the ULJ calculated in step 4. Round this load up to the next higher load located in the left column of the table. Read right to the column containing the lumber material used as the joist. This is the member to be supported by the stringer. The value at this intersection is the on-center (OC) spacing of the stringer.

Step 7. Calculate the uniform load on the stringer (ULS_{str}).

$$ULS_{str}(lb/ft) = \frac{TL \times masimum\ stringer\ spacing\ (in)}{12\ in/ft}$$

Step 8. Determine the maximum shore spacing the following two ways:

- Determine the maximum shore spacing based on the stringer strength. Use table 4-3 or table 4-4, page 4-11, (depending on the type of stringer) and the ULS_{str}, rounded to the next higher load shown in the left column of the table. Read right to the stringer material column. This intersection is the maximum OC spacing of the shore required to ensure the stringer is properly supported.
- Determine the maximum shore spacing based on the allowable load. This determination is dependent on both the shore strength and the end bearing of the stringer on the shore. Using tables 4-7 and 4-8, pages 4-22 and 4-23, select the allowable load on the shore both ways as follows:

Note. The unsupported length is equal to the height above the sill-sheathing thickness, joist thickness, and stringer thickness. This length is then rounded up to the next higher table value.

Table 4-7. Allowable load (in pounds) on wood shores, based on shore strength

Nominal Lumber Size, in Inches	4 x 4		4 x 6		6 x 6	
Unsupported Length, in Feet	R	S4S	R	S4S	R	S4S
4	9,900	9,200	15,300	14,400	23,700	22,700
5	9,900	9,200	15,300	14,400	23,700	22,700
6	9,900	9,200	15,300	14,400	23,700	22,700
7	8,100	7,000	12,500	11,000	23,700	22,700
8	6,200	5,400	9,600	8,400	23,700	22,700
9	4,900	4,200	7,600	6,700	23,700	22,700
10	4,000	3,400	6,100	5,000	23,000	21,000
11	3,300	2,800	5,100	4,500	19,000	17,300
12	2,700	2,400	4,300	3,700	16,000	14,600
13	2,300	2,000	3,600	3,200	13,600	12,400
		l/d = 50		l/d = 50		
14	2,000	1,700	3,100	2,800	11,700	10,700
	l/d = 50		l/d = 50			
15	1,800		2,700		10,200	9,300
					9,000	8,200
					7,900	7,300
					7,100	6,500
					6,400	5,800
					5,700	5,200

Notes:
1. The above table values are based on wood members with the following characteristics strength: Compression parallel to gram = 750 psi.
E = 1,100,000 psi.
2. R indicates rough lumber.

Form Design and Construction

Table 4-8. Allowable load on specified shores, in pounds, based on bearing stresses where the maximum shore area is in contact with the supported member

Nominal Lumber Size, in Inches	4 x 4		4 x 6		6 x 6	
C $\perp$ of Member Supported	R	S4S	R	S4S	R	S4S
250	3,300	3,100	5100	4800	7900	7600
350	4600	4300	7100	6700	11100	10600
385	5100	4700	7800	7400	12200	11600
400	5300	4900	8200	7700	12700	12100

Notes:
1. When the compression perpendicular to the grain of the member being supported is unknown, assume the most critical C is $\perp$ to the grain.
2. R indicates rough lumber.
3. S4S indicates surfaced on four sides.

- First, determine the allowable load based on the shore strength. Select the shore material dimensions and determine the unsupported length in feet of the shore. See table 4-7. The allowable load for shore is given in pounds at the intersection. Read down the left column to the unsupported length of the shore, then read right to the column of the size material used as the shore.
- Second, determine the allowable load based on end bearing area. Select the size of the shore material and the compression C perpendicular ($\perp$) to the grain of the stringer. If the C to the grain is unknown, use the lowest value provided in table 4-8. Read down the left column to the C$\perp$ to the grain of the stringer material and then right of the column of the shore material. The allowable load between the stringer and the shore will be in pounds.
- Compare the two loads just determined and select the lower as the maximum allowable load on the shore to be used in the formula below.

Calculate shore spacing by the following formula:

$$\text{Shore spacing} = \frac{\text{allowable load on shore} \times 12 \text{ in/ft}}{\text{ULS}_{str}}$$

Step 9. Determine the most critical shore spacing. Compare the shore spacing based on the stringer strength in step 7, with the shore spacing based on the allowable load in step 7. Select the smaller of the two spacings.

Step 10. Check the shore bracing. Verify that the unbraced length (l) of the shore (in inches) divided by the least dimension (d) of the shore does not exceed 50. If l/d exceeds 50, the lateral and cross bracing must be provided. Table 4-7 indicates the l/d > 50 shore lengths and can be used if the shore material is sound and unspliced. It is good engineering practice to provide both lateral and diagonal bracing to all shore members if material is available.

Chapter 4

SAMPLE FORM DESIGN PROBLEM NUMBER THREE

4-34. Design the form for the roof of a concrete water tank to be 6 inches thick, 20 feet wide, and 30 feet long. The slab will be constructed 8 feet above the floor (to the bottom of the slab). Available materials are 3/4-inch plywood and 4 by 4 (S4S) (surfaced on four sides) lumber. Mechanical buggies will be used to place concrete. Use the following steps in design:

Step 1. Lay out available materials for construction.

    Sheathing: 3/4-inch plywood (strong way)

    Joists: 4 x 4 (S4S)

    Shores: 4 x 4 (S4S)

    Stringers: 4 x 4 (S4S)

Step 2. Determine the maximum TL.

$$DL = concrete\ load = 150\ lb\ sq\ ft\ x\ \frac{6\ in}{12\ in/ft} = 75\ lb/sq\ ft$$

$$LL = personnel\ and\ equipment = 75\ lb/sq\ f$$

$$TL = DL + LL$$

$$TL = 75\ lb/sq\ ft + 75\ lb/sq\ ft$$

$$TL = 150\ lb/sq\ ft$$

Step 3. Determine the maximum joist spacing using table 4-2, page 4-8.

$$3/4 - inch\ plywood\ (strong\ way)\ and\ TL = 150\ pounds\ per\ square\ foot$$

$$Joist\ spacing = 22\ inches$$

Step 4. Calculate the ULJ.

$$ULJ = TL\ x\ \frac{joist\ spacing\ (in)}{12\ in/ft}$$

$$= 150\ lb/sq\ ft\ x\ \frac{22\ in}{12\ in/ft} = 275\ lb/lin\ ft$$

Step 5. Determine the maximum stringer spacing using table 4-3, page 4-9.

$$Load = 275\ pounds\ per\ foot$$

$$Joist\ material = 4\ x\ 4$$

$$Maximum\ stringer\ spacing = 55\ inches$$

Step 6. Calculate the ULS_{str}.

$$ULS_{str} = TL\ x\ \frac{maximum\ stringer\ spacing\ (in)}{12\ in/ft}$$

$$= 150 \, lb/sq\,ft \times \frac{55\,in}{12\,in/ft} = 687.5 \, lb/ft$$

Step 7. Determine the maximum shore spacing based on stringer strength. Use the maximum shore spacing shown in table 4-3, page 4-10, single member stringers.

- Load = 687.5 pounds per foot (found up to 700 pounds per foot)
- Stringer material = 4x4 (S4S)
- Maximum shore spacing = 35 inches, bases on stringer strength

Step 8. Determine shore spacing based on allowable load. This determination is based on both the shore strength and end bearing stresses of the stringer to the shore. Determine both ways as follows:

- Allowable loads based on shore strength are shown in table 4-7, page 4-22.
- Unsupported length = 8 feet − 3/4 inch − 3 1/2 inches − 3 1/2 inches = 7 feet 4 1/4 inches (round up to 8 feet).
- For an 8-foot 4 by 4 (S4S), the allowable load = 5,400 pounds, based on shore strength.
- For an allowable load based on end bearing stresses see table 4-8, page 4-23. Since we do not know what species of wood we are using, we must assume the worst case. Therefore, the ($C\perp$) to the grain = 250, and the allowable load for a 4 by 4 (S4S) = 3,100 pounds based on end bearing stresses.

Select the most critical of the two loads determined.

Since the $C\perp$ is less than the allowable load on the shore perpendicular to the grain (II), 3,100 pounds is the critical load.

Calculate the shore spacing as follows:

$$Shore \, spacing = \frac{3100 \, lb \times 12 \, in/ft}{ULS_{str} \, (lb/ft)}$$

$$= \frac{3{,}100 \, lb}{687.5 \, lb/ft} \times 12 \, in/ft = 54.1 \, in, based\,on\,allowable\,load$$

Step 9. Select the most critical shore spacing. The spacing determined in step 7 is less than the spacing determined in step 8; therefore, the shore spacing to be used is 35 inches.

Step 10. Check shore deflection.

$$Unsupported \, length = 8 \, feet - 3/4 \, inch - 3\,1/2 \, inch - 3\,1/2 \, inches$$
$$= 7 \, feet \, 4\,1/4 \, inches$$

$$= 88.25 \, inches$$

$$l/d = \frac{88.25\,in}{3.5\,in} = 25.21 < 50$$

$$d = least \, dimension \, of \, 4 \times 4 \, (S4S) \, lumber = 3.5 \, inches$$

Note. Lateral bracing is not required. Cross bracing is always required.

4-35. Summary of materials needed for construction.
- Sheathing: 3/4-inch plywood (strong way)
- Joists: 4 x 4 (S4S) lumber spaced 22 inches OC

- Stringer: 4 x 4 (S4S) lumber spaced 55 inches OC
- Shores: 4 x 4 (S4S) lumber spaced 35 inches OC
- Lateral braces: not required

CONCRETE SLAB ON GRADE THICKNESS DESIGN

4-36. Concrete slabs on grade are the most often constructed concrete projects by engineer units. Many slabs that are constructed fail because the thickness of the slab is not adequate. In other cases, the slab thickness is so excessive that materials and personnel are wasted in the construction. Use the following procedure, tables, and figures for the design of concrete slab thickness in the field to eliminate possible failure or wasted materials.

BASIC ASSUMPTIONS

4-37. The following three points are for the design thickness:

- The minimum slab thickness will be 4 inches for class 1, 2, and 3 floors, as listed in table 4-9.
- Only the load area and the flexural strength required will be considered using this method. Whenever the loaded area exceeds 80 square inches, the soil-bearing capacity must be considered. When the load will be applied at the edges of the concrete slab, either thicken the edge by 50 percent and taper back to normal slab thickness at a slope of not more than 1 in 10 or use a grade beam to support the edge.
- The controlling factor in determining the thickness of a floor on ground is the heaviest concentrated load the slab will carry. The load is usually the wheel load plus impact loading caused by the vehicle.

Table 4-9. Concrete floor classifications

Class	Usual Traffic	Use	Special Consideration	Concrete Finishing Technique
1	Light foot	Residential or tile-covered	Grade for drainage; make plane for tile	Medium steel trowel
2	Foot	Offices, churches, schools, hospitals	Nonslip aggregate; mix in surface	Steel trowel; special finish for nonslip
		Ornamental residential	Color shake, special	Steel trowel, color, exposed aggregate; wash if aggregate is to be exposed
3	Light foot and pneumatic wheels	Drives, garage floors, sidewalks for residences	Crown; pitch joints Air entrainment	Float, trowel, and broom
4	Foot and pneumatic wheels*	Light industrial commercial	Careful curing	Hard steel trowel and brush for nonslip
5	Foot and wheel abrasive wear*	Single-course industrial, integral topping	Careful curing	Special hard aggregate float and trowel
6	Foot and steel-tire vehicles - severe	Bonded two-course, heavy	Base / Textured surface and bond	Surface leveled by screeding

Table 4-9. Concrete floor classifications

Class	Usual Traffic	Use	Special Consideration		Concrete Finishing Technique
	abrasion	industrial	Top	Special aggregate and/or mineral or metallic surface treatment	Special disc-type power floats with repeated steel troweling
7	Same as classes 3, 4, 5, 6	Unbonded topping	Mesh reinforcing; bond breaker on old concrete surface; minimum thickness 2 1/2 inches		

*Under abrasive conditions on floor surface, the exposure will be much more severe and a higher quality surface will be required for class 4 and 5; for a class 6, two-course floor, a mineral or metallic aggregate monolithic surface treatment is recommended.

DESIGN PROCEDURE

4-38. Use the following steps in design procedures for concrete slabs.

   Step 1. Determine the floor classification. Use table 4-9 with the usual traffic and use of the slab.

   Step 2. Determine the minimum compressive strength. Using table 4-10, read down the class of floors column until you find the floor classification selected. Read to the right and select the f'_c in psi. Note that this table gives the recommended slump range for the plastic concrete.

Table 4-10. Recommended slumps and compressive strengths

Class of Floors	Slump Range, in inches	Minimum Compressive Strength, f'_c in Pounds per Square Inch*
1	2-4	3,000
2	2-4	3,500
3	2-4	3,500
4	1-3	4,000
5	1-3	4,500
6 base	2-4	3,500
6 topping**	0-1	5,000 - 8,000
7	1-3	4,000

Table 4-10. Recommended slumps and compressive strengths

Class of Floors	Slump Range, in inches	Minimum Compressive Strength, f'_c in Pounds per Square Inch*
Note. These recommendations are specially for concrete made with normal-weight aggregate. For structural slabs, the requirements of ACI 318 and the contract documents should be met.		
*Refer to the minimum compressive strength of cylinders made and tested according to applicable ASTM standards for 28 days. The average of any five consecutive strength tests of continuously moist-cured specimens representing each class of concrete should be equal to or greater than the specified strength.		
**The cement content of the heavy-duty floor topping depends upon the severity of the abrasion. The minimum is 846 pounds per cubic yard.		

Step 3. Determine the allowable flexural tensile stress f'_t after the concrete f'_c is determined. Use the formula: $f'_t = 4.6 \sqrt{f'_c}$

Step 4. Determine the equivalent static load (ESL). The expected impact loading is needed for this step. The impact loading is 25 percent more than the static load (SL) for the vehicles.

$$ESL = (1 + 25\%) \times SL$$

Step 5. Evaluate and correct the ESL if f'_t is not equal to 300 psi. When calculating the allowable flexural tensile f'_c in step 4 above, an f'_t was determined. If this f'_t is not equal to 300 psi, then the ESL must be corrected based on a ratio between the standard (300 psi) and the actual f'_t. This correction is necessary so that the standard thickness (see figure 4-6) may be used to determine the required slab thickness. The procedure for correction is as follows:

$$CESL = \frac{ESL \times 300}{f'_t}$$

Step 6. Determine the slab thickness. Using figure 4-6 read up the left side until the TL is the same as the design CESL from 5 above. Read to the right until the loaded area (in square inches) for the slab design is intersected. The slab thickness can then be determined by interpolating between the slanted slab thickness lines on the figure. Round up to the next 1/4 inch.

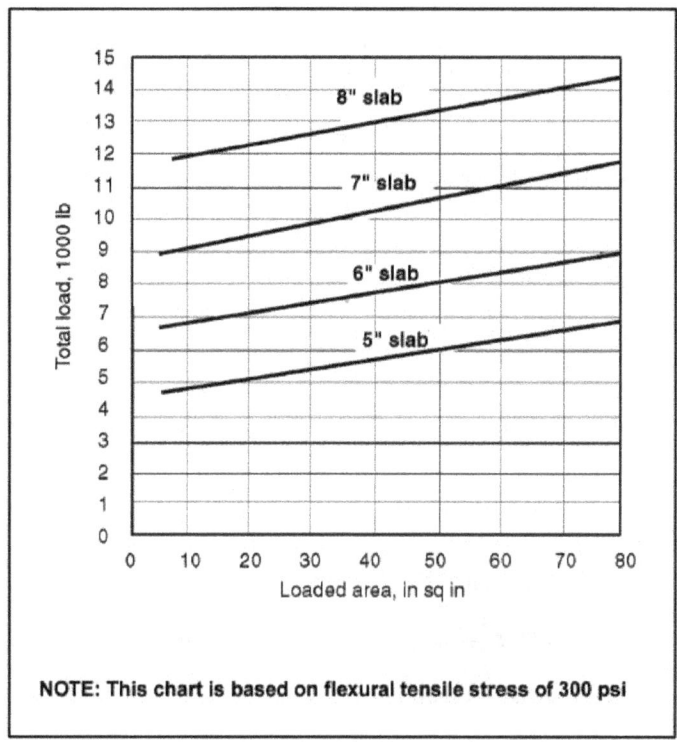

Figure 4-6. Maximum wheel loads for industrial floors

Step 7. Determine the minimum cement content and recommended air content for structures subjected to freeze-thaw cycles, depending upon the maximum size of CA to be used.

4-39. The slab thickness design procedure is now complete, and with the accumulated information you should proceed to the mix design procedure as detailed in Chapter 3 of this manual.

SLAB CONSTRUCTION

4-40. Slab construction is often spoiled by improper construction practices that can be easily prevented. Forms for slabs on grades are relatively simple to construct. See Section III of this chapter. The following practices should always be employed for ensuring that the slab will hold up under service load.

- Always use nonfrost susceptible (free-draining) material under the slab. Material such as silts and very fine sand expands under frost action if groundwater is present. Removing these types of soil and replacing them with a free-draining material (gravel or coarse sand) to one-half of the expected frost penetration, (in addition to providing adequate drainage at the edges of the structure), will prevent frost heave.
- Always use some form of steel reinforcement. Welded-wire mesh is very desirable for preventing excessive cracking and preventing widening of cracks that do develop.
- Always keep the slab moist and protect it from excessive heat and freezing for the required curing time. This will ensure the concrete is cured properly.

Chapter 4

SAMPLE SLAB CONSTRUCTION PROBLEM

4-41. Determine the slab thickness for the floor of a wheeled-vehicle shop. The total weight on one wheel is 7,500 pounds and the loaded area equals 40 square inches. The surface will be exposed to foot and wheel traffic. Abrasive wear is expected. The maximum-size aggregate will be 1 inch.

Step 1. Determine the floor classification, see class 5, table 4-9, page 4-26.

Step 2. Determine the minimum compressive strength from table 4-10, page 4-27.

$$f'_c = 4,500 \; psi$$

Step 3. Determine the allowable flexural tensile stress.

$$f'_t = 4.6\sqrt{f'_c} = 308.57 \; (use \; 309 \; lbs)$$

Step 4. Determine the ESL.

$$ESL = (1 + .25) \; x \; SL$$
$$= 1.25 \; x \; 7,500 = 9375 \; lbs$$

Step 5. Evaluate and correct the ESL if $f'_t \neq 300 \; psi$.

$$CESL = ESL \; x \frac{300}{f'_t} = 9,375 \; x \frac{300}{309} = 9,101.9 \; (use \; 9,102 \; lbs)$$

Step 6. Determine the slab thickness from figure 4-6.

$$Loaded \; area = 40 \; sq \; i$$
$$Slab \; area = 6.6 \; in$$

Note. Always round up to the next higher 1/4-inch thickness for convenience, therefore thickness is rounded up to 6 3/4-inches

4-42. The minimum cement content per cubic yard from, table 4-11, with 1-inch maximum-size aggregate (MSA) is 520 pounds. Because the structure is a vehicle-shop floor, the air content may be less than the table allowance. The slab design information is now:

$$f'_c = 4,500 \; psi.$$
$$Slab \; thickness = 6 \; 3/4 \; inches$$
$$Minimum \; cement \; content = 520 \; lb/cu \; yd$$
$$Air \; content = 6\%$$

Table 4-11. Minimum cement contents and percentages of entrained air

Maximum Size of Coarse Aggregate	Minimum Cement Content, in Pounds per Cubic Yard	Air Content for Freeze-Thaw Resistance, Shown in Percentage*
1 1/2	470	5 ± 1
1	520	6 ± 1
3/4	540	6 ± 1
1/2	590	7 1/2 ± 1
3/8	610	7 1/2 ± 1
Notes: These mixtures are specially for concrete made with normal-weight aggregate; different mixtures may be needed for lightweight aggregate concrete. For structural slabs, the requirements of ACI 318 and the contract document must be met.		
*Smaller percentages of entrained air can be used for concrete floors that will not be exposed to freezing and thawing or deicing. This may improve the workability of the concrete and the finish product.		

COLUMN FORM DESIGN

4-43. Use the following steps in determining design procedures of a wood form for a concrete column.

Step 1. Determine the materials you will use for sheathing, yokes, and battens. The standard materials for column forms are 2 by 4s and 1-inch sheathing.

Step 2. Determine the column height.

Step 3. Determine the largest cross-sectional column dimension.

Step 4. Determine the maximum yoke spacing by referring to table 4-12, page 4-32. First, find the column height in feet in the first column. Then move right horizontally to the column heading of the largest cross-sectional dimension in inches of the column you are constructing. The center-to-center spacing between the second yoke and the base yoke is the lowest value in the interval that falls partly in the correct column height line. You can obtain all subsequent yoke spacing by reading up this column to the top. These are maximum yoke spacing; you can place yokes closer together. Adjust final spacing to be at the top of the column.

Table 4-12. Column yoke spacing using 2 by 4s and 1 inch sheathing

Largest Cross-Sectional Dimension, in inches (L)

Column height in ft	16	18	20	24	28	30	32	36
1								
2								
3								
4								
5								
6								
7								
8								
9								
10								
11								
12								
13								
14								
15								
16								
17								
18								
19								
20								

SAMPLE COLUMN FORM DESIGN PROBLEM

4-44. Determine the yoke spacing for a 9-foot column whose largest cross-sectional dimension is 36 inches. Construction materials are 2 by 4s and 1-inch sheathing.

Solution Steps:

Step 1. Lay out materials available.

  2 x 4s and 1-inch sheathing

Step 2. Determine column height.

  9 feet

Step 3. Determine the largest cross-sectional dimension.

  36 inches

Step 4 and Step 5. Determine the maximum yoke spacing, refer to table 4-12. Starting from the base, the yokes are: 8, 8, 10, 11, 12, 15, 17, 17, and 10-inches. The spacing between the top

Form Design and Construction

two yokes are reduced due to the limits of the column height. Adjust the final spacing to be at the top of the column.

SECTION III - CONSTRUCTION

FOUNDATION FORMS

4-45. The portion of a structure that extends above the ground level is called the *superstructure*. The portion below ground level is called the *substructure*. The parts of the substructure that distribute building loads to the ground are called *foundations*. Footings are installed at the base of the foundations to spread the loads over a larger ground area to prevent the structure from sinking into the ground. Forms for large footings such as bearing wall footings, column footings, and pier footings are called foundation forms. Footings or foundations are relatively low in height since their primary function is to distribute building loads. Because the concrete in a footing is shallow, pressure on the form is relatively low. Therefore, it is generally not necessary to base a form design on strength and rigidity.

SIMPLE FOUNDATION FORMS

4-46. Whenever possible, excavate the earth and use it as a mold for concrete footings. Thoroughly moisten the earth before placing the concrete. If this is not possible, you must construct a form. Because most footings are rectangular or square, it is easier to erect the four sides of the form in panels. The following steps and figures helps explain this process:

Step 1. Build the first pair of opposing panels (see [a] in figure 4-7, page 4-34) to the footing width. Nail vertical cleats to the exterior sides of the sheathing. Use 2-inch dressed lumber for the cleats and space them 2 1/2 inches from each end of the exterior sides of the (a) panels and on 2-foot centers between the ends. Use forming nails driving them only part way for easy removal of the forms.

Step 2. Nail two cleats to the ends of the interior sides of the second pair of panels (see [b] in figure 4-7). The space between them should equal the footing length, plus twice the sheathing thickness. Nail cleats on the exterior sides of the (b) panels, spaced on 2-foot centers. Erect the panels into either a rectangle or a square, and hold them in place with form nails. Make sure all reinforcing bars are in place.

Step 3. Drill small holes on each side of the center cleat on each panel less than 1/2 inch in diameter to prevent paste leakage. Pass number 8 or number 9 black annealed iron wire through these holes and wrap it around the center cleats of opposing panels to hold them together as shown in figure 4-7.

Step 4. Mark the top of the footing on the interior side of the panels with grade nails. For forms 4 feet square or larger, drive stakes against the sheathing as shown in figure 4-7. Both the stakes and the 1- by 6-inch tie braces nailed across the top of the form keep it from spreading apart. If a footing is less than 1 foot deep and 2 feet square, you can construct the form from 1-inch sheathing without cleats. Simply make the side panels higher than the footing depth, and mark the top of the footing on the interior sides of the panels with grade nails. Cut and nail the lumber for the sides of the form as shown in figure 4-8, page 4-34.

Chapter 4

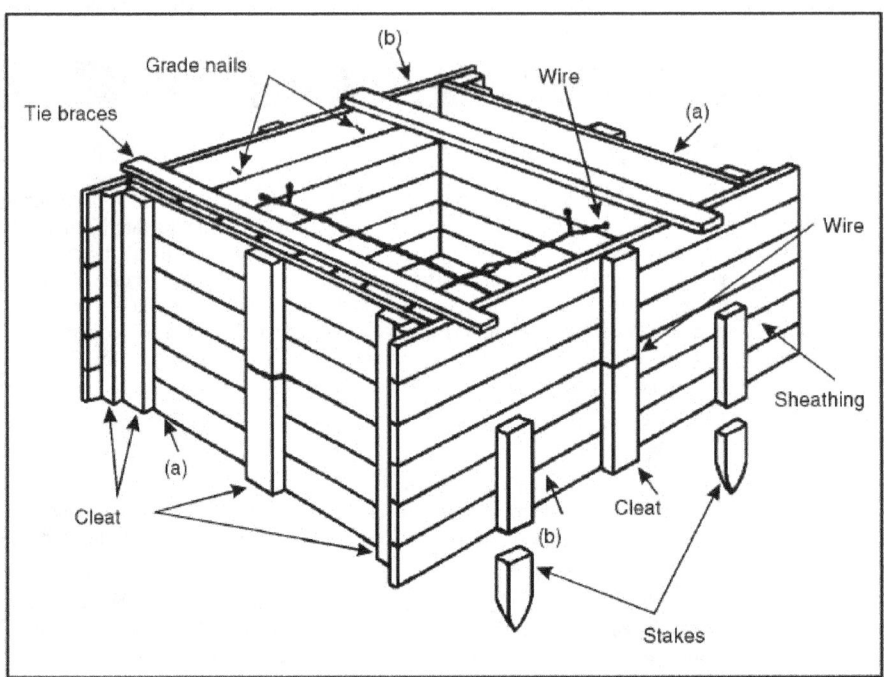

Figure 4-7. Typical foundation

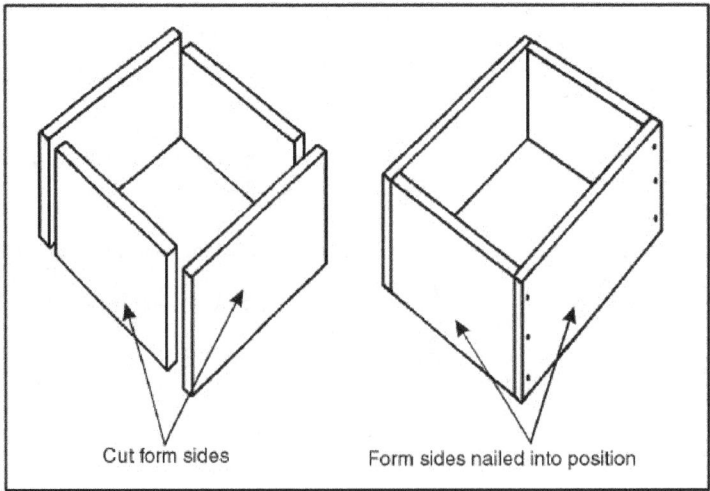

Figure 4-8. Small footing form

FOUNDATION AND PIER FORMS COMBINED

4-47. It is best to place the footing and a small pier at the same time. A pier is a vertical member, either rectangular or round, that supports the concentrated loads of an arch or bridge super-structure. The form that is build for this type of construction is shown in figure 4-9. The footing form should look like the one in figure 4-7. You must provide support for the pier form while not interfering with concrete placement in the footing form. This can be accomplished by first nailing 2 by 4s or 4 by 4s across the footing form, as shown in figure 4-9, that serve as both supports and tie braces. Nail the pier form to these support pieces.

BEARING WALL FOOTING

4-48. Figure 4-10 shows footing formwork for a bearing wall and figure 4-11, page 4-36, shows bracing methods for a bearing wall footing. A bearing wall, also called a load-bearing wall, is an exterior wall that serves not only as an enclosure, but also transmits structural loads to the foundation. The form sides are 2 inch lumber whose width equals the footing depth. Stakes hold the sides in place while spreaders maintain the correct distance between them. The short braces at each stake hold the form in line.

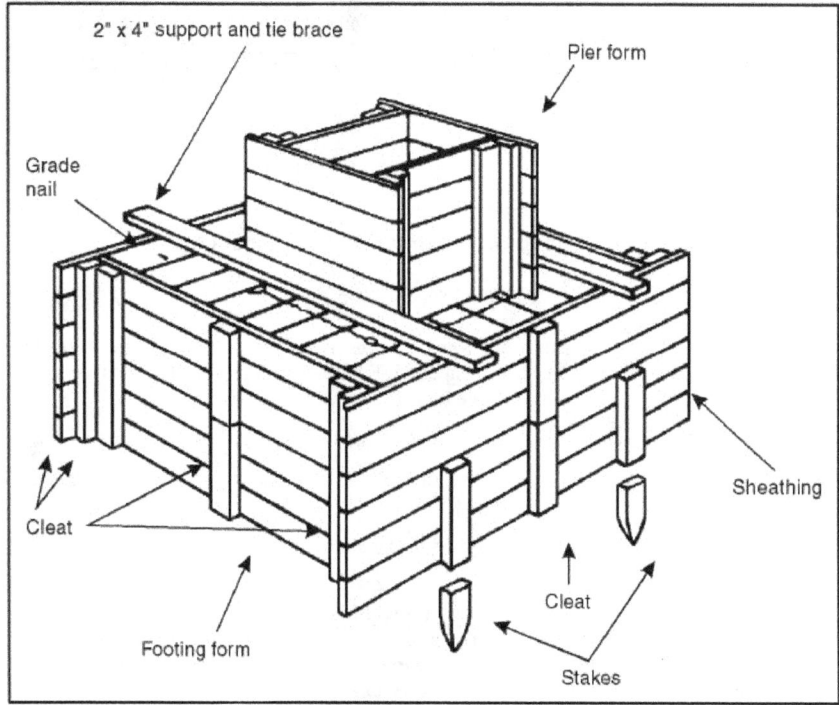

Figure 4-9. Footing and pier form

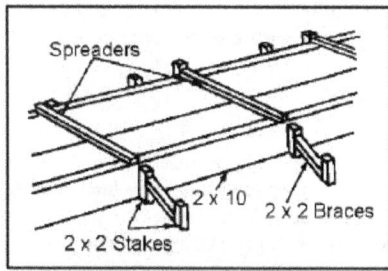

Figure 4-10. Typical footing form

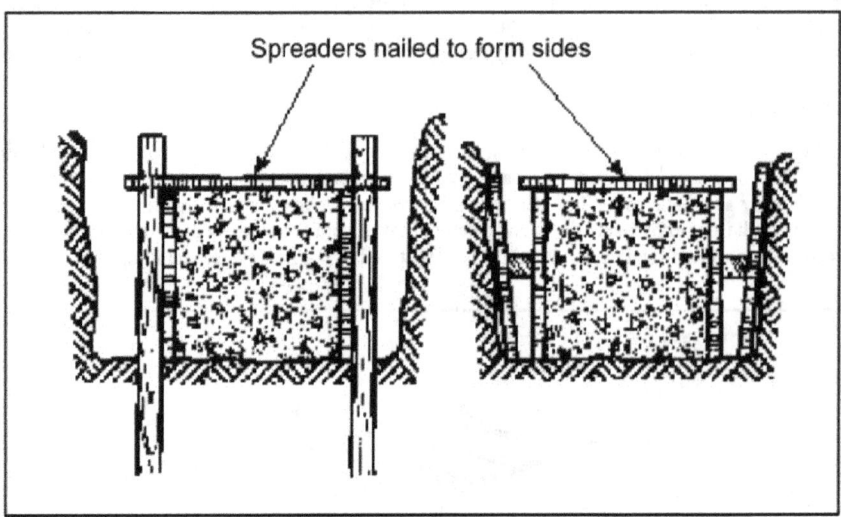

Figure 4-11. Methods of bracing bearing wall footing form

COLUMN AND FOOTING FORMS

4-49. Square column forms are made from wood whereas round column forms are made from steel or cardboard impregnated with waterproofing compound. Figure 4-2, page 4-4, shows an assembled column and footing form. After constructing the footing form follow the steps below:

Step 1. Build the column form sides in units first and then nail the yokes to them.

Step 2. Determine the yoke spacing from table 4-12, page 4-32, according to the procedure described in paragraph 4-43.

Step 3. Erect the column form as a unit after assembling the steel reinforcement and tying it to the dowels in the footing.

Step 4. Place reinforcing steel in columns as described in Chapter 6. The column form should have a clean-out hole in the bottom to remove construction debris. Be sure to nail the pieces of lumber that you remove for the clean-out hole to the form. This way you can replace them exactly before placing concrete in the column.

PANEL-WALL FORMS

4-50. A panel wall is an exterior or curtain wall made up of a number of units that enclose a structure below the roof line. It is built only strong enough to carry its own weight and withstand wind pressure on its exterior face.

PANEL-WALL UNITS

4-51. Panel-wall units should be constructed in lengths of about 10 feet for easy handling. The sheathing is normally a 1-inch (13/16-inch dressed) tongue-and-groove lumber or 3/4-inch plywood. Make the panels by first nailing the sheathing to the studs. Then connect the panels as shown in figure 4-12. Figure 4-13 shows the form details at the wall corner. When placing concrete panel-walls and columns at the same time, construct the wall form as shown in figure 4-14, page 4-38. Make the wall form shorter than the clear distance between the column forms to allow for a wood strip that acts as a wedge. When stripping the forms, remove the wedge first to facilitate form removal.

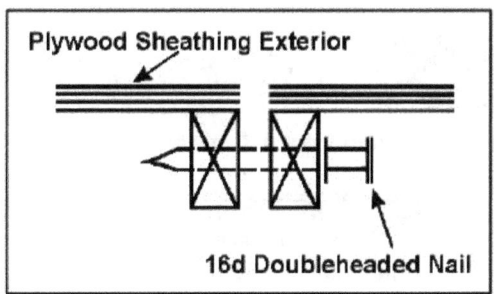

Figure 4-12. Method of connecting panel wall unit form together

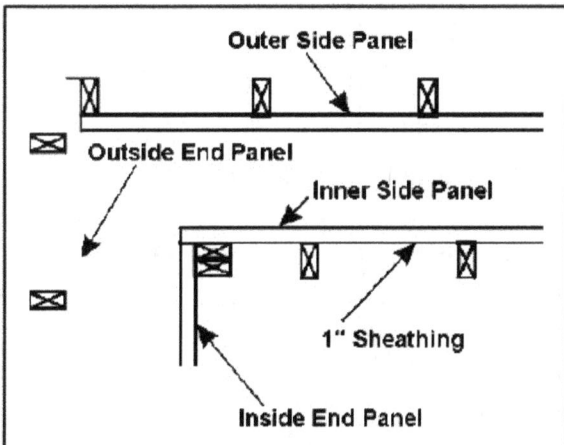

Figure 4-13. Detail at corner of panel wall form

SPREADERS AND TIES

4-52. Spreaders keep the sides of the wall form at the proper distance from each other until you place the concrete. Tie wires or tie-rods hold the sides firmly in place so that the weight of the freshly placed concrete does not push them apart. Figure 4-15, page 4-38 and figure 4-16, page 4-39, show the installation of tie wires and tie-rods. Use the following steps to install them:

Step 1. Use tie wires (see figure 4-15) only for low walls or when tie-rods are not available. Number 8 or number 9 black annealed iron wire is recommended, but you can use barbwire in an emergency.

Chapter 4

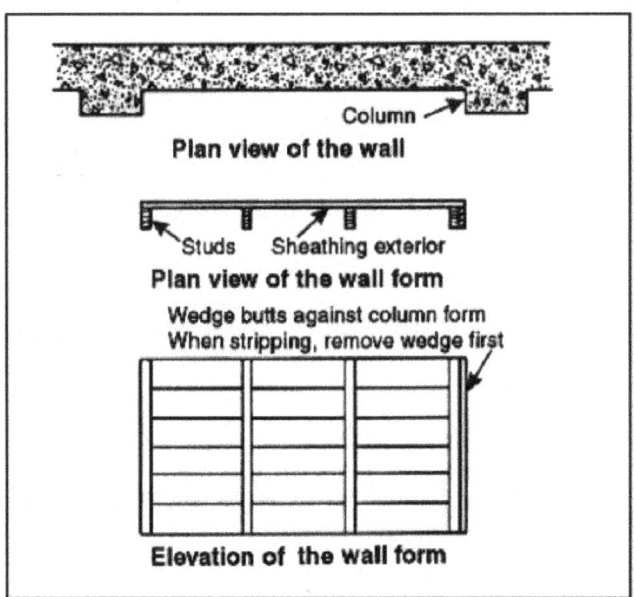

Figure 4-14. Form for panel walls and columns

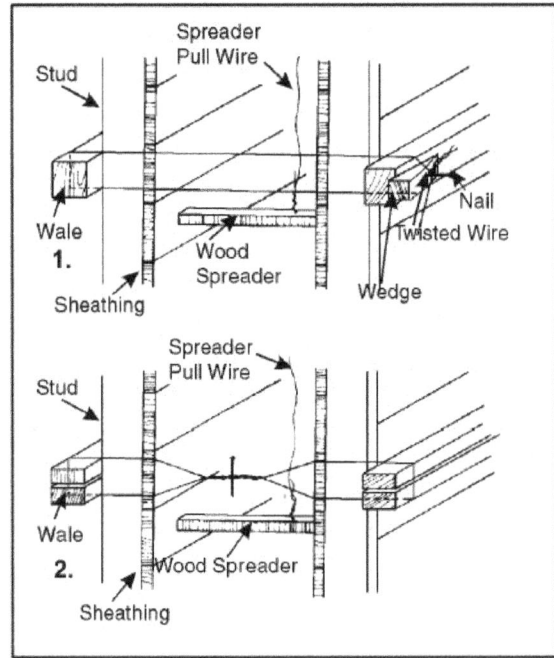

Figure 4-15. Wire ties for wall forms

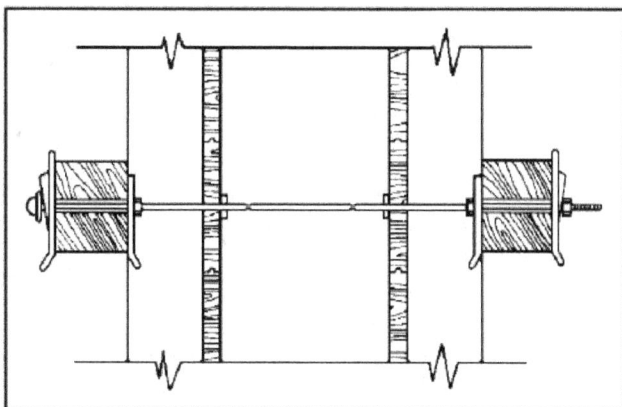

Figure 4-16. Tie-rod and spreader for wall form

Step 2. Space the individual ties the same as the stud spacing but never more than 3 feet apart.

Step 3. Place a spreader near the location of each tie wire or tie-rod, as shown in view 2 of figure 4-15. Form each tie by looping a wire around the wale on one form side, bringing, it through the form and looping it around the wale on the opposite form side.

Step 4. Make the wire taut by inserting a nail between the two strands and twisting. When using the wedge method shown in view 1 of figure 4-15, twist the ends of the wire strands around a nail driven part way into the point of a wedge. Remove the nail and drive the wedge until it draws the wire taunt.

Step 5. Use pull wires to remove the spreaders as you fill the form so that they are not embedded in the concrete. Figure 4-17, page 4-40, shows an easy way to remove spreaders. Loop a wire around the bottom spreader and pass it through off-center holes drilled in each succeeding spreader.

Step 6. Pull on the wire to remove one spreader after another as the concrete level rises in the form. Use tie-rods if rust stains are unacceptable or if the wall units must be held in an exact position.

Step 7. Space spreaders as shown in figure 4-17, the same as studs. After stripping the form, break off the rod at the notch located slightly inside the concrete surface. For a better appearance, grout in the tie-rod holes with a mortar mix.

STAIR FORMS

4-53. Various types of stair forms, including prefabricated ones, can be used. A design based on strength is not necessary for moderate-width stairs joining typical floors. Figure 4-18, page 4-40, shows one way to construct forms for stair widths up to and including 3 feet. Make the sloping wood platform that serves as the form for the underside of the steps from 1-inch tongue-and groove-sheathing.

Chapter 4

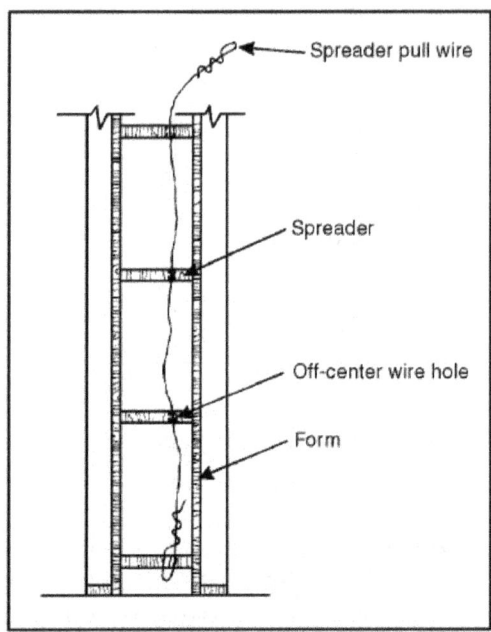

Figure 4-17. Removing wood spreader

4-54. The platform should extend about 12 inches beyond each side of the stairs to support the stringer bracing blocks. Support the back of the platform with 4 by 4 posts. Space the 2 by 6 cleats nailed to the post supports on 4-foot centers. The post supports should rest on wedges for easy adjustment and removal. Cut the 2 by 12 planks for the side stringers to fit the tread and risers. Bevel the 2 by 12 risers as shown.

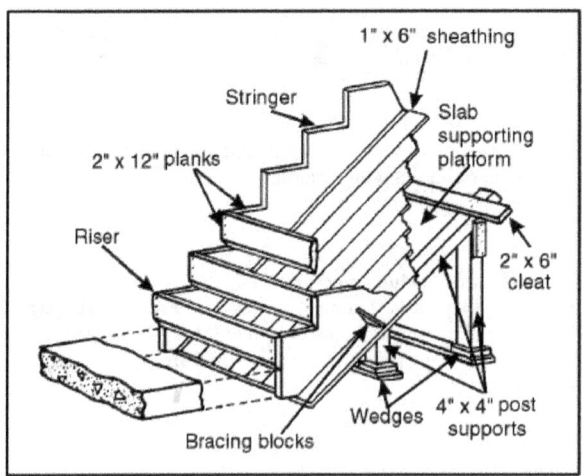

Figure 4-18. Stairway form

STEEL PAVEMENT FORMS

4-55. When possible, use standard steel pavement forms that can be set fast and accurately. They contain the concrete and provide a traction surface for form-riding equipment. These steel plate sections are 10 feet long and are normally available in slab thicknesses of 6, 7, 8, 9, 10, and 12 inches. The usual base plate width is 8 inches. Locking plates rigidly join the end sections. Three 1-inch diameter steel stakes per 10-foot section of form are required.

4-56. After driving the stakes, set the form true to line, them grade and clamp it to the stakes using special splines and wedges. Set the forms on the subgrade ahead of the paver and leave them in place for at least 24 hours after placing the concrete. Oil the forms immediately after setting them and before depositing the concrete. After oiling the form completely, wipe the top side or traction edge free from oil. Remove the forms as soon as practical, then clean, oil, and stack them in neat piles ready for reuse.

OILING AND WETTING FORMS

4-57. Make sure that any oil or other form coating that is used on forms does not soften or stain the concrete surface, prevent the wet surfaces from water curing, or hinder the proper functioning of sealing compounds used for curing. If standard form oil or other form coating is not obtainable, wet the forms to prevent sticking. Use water only in a emergency.

OIL FOR WOOD FORMS

4-58. Before placing concrete in wood forms, treat the forms with a suitable oil or other coating material to prevent the concrete from sticking to them. The oil should penetrate the wood and prevent water absorption, making it environmentally safe. Almost any light-bodied petroleum oil meets these specifications. On plywood, shellac works better than oil in preventing moisture from raising the grain and detracting from the finished concrete surface. Several commercial lacquers and similar products are also available for this purpose. A coat of paint or sealing compound helps to preserve the wood if you plan to reuse wood forms repeatedly. Sometimes lumber contains enough tannin or other organic substance to soften the surface concrete. To prevent this condition, treat the form surfaces with whitewash or limewater before applying the form oil or other coating.

OIL FOR STEEL FORMS

4-59. Oil steel column and wall forms before erecting them. You can oil all other steel forms when convenient but before reinforcing steel is in place. Use specially compounded petroleum oils, not oils intended for wood forms. Synthetic castor oil and some marine engine oils are typical examples of compound oils that give good results on steel forms.

APPLYING OIL

4-60. The successful use of form oil depends on the method of application and the condition of the forms. Forms should be clean and have smooth surfaces. Therefore, do not clean forms with wire brushes which can mar their surfaces and cause concrete to stick. Apply the oil or coating with a brush, spray, or swab. Cover the form surfaces evenly, but do not allow the oil or coating to contact construction joint surfaces or any reinforcing steel in the formwork. Remove all excess oil.

OTHER COATING MATERIALS

4-61. Fuel oil, asphalt paint, varnish, and boiled linseed oil are also suitable coatings for forms. Plain fuel oil is too thin to use during warm weather, but mixing one part petroleum grease to three parts of fuel oil provides enough thickness.

Chapter 4

SAFETY PRECAUTIONS

4-62. Safety precaution should be practiced at all times. When work is progressing on elevated forms, take precautions to protect both workers on scaffolds and ground personnel.

FORM CONSTRUCTION

4-63. Protruding nails are the main source of formwork accidents. Observe the following safety precautions during construction:

- Inspect tools frequently—particularly hammers.
- Place mudsills under shoring that rests on the ground.
- Observe the weather, do not attempt to raise large form panels in heavy wind gusts, either by hand or by crane.
- Brace all shoring to prevent possible formwork collapse.

FORMS STRIPPING

4-64. Observe of the following safety precaution when stripping forms:

- Permit only those workers actually stripping forms in the immediate work area. Do not remove forms until the concrete sets.
- Pile stripped forms immediately to avoid congestion, exposed nails, and other hazards.
- Take caution when cutting wires under tension to avoid backlash. See chapter 5, section XI.

FORM FAILURE

4-65. Even when all formwork is adequately designed, many form failures occur because of some human error or improper supervision.

WALL-FORM FAILURES

4-66. The following items are the most common construction deficiencies in wall-form failures and should be considered by supervisory personnel when working with concrete:

- Failure to control the rate of placing concrete vertically without regard to drop in temperature.
- Failure to sufficiently nail members.
- Failure to allow for lateral pressures and forces.
- Failure to tighten or secure form ties.
- Use of old, damaged, or weathered form materials.
- Use of undersized form material.
- Failure to sufficiently allow for eccentric loading due to placement sequences.

OVERHEAD SLAB FAILURES

4-67. The following are common construction deficiencies in overhead slab failure:

- Failure to adequately diagonally brace the shores.
- Failure to stabilize soil under mudsills.
- Failure to allow for lateral pressures on formwork.
- Failure to plumb shoring thus inducing lateral loading as well as reducing vertical load capacity.
- Failure to correct deficiencies where locking devices on metal shoring are not locked, inoperative, or missing.
- Failure to prevent vibration from adjacent moving loads or load carriers.
- Failure to prevent premature removal of supports especially under long continuous sections.

Form Design and Construction

- Lack of allowance in design for such special loads as winds, power buggies, or placing equipment.
- Failure to provide adequate reshoring.
- Failure to sufficiently allow for eccentric loading due to placement sequence.

OTHER GENERAL CAUSES OF FAILURE

4-68. The following are miscellaneous causes of form failure:

- Failure to inspect formwork during and after concrete placement to detect abnormal deflections or other signs of imminent failure which could be corrected.
- Failure to prevent form damage in excavated area caused by embankment failure.
- Lack of proper supervision and field inspection.
- Failure to ensure adequate anchorage against uplift due to battered form faces.

4-69. This list is not all inclusive; there are many reasons why forms fail. It is the responsibility of the form designer to ensure that the concrete will not fail when the designs are properly interpreted and constructed. It is the responsibility of the builder to make sure that the design is correctly constructed and that proper techniques are followed.

SECTION IV - JOINTS AND ANCHORS

JOINTS

4-70. Concrete structures are subjected to internal and external stresses caused by volume shrinkage during hydration, differential movement due to temperature changes, and differential movement caused by different loading conditions. These stresses can cause cracking and scaling of concrete surfaces and, in extreme cases, result in failure of the concrete structure. These stresses can be controlled by the proper placement of joints in the structure. Three basic types of joints will be addressed—the isolation joint, the control joint, and the construction joint.

ISOLATION JOINTS

4-71. Isolation joints are used to separate or isolate adjacent structural members. An example would be the joint that separates the floor slab from a column (to allow for differential movement in the vertical plane due to loading conditions or uneven settlement). Isolation joints are sometimes referred to as expansion or contraction joints to allow for differential movement as a result of temperature changes, as in two adjacent slabs. All isolation joints extend completely through the member and do not have any load transfer devices built into them. See figures 4-19 through 4-21, pages 4-44 and 4-46.

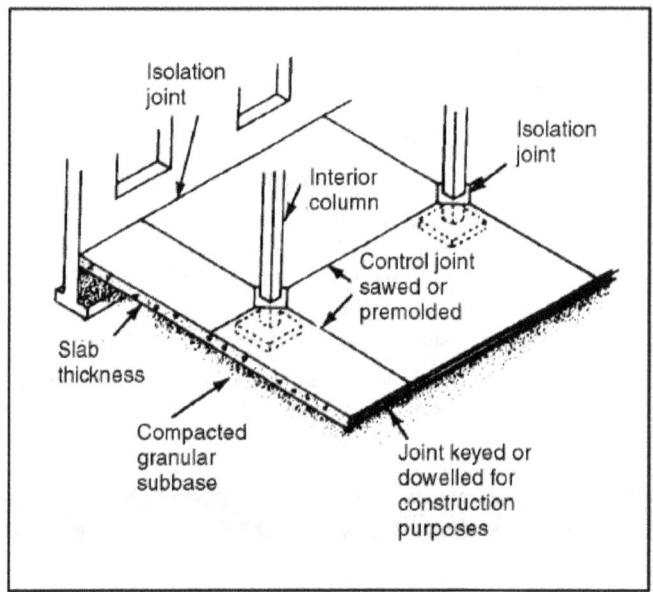

Figure 4-19. Typical isolation and control joints

Form Design and Construction

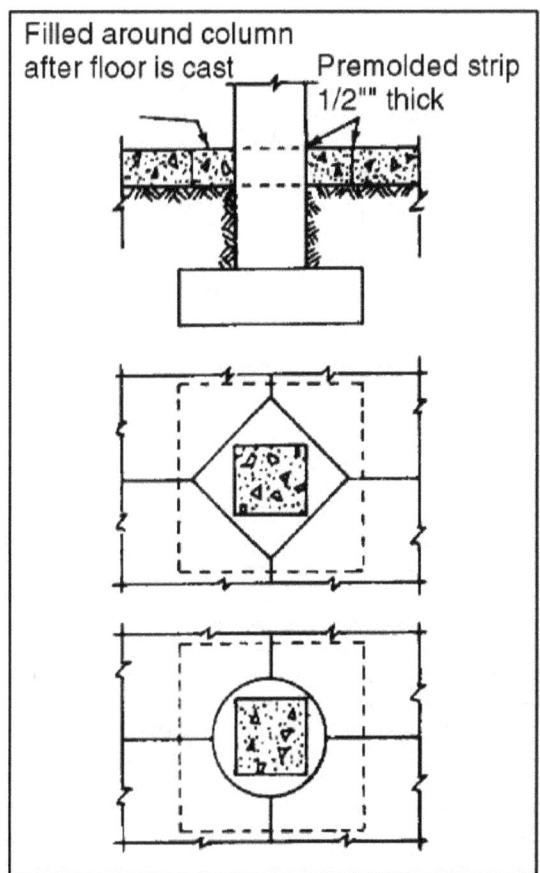

Figure 4-20. Joints at columns and walls

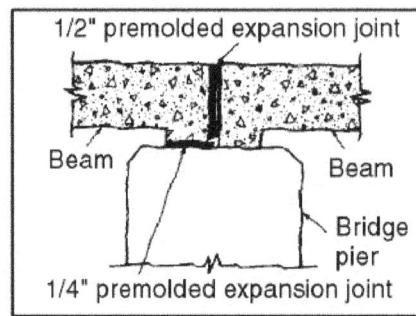

Figure 4-21. Expansion/contraction joint for a bridge

CONTROL JOINTS

4-72. Movement in the plane of a concrete slab is caused by drying shrinkage and thermal contraction. Some shrinkage is expected and can be tolerated, depending on the design and exposure of the particular structural elements. In a slab, the shrinkage occurs more rapidly at the exposed surfaces and causes upward curling at the edges. If the slab is restrained from curling, cracking will occur whenever the restraint imposes stress greater than the tensile strength. Control joints (see figure 4-22) are cut into the concrete slab to create a plane of weakness that forces cracking (if it happens) to occur at a designated place rather than randomly. These joints run in both directions at right angles to each other. Control joints in interior slabs are typically cut 1/3 to 1/4 inch of the slab thickness and then filled with lead or joint filler (see table 4-13 for suggested control joint spacing). Temperature steel (welded-wire fabric) may be used to restrict crack width for sidewalks, driveways, and tooled joints spaced at intervals equal to the width of the slab but not more than 20 feet (6 meters) apart. Surface irregularities along the plane of the cracks are usually sufficient to transfer loads across the joint in slabs on grade, the joint should be 3/4 to 1 inch deep.

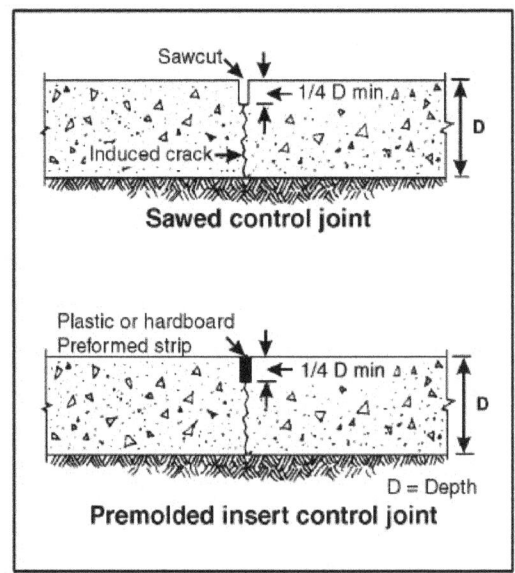

Figure 4-22. Control joint

Table 4-13. Spacing of control joints

Slab Thickness, in Inches	Less Than 3/4 inch Aggregate Spacing, in Feet	Larger Than 3/4 inch Aggregate Spacing, in Feet	Slump Less Than 4 Inches Spacing, in Feet
5	10*	13	15
6	12	15	18
7	14	18	21
8	16	20	24
9	18	23	27
10	20	25	30

Note. Spacing also applies to the distances from the control joints to the parallel isolation joints or to the parallel construction joints.

CONSTRUCTION JOINTS

4-73. Construction joints (see figure 4-23 and figures 4-24 through 4-26, page 4-48) are made where the concrete placement operations end for the day or where one structural element will be cast against previously placed concrete. These joints allow some load to be transferred from one structural element to another through the use of keys or (for some slabs and pavement) dowels. Note that the construction joint extends entirely through the concrete element.

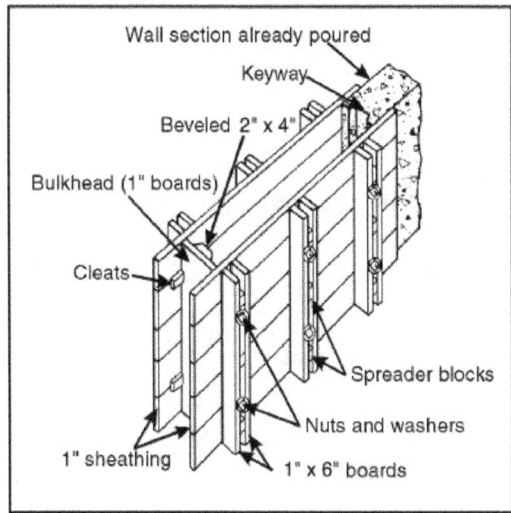

Figure 4-23. Keyed, wall construction joint (perspective view)

Chapter 4

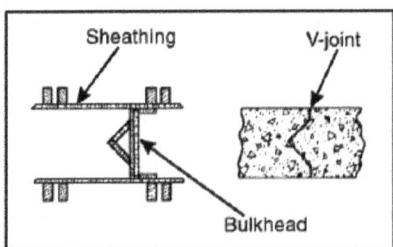

Figure 4-24. Keyed, wall construction joint (plan view)

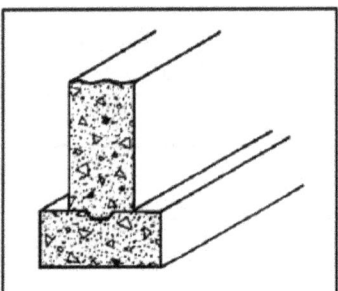

Figure 4-25. Construction joint between wall and footing showing keyway

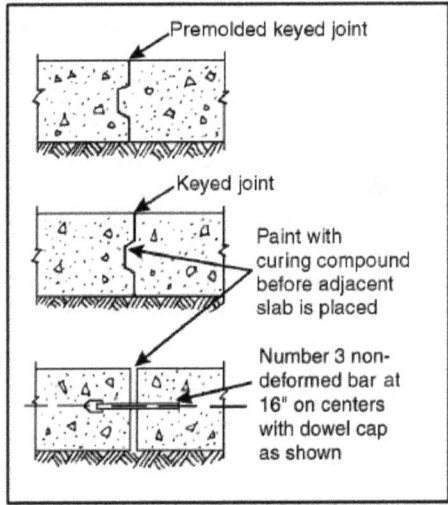

Figure 4-26. Types of construction joints

ANCHOR BOLTS

4-74. The anchor bolt (see figure 4-27) is used to anchor either machinery or structural steel. The diameter of the sleeve should be at least 1 inch larger than the bolt. The sleeve allows the bolt to shift to compensate for any small positioning error. The pipe sleeves are packed with grease to keep concrete out during paving operations. Do not use the sleeve if machinery mounts are on a raised base that is 4 or more inches above the floor. Instead set the anchor bolts in the floor so they extend above the base. Place the machine—

blocked up on scrap steel and leveled—and tighten the anchor bolts. Finally pour the base by packing the concrete into place from the sides.

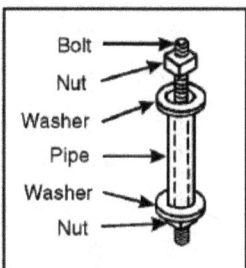

Figure 4-27. Anchor bolt with pipe sleeve

THE HOOKED ANCHOR BOLT AND THE SUSPENDED ANCHOR BOLT

4-75. The hooked anchor bolt (see figure 4-28) and the suspended anchor bolt (see figure 4-29, page 4-50) are both used to fasten a wood sill to either a concrete or masonry wall. Be careful not to disturb suspended bolt alignment when placing the concrete. Make the bolt holes in the board 1/16 inches larger than the bolt to permit adjustment.

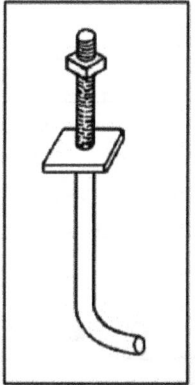

Figure 4-28. Hooked anchor bolt

Chapter 4

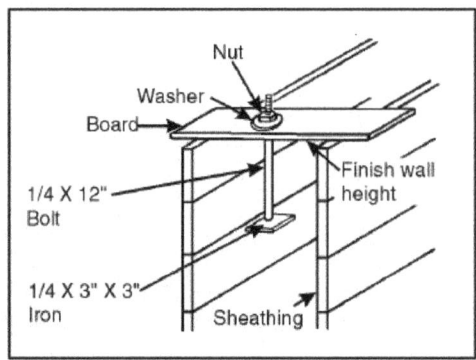

Figure 4-29. Suspended anchor bolt

A TEMPLET

4-76. (See figure 4-30). A templet is used to hold anchor bolts in place while placing concrete. When using sleeves in a templet adjust them so that the top of the sleeve is level with the top of the finished concrete.

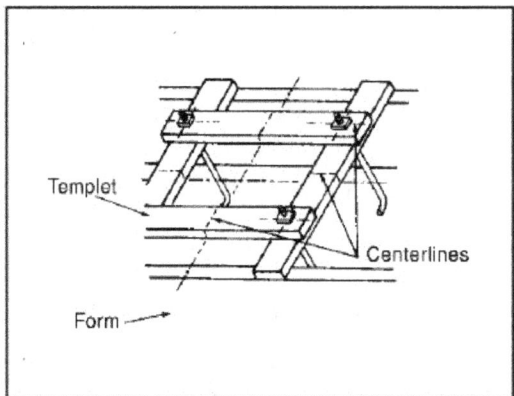

Figure 4-30. Anchor bolts held in place by a template

Chapter 5
Construction Procedures

Reconnaissance is primarily the first step taken for any construction project. Specific consideration should be given to—route selection, water and aggregate location, and time estimation. Site preparation includes—
- Clearing and draining the site.
- Establishing the location of the structure.
- Stockpiling construction materials.
- Establishing the location of the batch plants.

The types and amounts of excavation must be planned for equipment utilization and availability. Mixing, handling, and transporting must be strictly followed to maintain quality control measures. As concrete is placed in its final location, various finishing operations may be done. Initially the concrete should be screeded to the specified elevation. If a smoother surface is required, this may be accomplished by floating with a steel trowel. Curing methods, such as sprinkling the concrete with water after the initial set and placing plastic over the surface, provide effective moisture barriers that enhance the curing process.

SECTION I - RECONNAISSANCE

DETERMINING POSSIBLE DIFFICULTIES

5-1. The first step in any construction procedure is to make a thorough and efficient reconnaissance of the construction site. Note possible problems in clearing and draining the site and in transporting and storing materials. Investigate the site for any unusual characteristics that can cause construction problems such as undesirable soil or rock base. You can avoid construction delays by anticipating and considering such problems beforehand.

SPECIFIC CONSIDERATIONS

5-2. Special consideration such as the following must be given to any planning procedures:
- Route selection. Local traffic patterns, the quality of existing roads and bridges, and the equipment used all affect the selection of the best route to the construction site. Make maximum use of the existing road network; this will save time and effort by repairing or improving an existing road rather than constructing a new one. Select an alternate route when possible.
- Water and aggregate location. Locate the nearest or most convenient source of suitable mixing water. Note any alternate sources in case subsequent tests show that your first choice is unsuitable. Whenever possible, use local sand and gravel sources. Locate these sources and specify any necessary tests.
- Time estimation. Estimate the time for site preparation carefully during reconnaissance. This assures the proper equipment is available at both the place and time of need.

SECTION II - SITE PREPARATION

BUILDING APPROACH ROADS

5-3. Most new construction takes place on undeveloped land. Therefore, the approach roads need building up to deliver materials to the site. Although these are temporary roads, construct them carefully to withstand heavy loads. Build enough lanes to permit free traffic flow to and from the construction site because the routes may become permanent roads later.

CLEARING AND DRAINING THE SITE

5-4. Land clearing consists of removing all trees, downed timber, brush, and other vegetation and rubbish from the site to include the remains of previous construction efforts; digging up surface boulders and other materials embedded in the ground; and disposing of all materials cleared. Heavy equipment, hand equipment, explosives, and burning by fire may all be needed to clear the site of large timbers and boulders. The methods to be used depends on—

- The acreage to be cleared.
- The type and density of vegetation.
- The terrain affect upon equipment operations.
- The availability of equipment and personnel.
- The time available until completion.

ENSURING ADEQUATE DRAINAGE

5-5. Drainage is important in areas having high groundwater tables and for carrying off rainwater during actual construction. Use either a well-point system or mechanical pumps to withdraw surface and subsurface water from the building site.

LOCATING THE BUILDING SITE

5-6. Stake out the building site after clearing and draining the land. The batter-board layout is satisfactory in the preliminary construction phases. This method consists of placing batter boards about 2 to 6 feet outside of each corner of the site, driving nails into the boards, and extending strings between them to outline the building area.

STOCKPILING CONSTRUCTION MATERIALS

5-7. It is important to build up and maintain stockpiles of aggregate both at the batching plant and at the crushing and screening plant.

CONCRETE MATERIALS

5-8. Both aggregate and cement batching plants are essential in operations requiring large quantities of concrete. The batching-plant stockpiles prevent shortages caused by temporary production or transportation difficulties that allows the FA to reach a fairly stable and uniform moisture content and bulking factor. Large stockpiles are usually rectangular for ease in computing volumes. They are flat on top to retain gradation uniformity and to avoid segregation caused by dumping aggregate so that it runs downs a long slope; enough cement must be maintained at the cement-batching plant. The amount of concrete required by the project and the placement rate determine the size of the stockpiles. If admixtures are used, make sure that enough are on hand.

Lumber At Construction Site

5-9. Stockpile plenty of formwork and scaffolding materials at the construction site. The size and quantity of lumber stored depends on the type of forms and/or scaffolding used.

LOCATING BATCHING PLANTS

5-10. The initial location of the aggregate, cement, and water; the aggregate quality; and the location of the work all affect where the cement batching plant is positioned. Depending on these conditions, you can operate it at the same place as the aggregate batching plant or closer to the mixer. After developing a layout, position the batching plant within crane reach of the aggregate stockpiles and astride the batch truck routes. Although the crushing and screening plant is normally located at the pit, it can be operated at the batching plant or at a separate location. A hillside location permits gravity handling of materials without excessive new construction. This may eliminate the need for cranes or conveyors if the road is good.

CONSTRUCTING SAFETY FACILITIES

5-11. Plan and construct the safety facilities during site preparation. This includes overhead canopies and guardrails both to protect personnel from falling debris and to prevent anyone from falling into open excavations. Certain sites, such as those where landslides may occur, require additional safety facilities.

SECTION III - EXCAVATION

EXCAVATION AND SHORING CONSIDERATIONS

5-12. When the building site is cleared, drained, and outlined, cut the land to the proper elevation for placing footings. Use suitable equipment for the initial excavation, but excavate to the final depths by hand. The excavation should extend beyond the exterior wall edges to allow for placing forms and applying waterproofing material. Even if you excavate too much earth, place the concrete to the actual excavation depth. Attempts to refill an excavation to the depth specified are not recommended unless an elephant-foot tamper is used to properly compact the fill, because it is difficult to compact the fill surface properly. Some type of lateral provides support for both safety and economy whenever excavation is at such a depth that the slopes become unstable. Good engineering practice dictates using shoring whenever slope stability is questionable. The type of shoring varies with the depth and size of the excavation, the physical characteristics of the soil, and the fluid pressure under saturated conditions. Sandy soils and wet earth generally require more extensive shoring than firmer soils.

Machine Excavation

5-13. Machines are a necessity for large projects requiring substantial excavation. The most suitable types of equipment include power shovels, dragline buckets, and a backhoe. When selecting equipment consider:

- The total yardage to be move.
- The working time available.
- The type of excavation.
- The nature of the area.

5-14. Due to the many variables, it is not possible to give generalized rates of excavation for various types of equipment. However, table 5-1, page 5-4, gives some typical rates of excavation for specific conditions that still will vary considerably in practice.

Chapter 5

Table 5-1. Machine excavation

Equipment	Type of Material	Average Output, in Cubic Yards per Hour
Power shovel (1/2 cu yd capacity)	Sandy loam Common earth Hard clay Wet clay	70 60 45 25
Short-boom dragline (1/2 cu yd capacity)	Sandy loam Common earth Hard clay Wet clay	65 50 40 20
Backhoe (1/2 cu yd capacity)	Sandy loam Common earth Hard clay Wet clay	55 45 35 25

HAND EXCAVATION

5-15. Table 5-2 gives hand excavation rates which vary with soil types and excavation depth. Clear out and shape the last 6 inches of bottom excavation by hand; it is extremely difficult to excavate that closely with a machine.

Table 5-2. Hand excavation

Types of Material	Average Output Yards Per Hour					
	Excavation With Pick and Shovel to Depth Indicated				Loosening Earth - One Worker With Pick	Loading in Trucks or Wagons - One Worker With Shovel, Loose Soil
	0 to 3 ft	0 to 6 ft	0 to 8 ft	0 to 10 ft		
Sand	2.0	1.8	1.4	1.3	-	1.8
Silty sand	1.9	1.6	1.3	1.2	6.0	2.4
Loose gravel	1.5	1.3	1.1	1.0	-	1.7
Sandy silt-clay	1.2	1.2	1.0	.9	4.0	2.0
Light clay	.9	.7	.6	.7	1.9	1.7
Dry clay	.6	.6	.5	.5	1.4	1.7
Wet clay	.5	.4	.4	.4	1.2	1.2
Hardpan	.4	.4	.4	.3	1.4	1.7

Construction Procedures

SECTION IV - FORM WORK

MANAGEMENT ASPECTS

5-16. To perform a proper analysis, you must have a working knowledge of the equipment necessary for the form work job and a good idea of how much work the form builders can turn out per unit of time. It is important to be knowledgeable of—

- Equipment. The average form work job requires claw hammers, pinch bars, hand saws, portable electric saws, table saws, levels, plumb lines, and carpenter's squares. These tools should be readily available.
- Techniques. Break large projects into smaller, nearly identical units. Develop standardized methods for constructing, erecting, and stripping forms to the maximum extent possible. This saves time and material, and simplifies design problems.

TIME ELEMENT

5-17. A carpenter of average skill can build and erect 10 square feet of wood forms per hour. This figure increases as the worker becomes more skilled in form construction. It also varies with the tools and materials available and the type of form. Some forms, such as those for stairways, require considerable physical support from underneath. Some forms take more man-hours and materials to build than simpler forms. For carpenters to move from one level to another frequently requires additional time. Therefore, increased man power support at the ground level increases efficiency.

SECTION V - MIXING

PRINCIPLES

5-18. Established and well-defined concrete-mixing procedures must be followed to produce good quality finished concrete. Never become overconfident in this phase of concrete construction, whether caused by lack of competent and conscientious supervision or inattention to detail. Whoever is in charge of construction must know the concrete-mixing procedures and ensure that they are followed. The extra effort and care this requires are small in relation to the benefits in terms of strength consistency and finished appearance.

MEASURING MIX MATERIALS

5-19. Concrete of uniform quality requires measuring the ingredients accurately for each batch within these percentages: cement, 1 percent; aggregate, 2 percent; water, 1 percent; and admixtures, 3 percent. Equipment should be capable of measuring quantities within these tolerances for the smallest batch used, as well as for larger batches. Periodically check equipment for accuracy and adjust when necessary. Check admixture dispensers daily for errors in admixture measurements—particularly over-dosages; this can cause serious problems in both fresh and hardened concrete. Always measure the following:

- Cement. Concrete mixes normally call for sacked cement as the unit of measure, although bulk cement is common practice in commercial construction requiring large quantities. Bulk cement is stored in bins directly above a weighing cement hopper and discharged from the hopper. However, because bulk cement requires special equipment for transport, sacked cement is used almost exclusively in troop construction, particularly in the theater of operations.
- Aggregate. Measure aggregate for each batch accurately either by weight or by volume. Measurement by weight is the most reliable because the accuracy of volume measurement depends on an exact knowledge of the amount of moisture in the sand. Nevertheless, sometimes measurement by volume is more practical.

Chapter 5

- Measurement by weight. On comparatively small jobs, you can use platform scales placed on the ground to weigh aggregate. Construct runways as shown in figure 5-1, so that a wheelbarrow can run onto one side of the scale and off the other easily. With practice, you can fill a wheelbarrow so accurately that adding or removing material to obtain the correct weight is seldom necessary. Always place the same weight of aggregate on each wheelbarrow so that the quantity per batch equals the same number of wheelbarrow loads. Do not load a wheelbarrow to capacity.

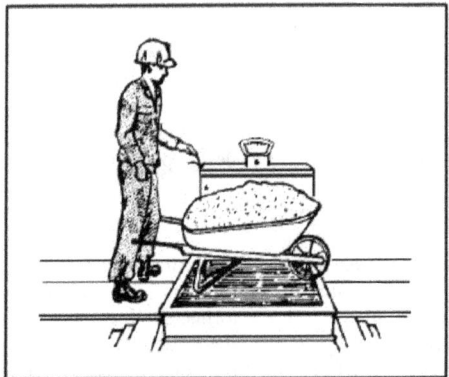

Figure 5-1. Measuring aggregate by weight

- Measurement by volume. Measure aggregate by volume using a 1-cubic foot measurement box built on-site or a wheelbarrow. Wheelbarrows having from 2 to 3-cubic feet capacity are also available in the engineering units. A simple way to mix batches is the 1:2:3 method. This means that each batch contains 1 part cement, 2 parts sand, and 3 parts aggregate, regardless of the units of measure used (shovels, cubic feet, wheelbarrows, and so forth).
- Water. Measure mixing water accurately for every batch. If the aggregate contains too much moisture, be sure to take this into account when adding mixing water. The water tanks on machine mixers often have automatic measuring devices that operate like a siphon.

HAND MIXING

5-20. Although a machine generally does the mixing, hand mixing sometimes may be necessary. A clean surface is needed for this purpose, such as a clean, even, paved surface or a wood platform having tight joints to prevent paste loss like the one shown at the top of figure 5-2. Moisten the surface and level the platform, spread cement over the sand, and then spread the CA over the cement as shown at the top of figure 5-2. Use either a hoe (see the middle of figure 5-2) or a square-pointed D-handled shovel to mix the materials. Turn the dry materials at least three times until the color of the mixture is uniform. Add water slowly while you turn the mixture again at least three times, or until you obtain the proper consistency. Although one worker can mix 1 cubic yard of concrete by hand in about 1 hour, this is not economical for batches of more than 1 cubic yard. Instead, two workers facing each other should work their way through the pile, and keep their shovels close to the platform surface while turning the materials. You can also mix in a hoe box shown at the bottom of figure 5-2.

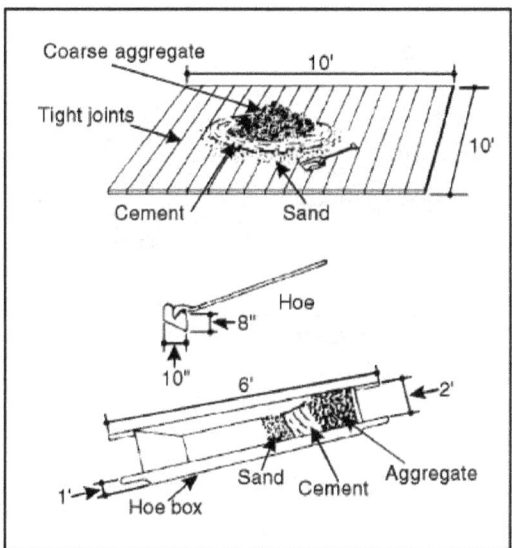

Figure 5-2. Hand mixing of concrete

MACHINE MIXING AND DELIVERY

5-21. The methods of mixing and delivering concrete ingredients and the types and sizes of equipment available vary greatly. Power-concrete mixers normally produce one batch about every 3 minutes including charging and discharging. Actual hourly output varies from 10 to 20 batches per hour. A mixer's cubic foot rating usually reflects the number of cubic feet of usable concrete that the machine mixes in one batch. Most mixers can handle a 10 percent overload. The stationary 16-cubic-feet mixer and the M919 Concrete-Mobile-Mixer unit (see figure 5-3, page 5-8) are table of organization and equipment (TOE) equipment in engineer construction battalions, and are well-suited for troop construction projects.

MIXING METHODS

5-22. Mixing methods include—

- Site mixed. Method used for delivering plastic concrete by chute, pump, truck, conveyor, or rail dump cars.
- Central-plant mixed. Method used for delivering plastic ready-mix in either open dump trucks or mixer trucks.
- Central-plant batched (weighed and measured). Method used for mixing and delivering "dry-batched" ready-mix by truck.
- Portable-mixing plant mixed. Method used for large building or paving projects distant from sources of supply.

Chapter 5

Figure 5-3. M919 concrete mobile mixer

MIXER TYPES

5-23. Mixer types include—

- Stationary mixers, which include both the on-site mixers and the central mixers in the ready-mix plants, are available in various sizes. They may be tilting or nontilting with open top, revolving blades or paddles.
- Mobile mixers include both truck- and trailer-mounted mixers. A truck mixer may pick up concrete from the stationary mixer in a partially or completely mixed state. In the latter case, the truck mixer functions as an agitator. Truck mixers generally deliver concrete from a centrally located stationary mixer to the construction site or pick up materials at a batching plant and mix the concrete en route to the job site. Trailer-mounted mixers are commonly used to patch concrete pavements and to form and widen curve during pavement construction. A battery of trailer-mounted mixers can serve either as a central mix plant for large scale operations or in conjunction with a central mix plant.

CENTRAL MIX PLANTS

5-24. Either stationary mixers or a battery of trailer-mounted mixers usually makes up a central-mix plant which is normally a gravity-feed operation. A clamshell bucket crane, conveyor belts, or elevators carry materials to a batching plant set high enough to discharge directly into dump trucks or other distribution equipment. Mixing time and mixing requirements do not differ much from those already discussed, but you must take special precautions to make sure that the concrete has the proper characteristics and workability upon arrival at the work site. Be especially careful to avoid segregation when using dump trucks.

SITE MIX PLANTS

5-25. The proper location of mixing equipment and materials at the site can yield large savings in time and labor. Figure 5-4 shows a typical on-site arrangement of mixer and materials. Always locate the mixer as close to the main section of the pour as possible. On a concrete wall project move the mixer to each wall in sequence to reduce transportation distance and time.

Construction Procedures

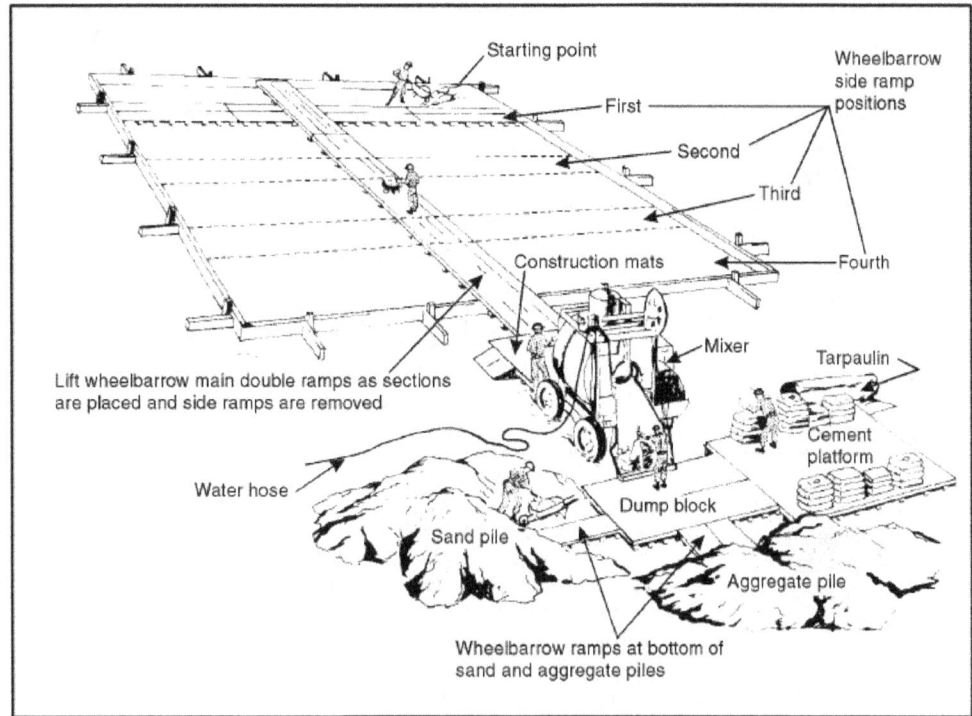

Figure 5-4. Typical on-site arrangement of mixing equipment and materials

5-26. Store aggregate and sand as close to the mixer as possible, without interfering with concrete transportation. The location of the mixing water depends on the type of water supply available. If water is piped in, use a hose to carry it to a barrel near the mixer. If you are using a water truck or trailer, park it next to the mixer.

OPERATING THE 16-CUBIC FOOT MIXER

5-27. Table 5-3, page 5-10, gives the physical characteristics of a typical 16-cubic-feet mixer. Normal operation of the mixer requires ten soldiers and one noncommissioned officer. The crew operates the mixer and handles the aggregate, sand, cement, and water. The noncommissioned officer, who must be competent, supervises the overall operation. The crew should produce about 10 cubic yards of concrete per hour, depending on their experience, the location of materials, and the mixer's discharge rate. You would need at least one platoon to operate an overall project like the one in figure 5-4.

Table 5-3. Physical characteristics of a typical 16-cubic feet mixer

Physical Characteristics	Concrete Mixer*	Physical Characteristics	Concrete Mixer*
Drum capacity • Hourly production • Rating, in sacks per batch	16 cu ft 10 cu yd 2	Drum Dimensions • Diameter (in) • Length (in)	57 46
Power unit • Horsepower • Fuel consumption (gal/hr)	26 0.5	Overall dimension (in) • Length • Width	158 96
Water tank • Supply • Measurement (gal)	none 26	Height (in) Weight (lb)	119 7,150
*The mixer is gas driven, liquid cooled, end discharge, trailer mounted, with 4-pneumatic-tired wheels.			

Charging the Mixer

5-28. Mixers can be charged in two ways: by hand or with a mechanical skip (see figure 5-5). When using the skip, deposit the aggregate, cement, and sand (in that order) into the skip and then dump it into the mixer while water runs into the mixing drum. Place the sand on top of the pile in the skip so that you do not lose too much cement as the batch dumps into the mixer. A storage tank on top of the mixer measures the water in the drum a few seconds before the skip dumps. This discharge also washes down the mixer between batches.

Discharging The Mixer

5-29. When the mix is ready for discharge from the mixer, move the discharge chute into place to receive the concrete from the drum. Concrete that is somewhat dry tends to cling to the top of the drum and not drop onto the chute in time. Very wet concrete may not carry up high enough on the drum to drop onto the chute. Correct these problems by adjusting the mixer speed. Increase the speed for very wet concrete and decrease the speed for dry concrete.

Mixing Time

5-30. The mixing time starts when water runs into the dry mixture. This is normally during the first quarter of the mixing period. The minimum mixing time per batch of concrete is 1 minute unless the batch exceeds 1 cubic yard. Each additional cubic yard of concrete, or fraction thereof, requires an additional 15 seconds of mixing time.

Construction Procedures

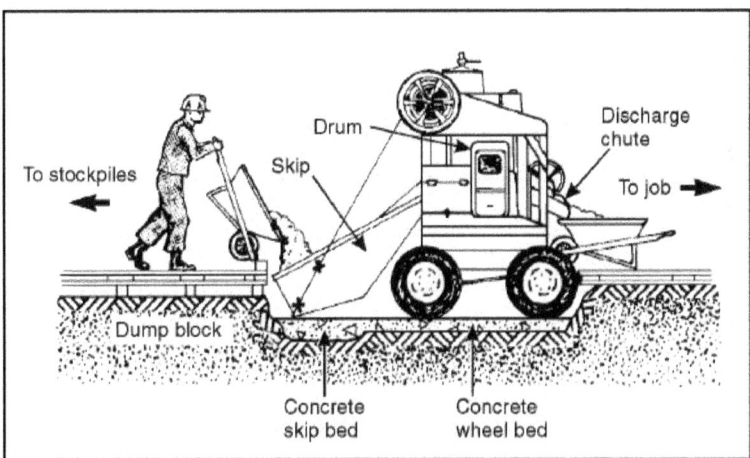

Figure 5-5. Charging a 16 cubic-foot mixer

Cleaning And Maintaining the Mixer.

5-31. Clean the mixer daily, if it operates continuously, or following each period of use if it operates less than 1 day. The exterior cleaning process goes faster if you coat the outside of the mixer with form oil before you use it. Knock off all accumulated concrete on the mixer exterior and wash it down with a hose. Mixer blades that are worn or coated with hardened concrete provide less efficient mixing action. Replace badly worn blades, and do not allow hardened concrete to accumulate in the mixer drum. Clean it out whenever you shut down for more than 1 1/2 hours. To do this, place a volume of CA equal to one-half the mixer capacity in the drum and allow it to revolve for about 5 minutes. Then discharge the aggregate and flush out the drum with water. Never strike the discharge chute, drum shell, or skip to remove aggregate or hardened concrete, because concrete adheres more readily to dents and bumps. Diligence in cleaning the drum is important because hardened concrete in a mixer absorbs mixing water during subsequent batches, diminishing the effectiveness and composition of the concrete.

OPERATING THE M919 CONCRETE-MOBILE-MIXER UNIT

5-32. The Concrete-Mobile-Mixer unit is a combination materials transporter and an on-site mixing plant. Table 5-4, page 5-12, gives its physical characteristics and overall dimensions. The special body is mounted on a model M919 truck chassis. The unit carries enough unmixed material to produce up to 8 cubic yards of fresh concrete. Because the unit is precisely calibrated, you can produce mixes that meet or exceed both the ACI and the American Association of State Highway and Transportation Officials (AASHTO) standards for design strength. The unit operates on either an intermittent or continuous basis, although continuous operation depends on raw material availability at the site. Certain control settings for the mix operations vary from truck to truck and from site to site.

Table 5-4. M919 Concrete-mobile-mixer unit

Physical Characteristics	
Capacity	8 cubic yards
Hourly production	16 cubic yards
Rating, in sacks per batch	Varies
Power unit	M915-series Cummins engine powered by power take-off from the truck transmission
Water tank	400 gallons
Cement bin	63 cubic feet
Sand bin	128 cubic feet
Gravel bin	182 cubic feet
Hi flow admix tank	42 gallons
Lo flow admix tank	12 gallons
Dry admix tank	2.35 cubic feet
Overall Dimensions and Weight	
Length	374 inches
Width	96 inches
Height	142 inches
Weight (empty)	37,540 pounds
Use the dials on the rear and sides of the M919 to control concrete mixes by regulating the amount of sand, gravel, water, and admixes.	

REMIXING CONCRETE

5-33. When fresh concrete is left standing and not poured, it tends to stiffen before the cement can hydrate to its initial set. Such concrete is still usable if remixing makes it sufficiently plastic to be compacted in the forms. To remix a batch, carefully add a small amount of water, and remix the concrete for at least one-half of the minimum required mixing time or number of revolutions. Then test the concrete to make sure that it does not exceed the maximum allowable W/C ratio, maximum allowable slump, or maximum allowable mixing and agitating time. Do not add water randomly because this lowers concrete quality. Remixed concrete tends to harden rapidly. Any fresh concrete placed adjacent to or above remixed concrete may cause a cold joint.

SECTION VI - HANDLING AND TRANSPORTING

PRINCIPLES

5-34. Concrete consistency depends on the conditions at placement. Handling and transporting methods can affect its consistency. Therefore, if placing conditions allow a stiff mix, choose equipment that can handle and transport such a mix without affecting its consistency. Carefully control each handling and transporting step to maintain concrete uniformity within a batch and from batch to batch so that the completed work is consistent throughout.

HANDLING TECHNIQUES

5-35. Figure 5-6, page 5-14, shows several right and wrong ways to handle concrete to prevent segregation of the aggregates and paste. Segregation causes honeycomb concrete or rock pockets. Segregation occurs because concrete contains aggregates of different particle sizes and specific gravities. When placed in a bucket, the coarser particles tend to settle to the bottom and the water rises to the top.

TRANSPORTATION REQUIREMENTS

5-36. The three main requirements for transporting concrete from the mixing plant to the job site are:

- Speed. Fast transportation does not allow concrete to dry out or lose workability or plasticity between mixing and placing.
- Minimum material segregation. To produce uniform concrete take steps to reduce segregation of the aggregates and paste to a minimum; this will help prevent the loss of fine material, cement, or water.
- No delays. Organize the transportation to eliminate delays in concrete placement that cause undesirable fill planes or construction joints.

Chapter 5

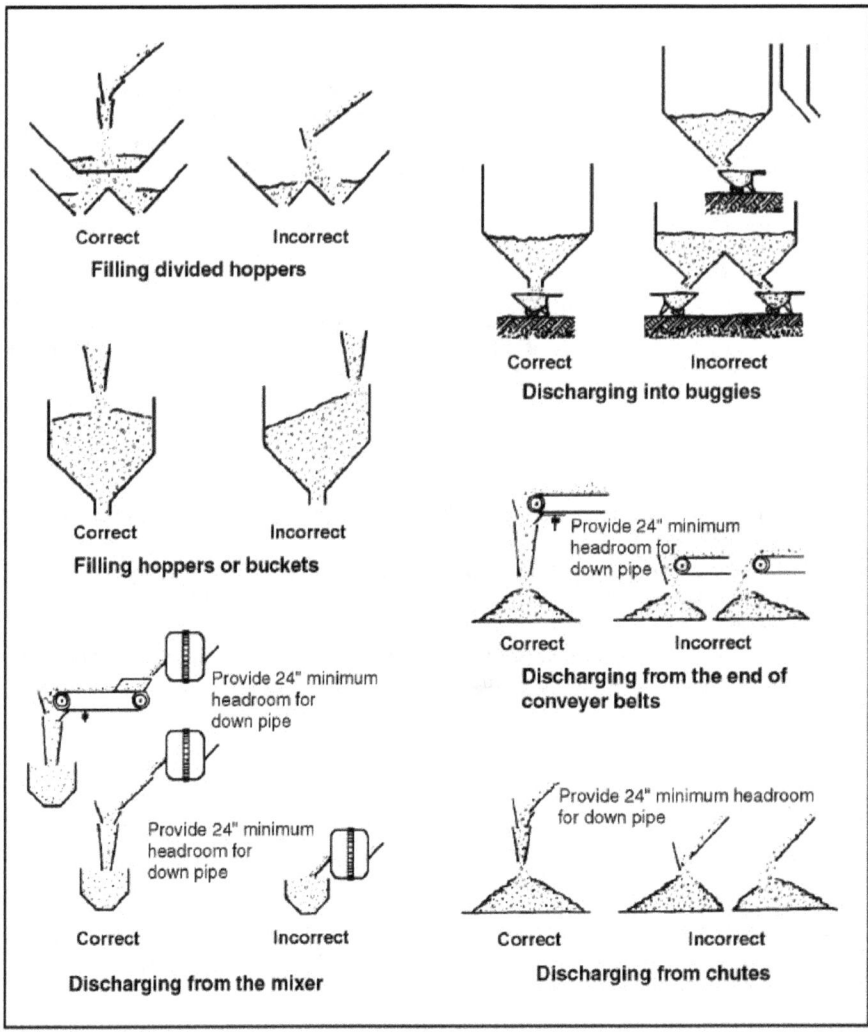

Figure 5-6. Concrete handling techniques to prevent segregation

DELIVERY METHODS

5-37. Concrete delivery equipment must be capable of handling 100 percent of the mixer capacity to meet peak demands.

SMALL JOBS

5-38. Wheelbarrows or buggies (see figure 5-7) are the most practical and economical to deliver concrete for foundations, foundation walls, or slabs on or below grade. If available, power buggies (see figure 5-8) are best for longer runs. Both wheelbarrows and buggies require suitable runways. If possible, arrange them so that the buggies or wheelbarrows do not need to pass each other on a runway at any time. When placing concrete below or at approximate grade, set 2-inch plank runways directly on the ground to permit pouring the concrete directly into the form. View 1 in figure 5-9, page 5-16, shows a runway along a wall form that is almost level with the top of the form sheathing. To provide room for the ledger, the top wale is at 1 foot below the top of the concrete. A runway can be made economically from rough lumber consisting of 2 by 10 planks, supported by 4 by 4s, and spaced on 6- foot centers. Nail the 1 by 6 ledger on the form side to the studs on the wall form. Always brace runways securely to prevent failure. A runway for placing concrete on a floor slab also consists of 2 by 10s supported by 4 by 4s placed on 6-foot centers. Such runways are built from either the formwork or the ground. When placing concrete about 5 to 6 feet above grade, economically construct inclined runways (see view 2 of figure 5-9) for buggy or wheelbarrow use. This will allows wheeling the concrete up inclined runways having a slope of 10:1. If a large quantity of concrete more than 5 to 6 feet is elevated, it is more economical to use elevating equipment, such as a bucket and a crane. If the difference in elevation from the runway to the bottom of the structure is large, use a hopper and chute like the ones shown in figure 5-10, page 5-17, to prevent segregation. The slope of the chute should normally be 2:1 or steeper for stiff mixes.

Figure 5-7. Handling concrete by buggy

Figure 5-8. Handling concrete by power buggy

Chapter 5

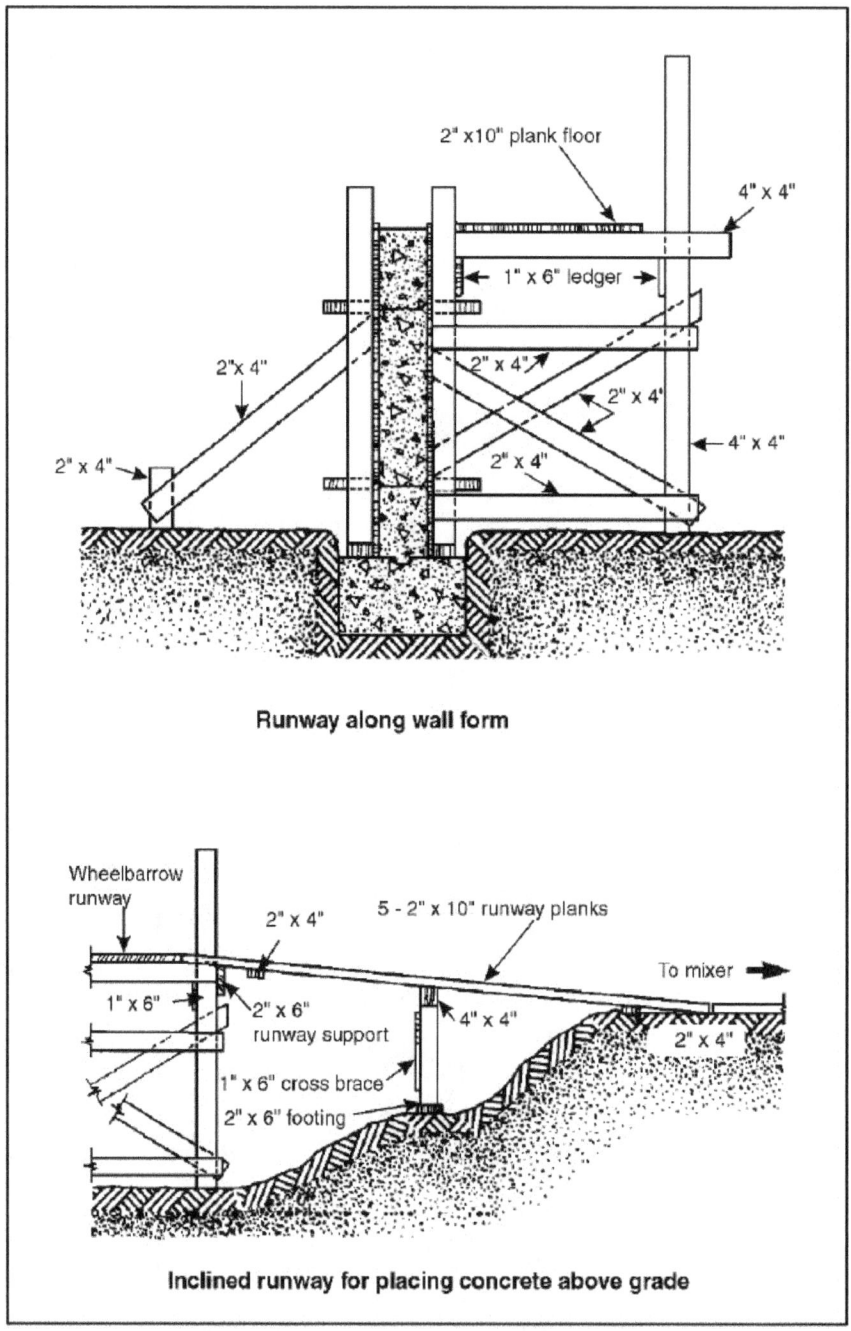

Figure 5-9. Runways for wheelbarrow or buggy use

Construction Procedures

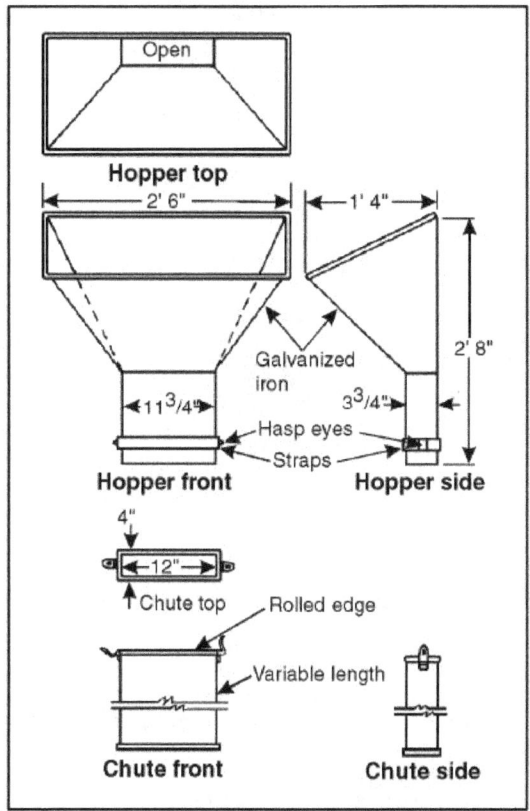

Figure 5-10. Hopper and chute for handling concrete

LARGE JOBS

5-39. Timely delivery of enough concrete on large projects requires careful planning and the selection of the right type of equipment for the purpose.

Dump trucks

5-40. Ordinary dump trucks are not designed as concrete carriers although they are commonly used to deliver concrete on large projects in the theater of operations. Exercise care in using them because no means of preventing segregation is provided as it is on mixers or agitator trucks. Even if you use a stiff mix and an air-entraining agent to reduce segregation, keep hauling distances as short as possible, maintain slow speed, and utilize smooth roads to reduce vibration.

Agitator trucks

5-41. You can use either transit-mix or ready-mix trucks as agitator trucks to deliver premixed concrete. A transit-mix truck is a concrete truck (M919) where materials are mixed after they arrive on the site. A ready-mix truck comes from the batch plant to the site already mixed, ready for placement. The load capacity of a ready-mix trucks is 30 to 35 percent greater than transit-mix trucks, but the operating radius of the ready-mix trucks is somewhat more limited. The trucks discharge concrete either continuously or intermittently from a spout or chute that moves from side to side.

Chapter 5

- Buckets. Figure 5-11 shows a crew using a bucket and crane to place concrete. Buckets in 2 yard capacities are standard TOE and Class IV items in combat heavy engineer battalions. They are either square or cylindrical with a clamshell door or gate at the bottom. The doors or gates are hand operated for flexibility in discharging the bucket.

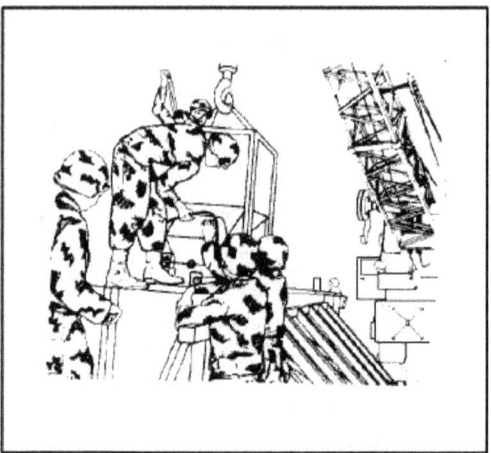

Figure 5-11. Placing concrete using a bucket and crane

- Pumps. When limited space prevent other more conventional means of delivery, use a heavy-duty piston pump to force concrete through 6-, 7-, or 8-inch pipeline as shown in figure 5-12. Pumps operated by a 25 horsepower (HP) gasoline engine have a rated capacity of 15 to 20 cubic yards per hour. Larger equipment with a double-acting pump has a rated capacity of 50 to 60 cubic yards per hour. Both machines can pump a mix having 2 or more inches of slump and force their rated capacities up to 800 feet horizontally, 100 feet vertically, or any equivalent combination of these distances. However, a 90-degree bend in the pipeline decreases horizontal delivery distance by 40 feet, and each foot of vertical lift decreases horizontal delivery distance by 8 feet. When starting the pump, lubricate the pipeline with a light cement grout first. Then make sure the pump receives an uninterrupted flow of fresh, plastic, unsegregated concrete having medium consistency. Maximum aggregate size is 3 inches for the 8-inch pipeline, 2 1/2 inches for the 7-inch pipeline, and 2 inches for the 6-inch pipeline. The discharge line should be as straight as possible, with a 5-foot-radius bend. Take appropriate steps to cool the pipe in hot weather for smooth concrete flow. An interrupted flow in the pipeline can seriously delay the concrete pour, causing undesirable joints in the structure. A deflector or choke (restricted section) can be used at the discharge end of the pipeline to direct and control the discharge flow. Be sure to thoroughly flush both the pump and line with water after each use. One disadvantage of using pumps is that the concrete mix is usually more expensive due to the smaller aggregate required to pass through the pump, resulting in a correspondingly higher requirement for expensive cement.

Construction Procedures

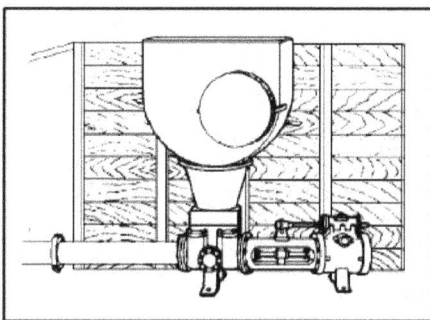

Figure 5-12 Piston pump and discharge pipeline

SECTION VII - PLACEMENT

IMPORTANCE OF PROPER PROCEDURES

5-42. You cannot obtain the full value of well-designed concrete without using proper placing and curing procedures. Good concrete placing and compacting techniques produce a tight bond between the paste and the CA and fills the forms completely, both contributes to the full strength and best appearance.

PRELIMINARY PREPARATION

5-43. Preparation prior to concrete placement includes compacting, moistening subgrade or placing vapor barrier, erecting forms, and setting reinforcing steel.

5-44. Moistening the subgrade is especially important in hot weather to prevent water extraction from the concrete. Preparation includes—

- Preparing rock surfaces. Cutting rocks out will make the surfaces either vertical or horizontal, rather than sloping. The rock surfaces should be roughened and thoroughly cleaned. Use stiff brooms, water jets, high-pressure air, or wet sandblasting. Remove all water from the depressions and coat the rock surfaces with a 3/4-inch-thick layer of mortar. Make the mortar with only FA using the same W/C ratio as the concrete. The mortar should have a 6-inch slump. Finally, work the mortar into the rock surfaces using stiff brushes.
- Moistening clay subgrades. Moisten a subgrade composed of clay or other fine-grained soils to a depth of 6 inches to help cure the concrete. Sprinkling the soil with water intermittently will saturate it without making it muddy. Clean the surface of debris and dry loose material before placing the concrete.
- Preparing gravel and sand subgrades. Cover a subgrade consisting of gravel, sand, or other loose material with tar paper or burlap before placing concrete. Compacted sand is not covered, but be sure it is moist before placement to prevent water absorption from the concrete. Lap tar paper edges not less than 1 inch and staple them together. Join burlap edges by sewing them together with wire. Sprinkle the burlap with water before placement.
- Preparing forms. Just before placement, check the forms for both tightness and cleanliness. Check the bracing to make sure the forms will not move during placing. Make sure that the forms are coated with a suitable form oil or coating material so that the concrete will not stick to them. Remember, in an emergency, moisten the forms with water to prevent concrete from sticking. Forms exposed to the sun for some time dry out and the joints tend to open up. Saturating such forms with water helps to close the joints.
- Planning placement. The joint between fresh and hardened concrete is never as strong as a continuous slab. Ideally, all the concrete required in a project is placed at once in a monolithic

placement. If subsequent placements are needed between placements, see figure 4-2, page 4-4. The size of each day's placement is planned based on where construction joints can be placed.
- Depositing fresh concrete on hardened concrete. To obtain a good bond and a watertight joint when depositing new concrete on hardened concrete, make sure that the hardened concrete is nearly level, is clean and moist, and that some aggregate particles are partially exposed. If the surface of the hardened concrete is covered by a soft layer of mortar or Laitance (a weak material consisting mainly of lime), remove it. Wet sand-blasting and washing is the best way to prepare old surfaces if the sand deposit can easily be removed Always moisten hardened concrete before placing new concrete; saturate dried-out concrete for several hours. Never leave pools of water on the old surface when depositing fresh concrete on it.

PLACING CONCRETE

5-45. The principles of proper concrete placement include—
- Segregation. Avoid segregation during all operations from the mixer to the point of placement, including final consolidation and finishing.
- Consolidation. Thoroughly consolidate the concrete, working solidly around all embedded reinforcement and filling all form angles and corners.
- Bonding. When placing fresh concrete against or upon hardened concrete, make sure that a good bond develops.
- Temperature control. Take appropriate steps to control the temperature of fresh concrete and protect the concrete from temperature extremes after placement.
- Maximum drop. To save time and effort, it is tempting to simply drop the concrete directly from its delivery point regardless of form height. However, unless the free fall into the form is less than 5 feet, use vertical pipes, suitable drop chutes, or baffles. Figure 5-6, page 5-14, suggests several ways to control concrete fall and prevent honey-combing and other undesirable results.
- Layer thickness. Try to place concrete in even horizontal layers; do not puddle or vibrate it into the form. Place each layer in one operation and consolidate it before placing the next one to prevent honeycombing or voids, particularly in wall forms containing considerable reinforcement. Use a mechanical vibrator or a hand spading tool for consolidation. Take care not to overvibrate, because segregation and a weak surface can result. Do not allow the first layer to take its initial set before adding the next layer. Layer thickness depends on the type of construction, the width of the space between forms, and the amount of reinforcement. When depositing from buckets in mass concrete, the layers should be from 15 to 20 inches thick. For reinforced concrete members, the layers should be from 6 to 20 inches thick.
- Compaction. Place concrete as near to its final position as possible. Work the concrete thoroughly around reinforcement and embedded fixtures, into the corners, and against the sides of the forms. Because paste tends to flow ahead of aggregate, avoid horizontal movements that result in segregation.
- Placement rate. To avoid too much pressure on forms for large projects, the filling rate should not exceed 5 vertical feet per hour, except for columns. Coordinate the placing and compacting so that the concrete is not deposited faster than it can be compacted properly. To avoid cracking during settlement, allow an interval of at least 4 hours, but preferably 24 hours, between placing columns and walls, and placing the slabs, beams, or girders they support.

WALL CONSTRUCTION

5-46. When constructing walls, beams, or girders, place the first batches of each layer at the ends of the section, then proceed toward the center to prevent water from collecting at the form ends and corners. For walls, stop off the inside form at the construction level. Overfill the form to about 2 inches and remove the excess just before the concrete sets to make sure of a rough, clean surface. Before placing the next lift of concrete, deposit a 1/2-to 1-inch thick layer of sand cement mortar. Make the mortar with the same water content as the concrete, with a slump of about 6 inches to prevent stone pockets this will help produce a watertight joint. View 1 of figure 5-13, page 5-22, shows the proper way to place concrete in the lower portion of high wall forms. Note the different kinds of drop chutes that can place concrete through port openings into the lower portion of the wall. Space the port openings at about 10-foot intervals up the wall. Show how to place concrete in the upper portion of the wall. (View 2 of figure 5-13.) When placing walls, be sure to remove the spreaders as you fill the forms.

SLAB CONSTRUCTION

5-47. When constructing slabs, place the concrete at the far end of the slab first, and then place subsequent batches against previously placed concrete, as shown in view 3 of figure 5-13. Do not place the concrete in separate piles, level the piles and work them together, or deposit the concrete in big piles and then move it horizontally to its final position. These practices would result in segregation.

PLACING CONCRETE ON SLOPES

5-48. View 4 of figure 5-13 shows how to place concrete on slopes. Always deposit the concrete at the bottom of the slope first, then proceed up the slope placing each new batch against the previous one. When consolidated, the weight of the new concrete increases the compacting of the previously placed concrete.

CONSOLIDATING CONCRETE

5-49. Concrete consolidation eliminates rock pockets and air bubbles.

PRINCIPLES

5-50. Except for concrete placed underwater, compact or consolidate all concrete after placement. Consolidation brings enough fine material both to the surface and against the forms to produce the desired finish. Mechanical vibrators are best for consolidating concrete, such hand tools such as spades, puddling sticks, or tampers, may also be used. Any compacting device must reach the bottom of the form and be small enough to pass between reinforcing bars. The process involves carefully working around all reinforcing steel with the compacting device to ensure proper embedding of reinforcing steel in the concrete. Be careful not to displace the reinforcing steel because the strength of the concrete member depends on proper reinforcement location.

VIBRATION

5-51. Mechanical vibrators consolidate concrete by pushing the CA downward away from the point of vibration. (See figure 5-14, page 5-23.) Vibrators allow placement of mixtures that are too stiff to place any other way, such as those having only a 1- or 2-inch slump. Stiff mixtures are more economical because they require less cement and present fewer segregation or excessive bleeding problems. However, do not use a mix so stiff that it requires too much labor to place it. The vibrators available in engineer construction battalions are called internal vibrators, because the vibrating element is inserted into the concrete. An external vibrator is applied to the form and is powered by either an electric motor, gasoline engine, or compressed air. When using an internal vibrator, insert it at about 18-inch intervals into air-entrained concrete for 5 to 10 seconds and into nonairentrained concrete for 10 to 15 seconds. The exact time period that you should leave a vibrator in the concrete depends on its slump. Overlap the vibrated areas at each insertion. Whenever possible, lower the vibrator into the concrete vertically and allow it to descend by gravity. The vibrator should not only pass through the layer just placed, but penetrate several inches into the layer underneath to make sure a good bond exists between the layers. Vibration does not normally damage the lower layers, as long as the concrete disturbed in these lower layers becomes plastic under the vibratory action. The concrete will be consolidated enough when a thin line of mortar appears along the form near the vibrator, the CA disappears into the concrete, or the paste just appears near the vibrator head. Withdraw the vibrator vertically at about the same gravity rate that it descended. Some hand spading or puddling should accompany all vibration. Do not vibrate mixes that you can consolidate easily by spading, because segregation may occur. Vibrated concrete has a slump of 5 or 6 inches. Use vibrators to move concrete any distance in the form.

Chapter 5

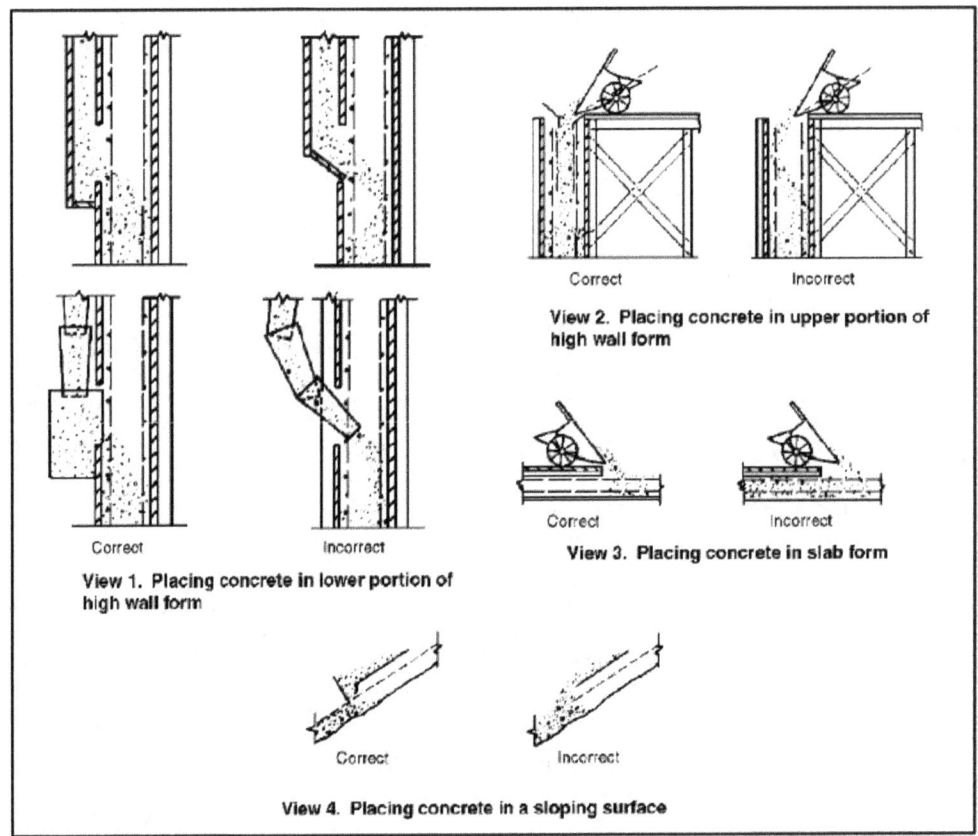

Figure 5-13. Concrete placing technique

Hand Methods

5-52. Manual consolidation methods require spades, puddling sticks, or various types of tampers. To consolidate concrete by spading insert the spade downward along the inside surface of the forms, as shown in figure 5-15, through the layer just placed into the layer underneath several inches. Continue spading or puddling until the CA disappears into the concrete.

Construction Procedures

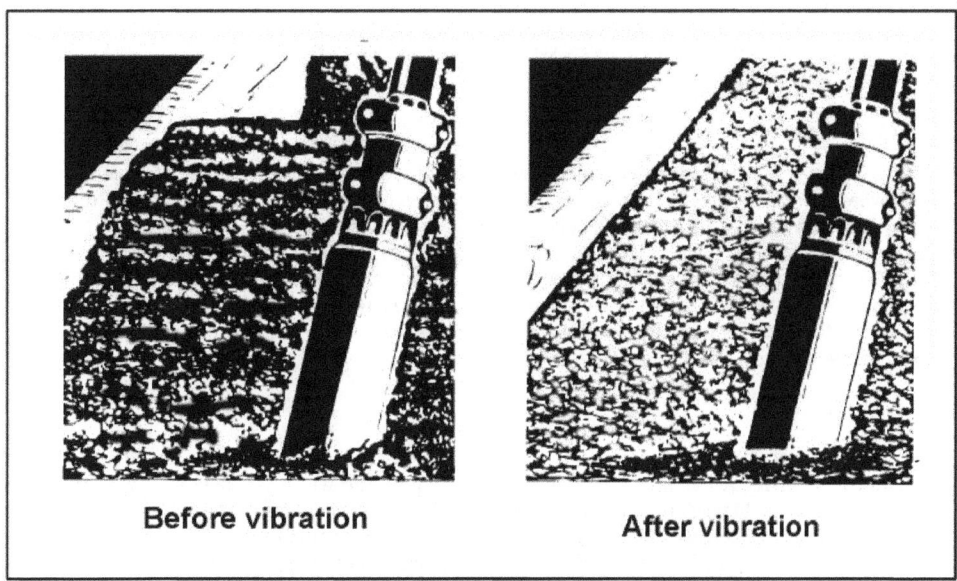

Figure 5-14. Using a vibrator to consolidate concrete

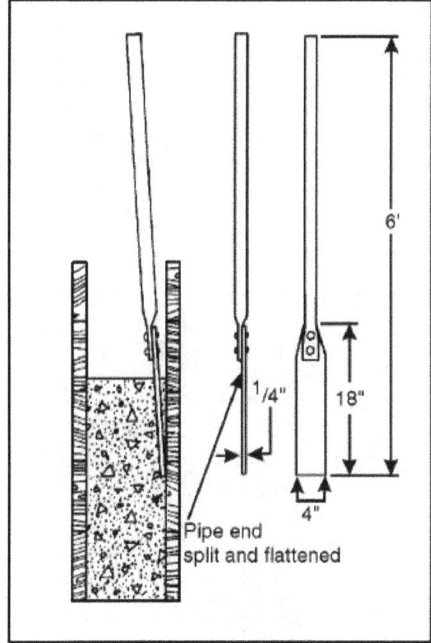

Figure 5-15. Consolidation by spading and a spading tool

Chapter 5

PLACING CONCRETE UNDERWATER

5-53. Whenever placing concrete underwater, make sure that the work is done under experienced supervision and that certain precautions are taken.

SUITABLE CONDITIONS

5-54. For best results, do not place concrete in water whose temperature is below 45°F, nor in flowing water whose velocity is greater than 10 feet per minute, even though sacked concrete can be placed in water having greater velocities. If concrete is placed in water temperature below 45°F, make sure the concrete temperature is above 60°F when deposited, but never above 80°F. If the velocity of flowing water is greater than 10 feet per minute, make sure that the cofferdams or forms are tight enough to reduce the velocity through the space to be concreted to less than 10 feet per minute. Do not pump water either while placing concrete or for 24 hours thereafter. Attempting to remove water during this time may cause water to mix inadvertently with concrete thus reducing strength.

TREMIE METHOD

5-55. You can place concrete underwater using several methods. The most common is with a tremie device shown in figure 5-16. A tremie is a long, funnel-shaped pipe into which you feed the concrete. The pipe must be long enough to reach from a working platform above the water level to the lowest point underwater at which you must place concrete. The discharge end of the pipe is usually equipped with a gate that you can close to fill the tremie before inserting it into the water. You can open the gate from above the water level at the proper time. When filled and lowered into position, the discharge end remains continually buried in newly placed concrete. By keeping the tremie constantly filled with concrete, you can exclude both air and water from the pipe. To start placement, lift the tremie a little bit—slowly—to permit the concrete to begin to flow out, taking care not to lose the seal at the discharge end. If the seal is lost, you must raise the tremie, close the gate, fill the tremie, and lower it into position again. Do not move the tremie horizontally through deposited concrete. When you must move it, lift it carefully out of the newly placed concrete and move it slowly to the new position, keeping the top surface of the new concrete as level as possible. Use several tremies to deposit concrete over a large area, spaced on 20- to 25-foot centers. Be sure to supply concrete to all tremies at a uniform rate with no interruptions. Pumping concrete directly from the mixer is the best way to do this. Suspend large tremies from a crane boom and easily raise and lower them with the boom. Concrete placed with a tremie should have a slump of about 6 inches and a cement content of seven sacks per cubic yard of concrete. About 50 percent of the total aggregate should be sand, and the maximum CA size should range from 1 1/2 to 2 inches.

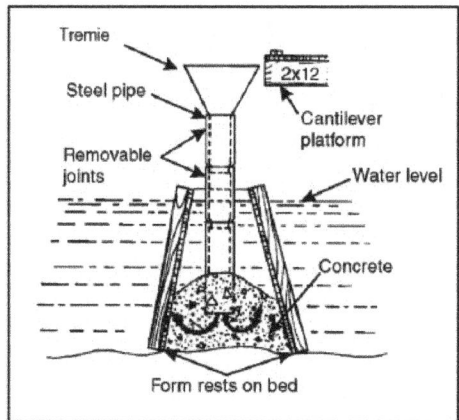

Figure 5-16. Placing concrete underwater with a tremie

Construction Procedures

CONCRETE BUCKET METHOD

5-56. Concrete can be placed at a considerable depth below the water surface using open-top buckets having a drop bottom. The concrete that is placed this way can be slightly stiffer than that placed by tremie, but should still contain seven sacks of cement per cubic yard. First, fill the bucket completely and cover the top with a canvas flap attached to one side of the bucket only. Then lower the bucket slowly into the water without displacing the canvas. Do not discharge the concrete before reaching the placement surface. Make soundings frequently to keep the top surface level.

SACK METHOD

5-57. In an emergency, place concrete underwater using partially filled sacks arranged in header and stretcher courses that interlock the entire mass together. Cement from one sack seeps into adjacent sacks and thus bonds them together. First, fill jute sacks of about 1-cubic foot capacity about two-thirds full. Then lower them into the water, placing the header courses so that the sack lengths are at right angles to the stretcher course sacks. Do not attempt compaction because the less concrete underwater is disturbed after placement, the better the concrete product

SECTION VIII - FINISHING

FINISHING OPERATIONS

5-58. The finishing process provides the desired final concrete surface. There are many ways to finish concrete surfaces depending on the effect required. Sometimes there is only a need to correct surface defects, fill bolt holes, or clean the surface. Unformed surfaces may require only screeding to proper contour and elevation, or a broomed, floated, or troweled finish may be specified.

SCREEDING

5-59. The top surface of a floor slab, sidewalk, or pavement is rarely placed at the exact specified elevation. Screeding brings the surface to the correct elevation by striking off the excess concrete. Using a tool called a screed, which is a template having a straight lower edge to produce a flat surface or a curved lower edge to produce a curved surface, move it back and forth across the concrete using a sawing motion as shown in figure 5-17, page 5-26. With each sawing motion, move the screed forward a short distance along the forms. (The screed rides on either wood or metal strips established as guides.) This forces the excess concrete built up against the screed face into the low spots. If the screed tends to tear the surface—as it may on air-entrained concrete due to its sticky nature—either reduce the rate of forward movement or cover the lower edge of the screed with metal, which will stop the tearing action in most cases. Hand-screed surfaces up to 30 feet wide, but expect diminished efficiency on surfaces more than 10 feet wide. Three workers (excluding a vibrator operator) can screed about 200 square feet of concrete per hour. Two of the workers operate the screed while the third pulls excess concrete from the front of the screed. Screed the surface a second time to remove the surge of excess concrete caused by the first screeding. This surge is when the concrete seems to grow out of the form into a convex shape.

Chapter 5

Figure 5-17. Screeding operation

FLOATING

5-60. If a surface smoother than that obtained by screeding is required, work the surface sparingly using either a wood or aluminum-magnesium float, or a finishing machine. The wood float is shown in use in view 2 of figure 5-18. Begin floating immediately after screeding while the concrete is still plastic and workable. Floating has three purposes: to embed aggregate particles just beneath the surface; to remove slight imperfections, high spots, and low spots; and to compact the concrete at the surface in preparation for other finishing operations.

5-61. Do not overwork the concrete while it is still plastic or an excess of water and paste will rise to the surface. This fine material will form a thin weak layer that will scale or wear off under use. To produce a coarse texture as the final finish, float the surface a second time after it partially hardens. Use a long-handled wood float for slab construction as shown in view 3 of figure 5-18. Use an aluminum float the same way as the wood float, to give the finished concrete a much smoother surface. To avoid cracking and dusting of the finished concrete, begin aluminum floating when the water sheen disappears from the freshly placed concrete surface. Do not use either cement or water as an aid in finishing the surface.

Construction Procedures

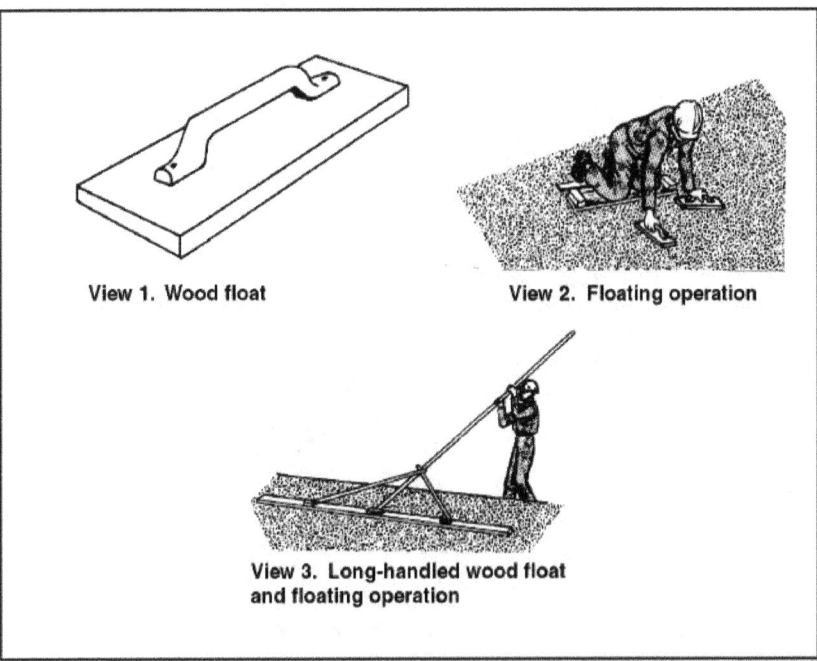

Figure 5-18. Wood floats and floating operations

TROWELING

5-62. For a dense, smoother finish, follow floating with steel troweling (see figure 5-19, page 5-28). Begin this procedure when the moisture film or water sheen disappears from the floated surface and the concrete has hardened enough to prevent fine material and water from working to the surface. Delay this operation as long as possible. Too much troweling too soon tends to produce crazing and reduces durability. However, too long a delay in troweling makes the surface hard to finish properly. Troweling should leave the surface smooth, even, and free from marks and ripples.

Avoid all wet spots if possible. When they do occur, do not resume finishing operations until the water has been absorbed, evaporated, or mopped up. When a wear resistant and durable surface is required, it is poor practice to spread dry cement on the wet surface to absorb excess water. Obtain a surface that is fine-textured, but not slippery, by a second light troweling over the surface with a circular motion immediately following the first regular troweling, keeping the trowel flat against the surface. When a hard steel-troweled finish is specified, follow the first regular troweling with a second troweling only after the concrete is hard enough that no paste adheres to the trowel, and passing the trowel over the surface produces a ringing sound. During this final troweling, tilt the trowel slightly and exert heavy pressure to compact the surface thoroughly. Hair cracks usually result from a concentration of water and fines at the surface due to overworking the concrete during finishing operations. Rapid drying or cooling aggravates such cracking. Close cracks that develop before troweling by pounding the concrete with a hand float.

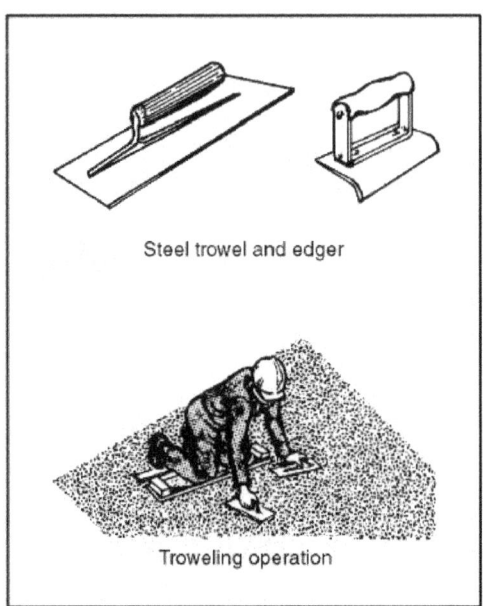

Figure 5-19. Steel finishing tools and troweling operation

BROOMING

5-63. Produce a nonskid surface by following the floating operation. After waiting 10-15 minutes, broom the concrete before it hardens thoroughly with a straw broom of fibers at least 4 1/2 inches long. The grooves cut by the broom should not be more than 3/16 inch deep. When severe scoring is not desirable, such as in some floors and sidewalks, produce the broomed finish using a hairbrush after troweling the surface once to a smooth finish. However, when rough scoring is specified, use a stiff broom made from either steel wire or coarse fiber. The direction of scoring when brooming should be at right angles to the direction of any traffic.

RUBBING

5-64. The most uniform and attractive surface requires a rubbed finish. A surface finish having a satisfactory appearance can be produced simply by using plywood or lined forms. As soon as the concrete hardens, rub the surface first with coarse carborundum stones so that the aggregate does not pull out. Allow the concrete to cure before the final rubbing with finer carborundum stones. Keep the concrete damp while rubbing. To properly cure any mortar used as an aid in this process and left on the surface, keep it damp for 1 to 2 days after it sets. Restrict the mortar layer to a minimum, because it is likely to scale off and mar the surface appearance.

FINISHING PAVEMENT

5-65. A machine finish is needed when the concrete takes its initial set, but still remains in workable condition. Technical Manual (TM) 5-331D describes the operation and maintenance of machine finishers.

MACHINE FINISHING

5-66. Set both the screeds and vibrators on the machine finisher to produce the specified surface elevation as well as a dense concrete. Generally, a sufficiently thick layer of concrete should build up ahead of the screed to fill all low spots completely. The sequence of the operation is: screed, vibrate, then screed again.

Construction Procedures

If forms are in good alignment and firmly supported, and if the concrete has the correct workability, only two passes of the machine finisher are needed to produce a satisfactory surface. Adjust the second screed to carry enough concrete ahead of it so that the screed continually contacts the pavement.

HAND FLOATING

5-67. Hand floating often does more harm than good; overworking can cause the wearing surface to deteriorate. Scaling is a good example. How-ever, sometimes there may be a need to use a longitudinal float to decrease variations running lengthwise in the surface. Such a float is made from wood, 6 by 10 inches wide and 12 by 18 feet long, fitted with a handle at each end. It is operated on form-riding bridges by two workers. The float oscillates longitudinally as it moves transversely (crosswise). A 10-foot straightedge pulled from the center of the pavement to the form removes any minor surface irregularities as well as laitance. The surface should have no coating of weak mortar or scum that will later scale off. Unless the workers use considerable care as the straightedge approaches the form, it will ride up on the concrete causing a hump in the surface, especially at construction and expansion joints. When the water sheen disappears, obtain the final surface finish by dragging a clean piece of burlap along the pavement strip longitudinally. This operation, called belting, requires two workers, one on each side of the form. Be sure to round all pavement corners with an edging tool, clean out expansion joints, and prepare them for filling.

FINISHING REPAIRS

5-68. As soon as possible after removing the forms, knock off all small projections, fill tie-rod holes, and repair any honeycombed areas. Section XII in this chapter covers concrete repair in detail.

CLEANING THE SURFACE

5-69. Concrete surfaces are not always uniform in color when forms are removed. If appearance is important, clean the surface using one of the methods described below.

CLEANING WITH MORTAR

5-70. Methods of cleaning with mortar includes—

- Using a solution. Clean the surface with a cement-sand mortar consisting of 1 part portland cement and 1 1/2 to 2 parts fine sand. Use white portland cement for a light-colored surface. Apply the mortar with a brush after repairing all defects. Then immediately scour the surface vigorously using a wood or cork float. Remove excess mortar with a trowel after 1 or 2 hours, allowing the mortar to harden enough so that the trowel will not remove it from the small holes. After the surface dries, rub it with dry burlap to remove any loose material. Mortar left on the surface overnight is very difficult to remove. Complete one section without stopping. There should be no visible mortar film remaining after the rubbing.
- Rubbing with burlap. An alternate method is to simply rub the mortar over the surface using clean burlap. The mortar should have the consistency of thick cream and the surface should be almost dry. Remove the excess mortar by rubbing the surface with a second piece of clean burlap. Delay this step long enough to prevent smearing, but complete it before the mortar hardens. Allow the mortar to set several hours and cure for 2 days. After curing, permit the surface to dry and then sand it vigorously with number 2 sandpaper. This removes all excess mortar remaining after the second burlap rubbing and produces a surface having a uniform appearance. For best results, clean concrete surfaces in the shade on a cool, damp day.

SANDBLASTING

5-71. Completely remove surface stains, particularly rust, by lightly sandblasting the surface. This method is more effective than washing with acid.

CLEANING WITH ACID

5-72. Use this method when surface staining is not severe. Precede acid washing by a 2-week period of moist curing. First, wet the surface and while it is still damp scrub it thoroughly using a stiff bristle brush with a 5 to 10 percent solution of muriatic acid. Remove the acid immediately and thoroughly by flushing with clean water. If possible, follow the acid washing with four more days of moist curing. When handling muriatic acid, be sure to wear goggles to protect your eyes and take precautions to prevent acid from contacting hands, arms, and clothing.

SECTION IX - CURING

IMPORTANCE OF CURING TO HYDRATION

5-73. Curing keeps concrete moist. The moisture is needed for any chemical reactions.

HYDRATION

5-74. Adding water to portland cement to form the water-cement paste that holds concrete together starts a chemical reaction that makes the paste into a bonding agent. This reaction, called hydration, produces a stone-like substance—the hardened cement paste. Both the rate and degree of hydration, and the resulting strength of the final concrete, depend on the curing process that follows placing and consolidating the plastic concrete. Hydration continues indefinitely at a decreasing rate as long as the mixture contains water and the temperature conditions are favorable. Once the water is removed, hydration ceases and cannot be restarted.

CURING FACTORS

5-75. *Curing* is the period of time from consolidation to the point where the concrete reaches its design strength. There are numerous facts which affect the curing process.

- Importance of moisture and temperature. During this period, take steps to keep the concrete moist and as near to 73°F as practicable. The properties of concrete, such as freeze and thaw resistance, strength, watertightness, wear resistance, and volume stability, cure or improve with age as long as moisture and temperature conditions are maintained and are favorable to continued hydration.
- Length of curing period. The length of time that concrete is protected against moisture loss depends on the type of cement used; mix proportions; the required strength, size, and shape of the concrete mass; the weather; and future exposure conditions. The period can vary from a few days to a month or longer. For most structural use, the curing period for cast-in-place concrete is usually 3 days to 2 weeks, depending upon such conditions as temperature, cement type, mix proportions, and so forth. Bridge decks and other slabs exposed to weather and chemical attack usually require more extended curing periods. Figure 5-20 shows how moist curing affects concrete's compressive strength.

Construction Procedures

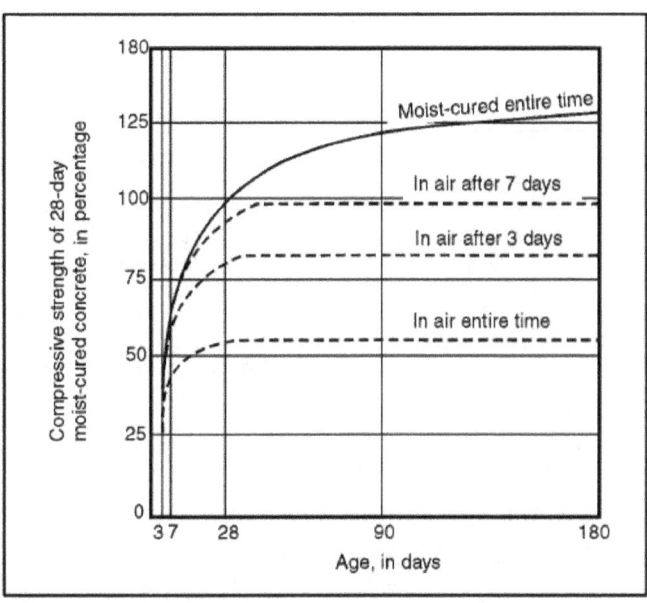

Figure 5-20. Moist curing effect on compressive strength of concrete

CURING METHODS

5-76. Several curing methods will keep concrete moist and, in some cases, at a favorable hydration temperature. They fall into two categories: those that supply additional moisture and those that prevent moisture loss. Table 5-5 lists several of these effective curing methods and their advantages and disadvantages.

Table 5-5. Curing methods

Method	Advantage	Disadvantage
Sprinkle with water or cover with wet burlap	Excellent results if constantly kept wet.	Likelihood of drying between sprinklings. Difficult on vertical walls
Strawing	Insulator in water.	Can dry out, blow away, or burn.
Moist earth	Cheap, but messy.	Stains concrete. Can dry out. Removal problem.
Pounding on flat surfaces	Excellent result, maintains uniform temperature.	Requires considerable labor; undersirable in freezing weather
Curing compounds	Inexpensive, easy to apply.	Sprayer needed. Inadequate coverage allows drying out. Film can be broken or tracked off before curing is completed. Unless

Table 5-5. Curing methods

Method	Advantage	Disadvantage
		pigmented, concrete can get too hot.
Waterproof paper	Excellent protection, prevents drying.	Heavy cost can be excessive. Must be kept in rolls; storage and handling problem.
Plastic film	Watertight, excellent protection. Light and easy to handle.	Pigmented for heat protection; requires reasonable care and tear; must be patched; must be weighted down to prevent blowing away.

METHODS THAT SUPPLY ADDITIONAL MOISTURE

5-77. Both sprinkling and wet covers add moisture to the concrete surface during the early hardening or curing period. They also provide some cooling through evaporation, which is especially important in hot weather. Methods that gives additional moisture are—

- Sprinkling continually with water. This is an excellent way to cure concrete. However, if sprinkling only at intervals, do not allow the concrete to dry out between applications. The disadvantages of this method are the expense involved and volume of water required.
- Covering with wet material. This type of covering includes wet burlap, cotton mats, straw, earth, and other moisture-retaining fabrics. These are used extensively in curing concrete. Figure 5-21, shows a typical application of wet burlap. Lay the wet coverings as soon as the concrete hardens enough to prevent surface damage. Leave them in place and keep them moist during the entire curing period.
- Flooding with water. If practical, horizontal placements can be flooded by creating an earth dam around the edges, and submerging the entire concrete structure in water.

Construction Procedures

Figure 5-21. Curing a wall with wet burlap sacks

METHODS THAT PREVENT MOISTURE LOSS

5-78. Moisture lost prevention methods include laying waterproof paper, plastic film, or liquid-membrane-forming compounds, and simply leaving forms in place. Moisture loss can be prevented by sealing the surface with—

- Waterproof paper. Use waterproof paper (see figure 5-22) to efficiently cure horizontal surfaces and structural concrete having relatively simple shapes. The paper should be large enough to cover both the surfaces and the edges of the concrete. Wet the surface with a fine water spray before covering. Lap adjacent sheets 12 inches or more and weigh their edges down to form a continuous cover having completely closed joints. Leave the coverings in place during the entire curing period.

Figure 5-22. Waterproof paper used for curing

- Plastic. Certain plastic film materials are used to cure concrete. They provide lightweight, effective moisture barriers that are easy to apply to either simple or complex shapes. However, some thin plastic sheets may discolor hardened concrete, especially if the surface was steel-troweled to a hard finish. The coverage, overlap, weighing down of edges, and surface wetting requirements of plastic film are similar to those of waterproof paper.
- Curing compounds. These are suitable not only for curing fresh concrete, but to further cure concrete following form removal or initial moist curing. Apply them with spray equipment, such as hand-operated pressure sprayers, to odd slab widths or shapes of fresh concrete, and to exposed concrete surfaces following form removal. See TM 5-337 for application details. Respray any concrete surfaces subjected to heavy rain within 3 hours of application. Use brushes to apply curing compound to formed surfaces, but do not use brushes on unformed concrete due to the risk of marring the surface, opening the surface to too much compound penetration, and breaking the surface film continuity. These compounds permit curing to continue for long periods while the concrete is in use. Do not use curing compounds if a bond is necessary because curing compounds can prevent a bond from forming between hardened and fresh concrete.
- Forms. Forms provide enough protection against moisture loss if the exposed concrete surfaces are kept wet. Keep wood forms moist by sprinkling, especially during hot, dry weather.

SECTION X - TEMPERATURE EFFECTS

HOT-WEATHER CONCRETING

5-79. Concreting in hot weather poses some special problems such as strength reduction and cracking of flat surfaces due to too-rapid drying.

PROBLEMS ENCOUNTERED

5-80. Concrete that stiffens before consolidation is caused by too-rapid setting of the cement and too much absorption and evaporation of mixing water. This leads to difficulty in finishing flat surfaces. Therefore, limitations are imposed on placing concrete during hot weather and on the maximum temperature of the concrete; quality and durability suffer when concrete is mixed, placed, and cured at high temperatures. During hot weather take steps to limit concrete temperature to less than 90°F, but problems can arise even with concrete temperatures less than 90°F. The combination of hot dry weather and high winds is the most severe condition, especially when placing large exposed slabs.

EFFECTS OF HIGH CONCRETE TEMPERATURES

5-81. Three common things affect high concrete temperatures.

- Water requirements. Because high temperatures accelerate hardening, a particular concrete consistency generally requires more mixing water than normal. Figure 5-23 shows a linear relationship between an increase in concrete temperature and the increase in mixing water required to maintain the same slump. However, in-creasing water content without increasing cement content results in a higher W/C ratio, which has a harmful effect on the strength and other desirable properties of hardened concrete.

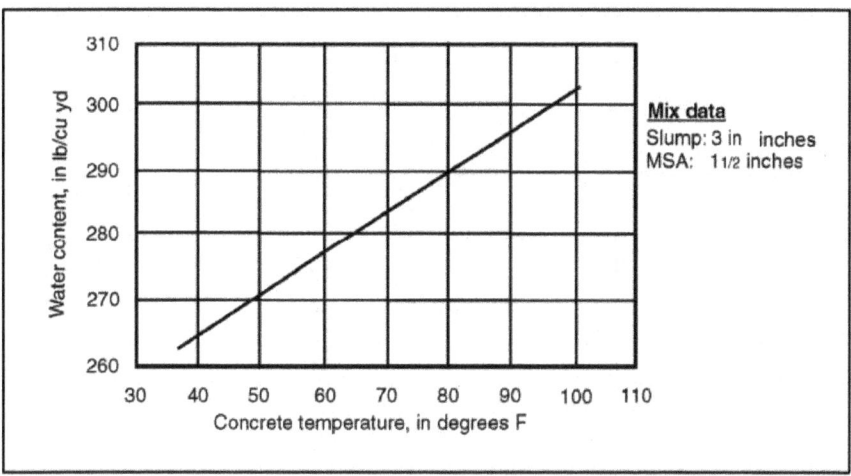

Figure 5-23. Concrete mix water requirements as temperature increases

- Compressive strength of concrete. Figure 5-24, page 5-36, demonstrates the effect of high concrete temperatures on compressive strength. Tests using identical concretes having the same W/C ratio show that while higher concrete temperatures increase early strength, the reverse happens at later ages. If water content is increased to maintain the same slump (without changing the cement content), the reduction in compressive strength is even greater than that shown in figure 5-24.
- Cracks. In hot weather the tendency for cracks to form increases both before and after hardening. Rapid water evaporation from hot concrete can cause plastic shrinkage cracks even before the surface hardens. Cracks can also develop in the hardened concrete because of increased shrinkage due to a higher water requirement, and because of the greater difference between the high temperature at the time of hardening and the low temperature to which the concrete later drops.

COOLING CONCRETE MATERIALS

5-82. The most practical way to obtain a low concrete temperature is to cool the aggregate and water as much as possible before mixing. Mixing water is the easiest to cool and is also the most effective, pound for pound, in lowering concrete temperature. However, because aggregate represents 60 to 80 percent of the concrete's total weight, the concrete temperature depends primarily on the aggregate temperature. Figure 5-25, page 5-36, shows the effects of the mixing water and aggregate temperatures on the temperature of fresh concrete. Lower the temperature of fresh concrete by—

- Using cold mixing water. In extreme cases, add slush ice to chill the water.
- Cooling CA by sprinkling, thereby avoiding too much mixing water.
- Insulating mixer drums or cooling them with sprays or wet burlap coverings.
- Insulating water supply lines and tanks or painting them white.
- Shading those materials and facilities not otherwise protected from the heat.
- Working only at night.
- Sprinkling forms, reinforcing steel, and subgrade with cool water just before placing concrete.

Chapter 5

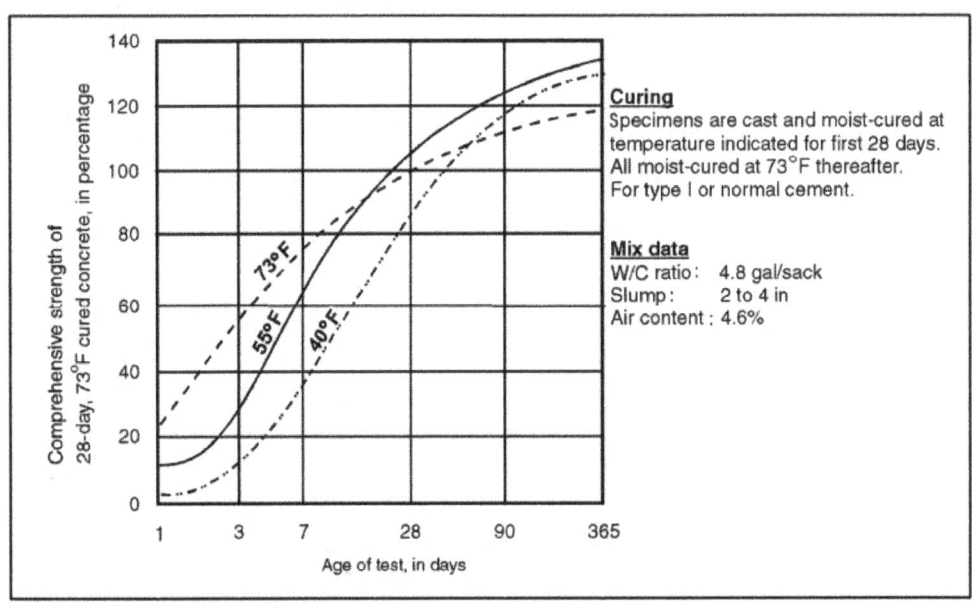

Figure 5-24. Effect of high temperature on concrete compressive strength at various ages

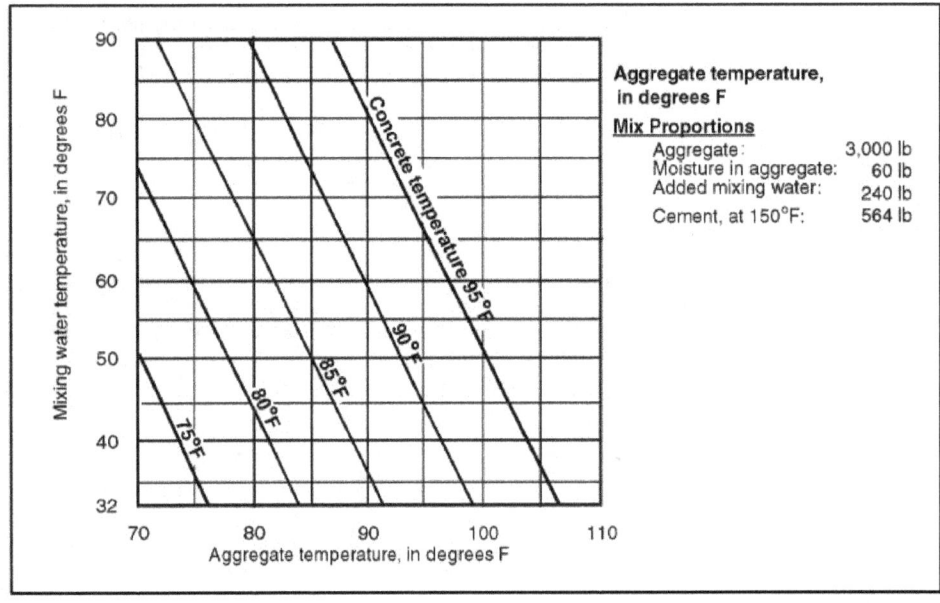

Figure 5-25. Mixing water temperatures required to produce concrete of required temperature

SPECIAL PRECAUTIONS

5-83. High temperatures increase the hardening rate, thereby shortening the length of time available to handle and finish the concrete. Concrete transport and placement must be completed as quickly as possible.

Take extra care to avoid cold joints when placing it. Proper curing is especially important in hot weather due to the greater danger of crazing and cracking. But curing is also difficult in hot weather, because water evaporates rapidly from the concrete and the efficiency of curing compounds is reduced. Leaving forms in place is not a satisfactory way to prevent moisture loss when curing concrete in hot weather. Loosen the forms as soon as possible without damaging the concrete and cover the concrete with water. Then frequent sprinkling, use of wet burlap, or use other similar means of retaining moisture for longer periods.

COLD WEATHER CONCRETING

5-84. Do not suspend concreting during the winter months. Take the necessary steps to protect the concrete from freezing in temperatures of 40°F or lower during placing and during the early curing period.

FREEZE PROTECTION

5-85. In your prior planning, include provisions for heating the plastic concrete and maintaining favorable temperatures after placement. The temperature of fresh concrete should not be less than that shown in lines 1, 2, and 3 of table 5-6. Note that lower temperatures are given for heavier mass sections than thinner sections, since less heat dissipates during the hydration period. Because additional heat is lost during transporting and placing, the temperatures given for the freshly mixed concrete are higher for colder weather. To prevent freezing, the concrete's temperature should not be less than that shown in line 4 of table 5-6 at the time of placement. To ensure durability and strength development, further thermal protection will need to be provided to make sure that subsequent concrete temperatures do not fall below the minimums shown in line 5 of table 5-6 for the time periods given in table 5-7, page 5-38. Concrete temperatures over 70°F are seldom necessary because they do not give proportionately longer protection from freezing, since the heat loss is greater. High concrete temperatures require more mixing water for the same slump and this contributes to cracking due to shrinkage.

Table 5-6. Recommended concrete temperatures for cold-weather construction

Line	Placing and Curing Conditions		Section, Size, Minimum Thickness, in Inches		
			< 12 inches	12-36 inches	36-72 inches
1	Minimum temperature, fresh concrete as mixed for weather indicated, °F.	above 30°F	60	55	50
2		0° to 30°F	65	60	55
3		below 0°F	70	65	60
4	Minimum temperature, fresh concrete as placed, °F.		55	50	45
5	Maximum allowable gradual drop in temperature throughout first 24 hours after end of protection, °F.		50	40	30
Adapted from recommended practice for cold weather concerning (ACI 306R-78).					

Chapter 5

Table 5-7. Recommended duration of protection for concrete placed in cold weather (air-entrained concrete)

Degree of Exposure to Freeze-Thaw	Normal Concrete*	High-Early-Strength Concrete**
No exposure	2 days	1 day
Any exposure	3 days	2 days

Note. Protection time for durability is at the temperature indicated in line 4, table 5-6.

*Made with Type I, II, or normal cement.

**Made with Type III or high-early-strength cement or an accelerator or an extra 100 lbs of cement.

EFFECTS ON LOW CONCRETE TEMPERATURES

5-86. Figure 5-26 demonstrates which temperature affects the hydration rate of cement; low temperatures retard hardening and compressive strength gain. The graph shows that the strength of concrete mixed, placed, and cured at temperatures below 73°F is lower than concrete cured at 73°F during the first 28 days, but becomes higher with age and eventually overtakes the strength of the concrete cured at 73°F. Concrete placed at temperatures below 73°F must be cured longer. Remember that strength gain practically stops when the moisture required for hydration is removed. Figure 5-27 shows that the early strengths achieved by Type III or high-early-strength cement are higher than those achieved by Type I cement.

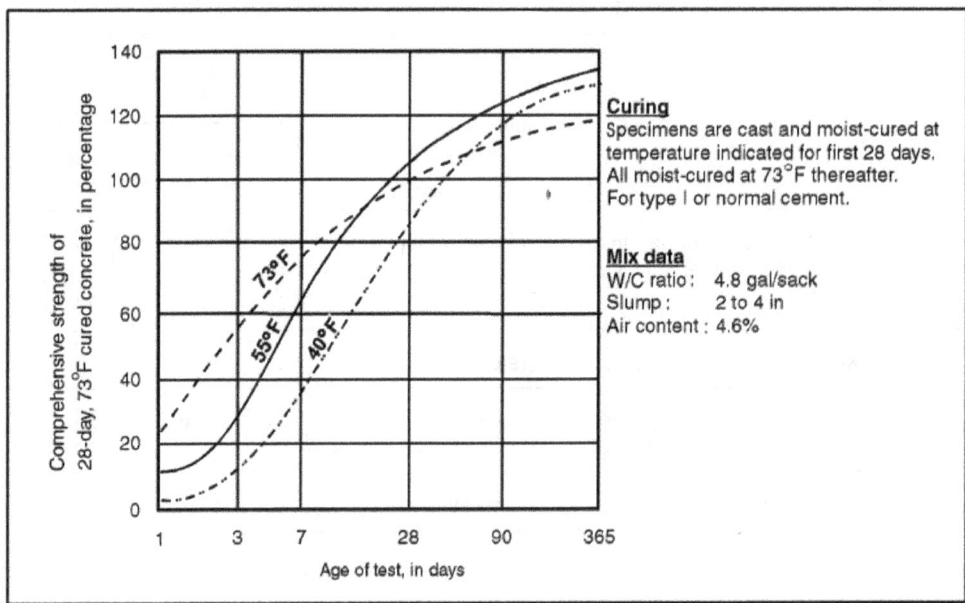

Figure 5-26. Effect of low temperature on concrete compressive strength at various ages

Construction Procedures

COLD WEATHER TECHNIQUES

5-87. When heating concrete ingredients, the thaw-frozen aggregate makes proper batching easier and avoids pockets of aggregate in the concrete after placement. If aggregate is thawed in the mixer, check for too much water content. Aggregate in temperatures above freezing seldom has to be heated, but at temperatures below freezing, the FA used to produce concrete may need to be heated.

- Heating aggregate. Use any of several methods to heat aggregate. One method for small jobs is to pile it over metal pipes containing fires or stockpile aggregate over circulating steam pipes. Cover the stockpiles with tarpaulins to both retain and distribute the heat. Another method is to inject live steam directly into a pile of aggregate, but the resulting variable moisture content can cause problems in controlling the amount of mixing water. The average temperature of the aggregate should not exceed 150°F.
- Heating water. Mixing water is easier to heat because it can store five times as much heat as solid materials having the same weight. Although aggregate and cement weigh much more than water, water's stored heat can be used to heat other concrete ingredients. When either aggregate or water is heated above 100°F, combine them in the mixer first before adding the cement. Figure 5-28, page 5-40, shows how the temperature of its ingredients affects the temperature of fresh concrete. This graph is reasonably accurate for most ordinary concrete mixtures. As shown in figure 5-28, mixing water should not be hotter than 180°F so that, in some cases, both aggregate and water must be heated. For example, if the weighted average temperature of aggregate is below 36°F, and the desired fresh concrete temperature is 70°F, heat the water to its maximum temperature of 180°F and heat the aggregate to make up the difference.

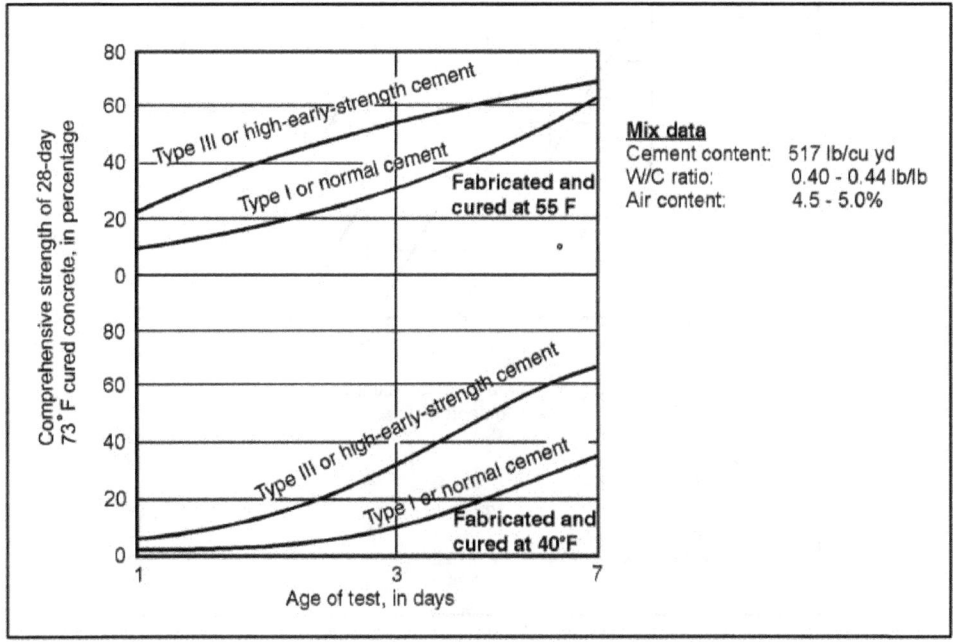

Figure 5-27. Relationship between early compressive strengths of portland cement types and low curing temperatures

- Using high-early-strength cement. High-early-strength cement produces much higher hydration temperatures which can offset some of the cold water effects. Other benefits include early reuse of forms and shore removal, cost savings in heating and protection, earlier flatwork finishing, and earlier use of the structure.

Chapter 5

- Using accelerators. Do not substitute accelerators for proper curing and frost protection. Also, do not try to lower the freezing point of concrete with accelerators (antifreeze compounds or similar products), because the large quantities required seriously affect compressive strength and other concrete properties. However, you may use smaller amounts of additional cement or such accelerators as calcium chloride to speed up concrete hardening in cold weather, as long as you limit it to no more than 2 percent of calcium chloride by weight of cement. But be careful in using accelerators containing chlorides where an in-service potential of corrosion exists, such as in prestressed concrete or where aluminum inserts are planned. When sulfate-resisting concrete is required, use an extra sack of cement per cubic yard rather than calcium chloride.

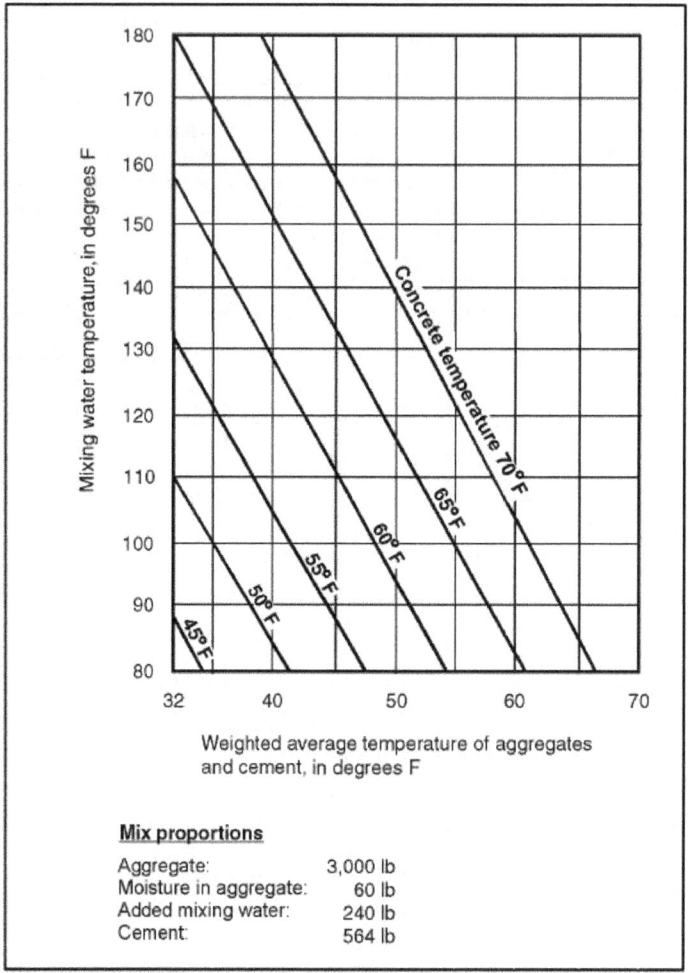

Figure 5-28. Effect of temperature of materials on temperature of fresh concrete

- Preparing for placement. Never place concrete on a frozen subgrade because severe cracks due to settlement usually occurs when the subgrade thaws. If only a few inches of the subgrade are frozen, thaw the surface by burning straw, by steaming or, if the grade permits, by spreading a layer of hot sand or other granular material. Be sure to thaw the ground enough to ensure that it will not refreeze during the curing period.

Curing

5-88. Concrete placed in forms or covered by insulation seldom loses enough moisture at 40°F to 55°F to impair curing. Forms distribute heat evenly and help prevent drying and overheating. Leave them in place as long as practicable. However, when using heated enclosures during the winter, moisten curing concrete to offset the drying effects. Keep the concrete at a favorable temperature until it is strong enough to withstand both low temperatures and anticipated service loads. Concrete that freezes shortly after placement is permanently damaged. If concrete freezes only once at an early age, favorable curing conditions can restore it to nearly normal, although it will neither weather as well nor be as watertight as concrete that has never frozen. Air-entrained concrete is less susceptible to freeze damage than nonair-entrained concrete. (See TM 5-349 for details of cold weather concreting.) Three methods for maintaining proper curing temperatures are described below.

- Live steam. When fed into an enclosure, live steam is an excellent and practical curing aid during extremely cold weather, because its moisture offsets the rapid drying that occurs when very cold air is heated. Use a curing compound after removing the protection if the air temperature is above freezing.
- Insulation blankets or bats. The manufacturers of these materials can usually provide information on how much insulation is necessary to protect curing concrete at various temperatures. Because the concrete's corners and edges are the most likely to freeze, check them frequently to determine the effectiveness of the protective covering.
- Heated enclosures. Use wood, canvas, building board, plastic sheets, or other materials to enclose and protect curing concrete at below-freezing temperatures. Build a wood framework and cover it with tarpaulins or plastic sheets. Make sure enclosures are sturdy and reasonably airtight, and allow for free circulation of warm air. Provide adequate minimum temperatures during the entire curing period. The easiest way to control the temperature inside the enclosure is with live steam. Unless enclosures are properly vented, do not use carbon monoxide-producing heaters (salamanders or other fuel-burning heaters) when placing concrete for 24 to 36 hours afterwards curing.

SECTION XI - FORM REMOVAL

FORM STRIPPING

5-89. Careless workers can cancel out the value of good detailing and planning by indiscriminate use of the wrecking bar. A pinch bar or other metal tool should never be placed against exposed concrete to wedge forms loose. If it is necessary to wedge between the concrete and the forms, use only wooden wedges.

Basic Considerations

5-90. Wall forms should not be removed until the concrete has thoroughly hardened, but specified curing should begin as early as possible in warm weather. Ties may be removed as early as 24 hours after casting to loosen forms slightly and permit entry of curing water between form and concrete. Ornamental molds must be left in place until they can be removed without damage to the concrete surface. In cold weather, removal of form work should be deferred or form work should be replaced with insulation blankets to avoid thermal shock and consequent crazing of the concrete surface.

5-91. When stripping forms in the vicinity of a belt course, cornice, or other projecting ornament, begin stripping some distance away from the ornament and work toward it. If there is any tendency for the forms to bind around the ornament, the pressure of the forms against projecting corners will be relieved so that there will be less chance of spalling sharp edges. Forms recessed into the concrete require special care in stripping. Wedging should be done gradually and should be accompanied by light tapping on the piece to crack it loose from the concrete. Never remove an embedded form with a single jerk. Embedded wood forms are generally left in place as long as possible so they will shrink away from the concrete. The embedded items should be separate from or loosely attached to the main form so that they will remain in place when the main form is stripped.

Chapter 5

STRIPPING TIME BASED ON CONCRETE STRENGTH

5-92. Since early form removal is usually desirable so that forms can be reused, a reliable basis for determining the earliest proper stripping time is necessary. When forms are stripped, there must be no excessive deflection or distortion and no evidence of cracking or other damage to the concrete, due either to removal of support or to the stripping operation. Supporting forms and shores must not be removed from beams, floors, and walls until these structural units are strong enough to carry their own weight and any approved superimposed load. Such approved load should not exceed the live load for which the member was designed unless provision has been made by the engineer architect to allow for temporary construction loads.

5-93. Forms for vertical members such as columns and piers may be removed before those for beams and slabs. The strength of the concrete necessary before form work is stripped and the time required to attain it vary widely with different job conditions, and the most reliable basis is furnished by test specimens cured under job conditions. In general, forms and supports for suspended structures can be removed safely when the ratio of cylinder test compressive strength to design strength is equal to or greater than the ratio of total dead load and construction loads to total design load with a minimum of 50 percent of design compressive strength being required.

5-94. For some applications, a definite strength must be obtained; for example, 2,500 psi or two-thirds of the design strength. However, even when concrete is strong enough to show no immediate distress or deflection under a load, it is possible to damage corners and edges during stripping and for excessive creep deflections to occur. If strength tests are to be the basis for the designers, instructions to the project officer on form removal, the type of test, the method of evaluating, and the minimum strength standards should be stated clearly in specification.

5-95. The number of test specimens, as well as who should take them and perform the tests, should also be specified. Ideally, test beams and cylinders should be job cured under conditions which are similar to those for the portions of the concrete structure which the test specimens represent. (The specimens must not be confused with those cured under laboratory conditions to evaluate 28-day strength of the concrete.) The curing record including the time, temperature, and method for both the concrete structure and the test specimens, as well as the weather record, will assist both the engineer and the project officer in determining when forms can be safely stripped. It should be kept in mind that specimens which are relatively small are more quickly affected by freezing or drying conditions than concrete in the structure.

5-96. On jobs where the engineer has made no provision for approval of shore and form removal based on strength and other considerations peculiar to the job, table 5-8 shows the minimum time forms and supports should remain in place under ordinary conditions.

FORM REMOVAL PROCEDURES

5-97. Forms are designed and constructed so that their removal does not harm the concrete. The form must be stripped carefully to avoid damaging the surface. Do not jerk the forms from the concrete after wedging at one end, or the edges will break. Withdraw all nails while stripping and immediately clean and oil all forms to be reused.

SECTION XII - REPAIRING

NEW CONCRETE

5-98. Although repairs are costly and time-consuming, field experience dictates that steps in the repair procedure cannot be omitted or performed carelessly without harming the serviceability of the repair work.

PRELIMINARY PROCEDURES

5-99. If they are not properly performed, repairs later loosen, crack at the edges, and allow water to permeate the structure. Preliminary procedures includes—

Construction Procedures

- Inspection. Following form removal, inspect the concrete for such surface defects as rock pockets, inferior quality, ridges at form joints, bulges, bolt holes, or form-stripping damage.
- Timely repair. On new work, when repairs are made immediately after form removal, while the concrete is quite green, the best bonds are developed and are more likely to be as durable and permanent as the original work. Therefore, make all repairs within 24 hours after removing forms.

Table 5-8. Recommended form stripping time

Forms	Setting Time	
Walls[1]	12-24 hours	
Columns[1]	12-24 hours	
Beams and Girders[1] (sides only)	12-24 hours	
Pan joist forms[3]		
• 30 inches wide or less	3 days	
• Over 30 inches wide	4 days	
	Where Design Live Load is	
	<DL	>DL
Joist, beam, or girder soffits[4]		
• Under 10 feet of clear span between supports	7 days[2]	4 days
• 10 to 20 feet of clear span between supports	14 days[2]	7 days
• Over 20 feet of clear span between supports	21 days[2]	14 days
Floor slab[4]		
• Under 10 feet of clear span between supports	4 days[2]	3 days
• 0 to 20 feet of clear span between supports	7 days[2]	4 days
• Over 20 feet of clear span between supports	10 days[2]	7 days

Note. These periods represent the cumulative number of days or fractions thereof not necessarily consecutive, during which the temperature of the air surrounding the concrete is above 50°F.

[1]Where such forms also support form work for slabs, beams, or soffits, the removal times of the latter should govern.
[2]Where form may be removed without disturbing shores, use half of values shown, but not less than 3 days.
[3]These are the type that can be removed without disturbing forming.
[4]The distances between supports refer to structural supports and not to temporary form work or shores.

Chapter 5

REMOVING RIDGES AND BULGES

5-100. Remove objectionable ridges and bulges by carefully chipping and rubbing the surface with a grinding stone.

PATCHING NEW CONCRETE

5-101. Before placing mortar or concrete into patch holes, keep the surrounding concrete wet for several hours. Then brush a grout made from cement and water mixed to a creamy consistency into the hole surfaces before applying the patch material. Start curing the patch as soon as possible to avoid early drying. Use damp burlap, tarpaulins, or membrane curing compounds. Patches are usually darker in color than the surrounding concrete. If appearance is important, mix some white cement into the mortar or concrete and use as a patch. Make a trial mix to determine the best proportion of white and gray cements to use.

Small Patches

5-102. Do not apply a single shallow layer of mortar on top of honey-combed concrete, because moisture will form in the voids and subsequent weathering will cause the mortar to spall off. Instead, chip out small defective areas, such as rock pockets or honeycomb, down to solid concrete. Cut the edges either as straight as possible at right angles to the surface, or undercut them slightly to make a key at the edge of the patch. Keep the surfaces of the resulting holes moist for several hours before applying the mortar. Fill shallow patch holes with mortar placed in layers not more than 1/2 inch thick. Make rough scratches in each layer to improve the bond with the succeeding layer, and smooth the surface of the last layer to match the adjacent surface. Allow the mortar to set as long as possible to reduce shrinkage and make a more durable patch. If absorptive form lining was used, make the patch match the adjacent surface by pressing a piece of the form lining against the fresh patch.

Large Patches

5-103. Patch large or deep holes with concrete held in place by forms. Reinforce such patches and dowel them to the hardened concrete as shown in figure 5-29.

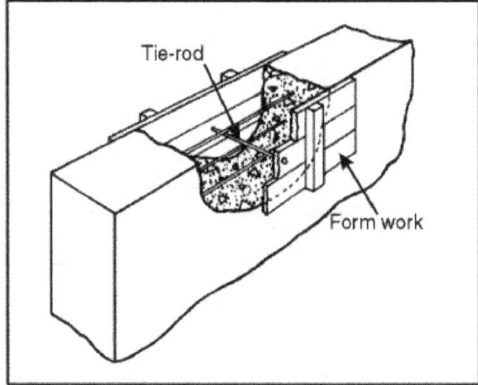

Figure 5-29. Repairing large and deep holes in new concrete

Bolt and Tie-Rod Holes

5-104. When filling bolt holes, pack small amounts of mortar into place carefully. Mix the mortar as dry as possible with just enough water to compact tightly when forced into place. Fill tie-rod holes extending through the concrete with mortar, using a pressure gun similar to an automatic grease gun.

Flat Surfaces

5-105. View 1 of figure 5-30 shows why feathered edges around a flat patch break down. First chip an area at least 1 inch deep with edges at right angles to the surface, as shown in view 2 of figure 5-30, before filling the hole. Then screed the patch as shown in view 3 of figure 5-30. The fresh concrete should project slightly beyond the surface of the hardened concrete. Allow it to stiffen before troweling and finishing to match the adjacent surfaces.

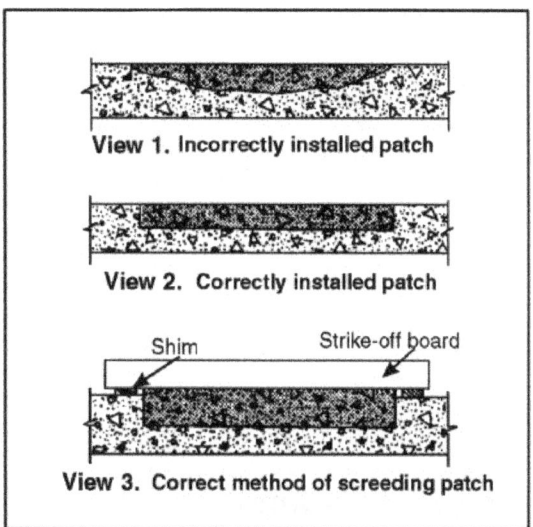

Figure 5-30. Patching flat surface in new concrete

OLD CONCRETE

5-106. When repairing old concrete, first determine how much material to remove by thoroughly inspecting the defect. Remember that it is far better to remove too much old material than not enough.

Preliminary Procedures

5-107. Remove all concrete of questionable quality unless nothing would be left. In this case, remove only the loose material. If the old and new concrete will join on a surface exposed to weathering or chemical attack, make sure the old concrete is perfectly sound. After removing weakened material and loose particles, thoroughly clean the cut surfaces using air or water or both. Keep the area surrounding the repair continuously wet for several hours, preferably overnight. Wetting is especially important in repairing old concrete because a good bond will not form without it.

Patching Old Concrete

5-108. The repair depth depends on many conditions. In large structures such as walls, piers, curbs, and slabs, the repair should be at least 6 inches deep, if possible. If the old concrete contains reinforcement bars, allow a clearance of at least 1 inch around each exposed bar. Use rectangular patches on small areas, cutting 1 to 2 inches vertically into the old concrete to eliminate thin or feathered edges. Following the wetting period, place the new concrete into the hole in layers and thoroughly tamp each layer. The patch concrete should be a low slump mixture allowed to stand for awhile to reduce shrinkage. Forms may be needed to hold the patch concrete in place. The design and construction of such forms often require a lot of cleverness, but well-designed and properly constructed forms are important in concrete repair. Reinforce deep patches to the hardened concrete. Good curing is essential. Begin curing as soon as possible to avoid early drying.

This page intentionally left blank.

Chapter 6
Reinforced-Concrete Construction

Concrete is strong in compression, but relatively weak in tension. The reverse is true for slender steel bars. When concrete and steel are used together, one makes up for the deficiency of the other. The most common type of steel reinforcement employed in concrete building construction consists of round bars, usually of the deformed type, with lugs or projections on their surfaces. The purpose of the surface deformations is to develop a greater bond between the concrete and the steel. The bars are made from billet steel, rail steel, or axle steel, conforming to rigid specifications. Welded-wire fabric is another type of reinforcement that consists of a series of parallel-longitudinal wires welded at regular intervals to transverse wires. It is available in sheets and rolls and is widely used as reinforcement in floors and walls.

SECTION I - DEVELOPMENT AND DESIGN

PRINCIPLES AND DEFINITIONS

6-1. From the structural engineering viewpoint, the three main elements that support live loads in a building are tension members, compression members, and bending members called beams. Beams command the most study, because the bending stress varies over the beam cross section, instead of being uniformly distributed.

REINFORCED CONCRETE

6-2. The concrete's beam strength is increased greatly by embedding steel in the tension area as shown in Figure 6-1, page 6-2. When steel reinforcements in concrete help carry imposed loads, the combination is reinforced concrete.

SHEAR STRENGTH

6-3. The concrete's shear strength is about one-third the unit compressive strength, whereas, tensile strength is less than one-half the shear strength. The failure of a concrete slab subjected to a downward concentrated load is due to diagonal tension. However, web reinforcement can prevent beams from failing in diagonal tension.

TENSILE STRENGTH

6-4. The concrete's tensile strength is such a small percentage of the compressive strength that it is ignored in calculations for reinforced-concrete beam. Instead, horizontal steel bars well embedded in the tension area provide tensile resistance.

Chapter 6

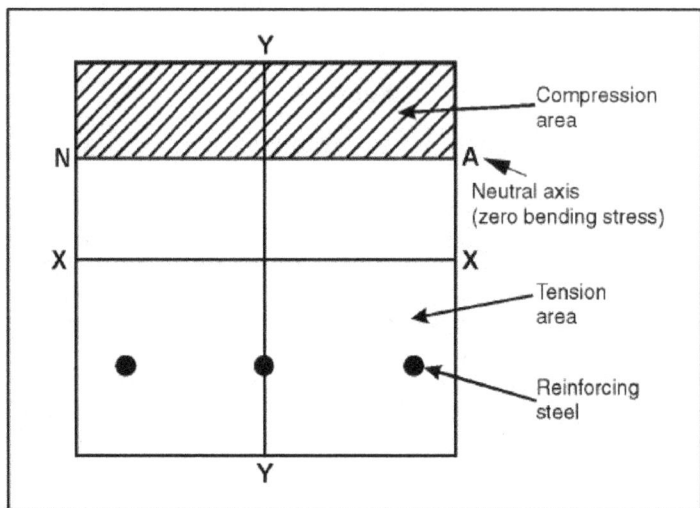

Figure 6-1. Cross section of a concrete beam showing location of reinforcement in tension

BOND STRENGTH

6-5. Bond Strength is the measure of effective grip between the concrete and the embedded-steel bar. The design theory of reinforced-concrete beam is based on the assumption that a bond develops between the reinforcement and the concrete that prevents relative movement between them as the load is applied. How much bond strength develops depends largely on the area of contact between the two materials. Because of their superior bonding value, bars having a very rough surface (deformed bars or rebars see figure 6-2) have replaced plain bars as steel reinforcement.

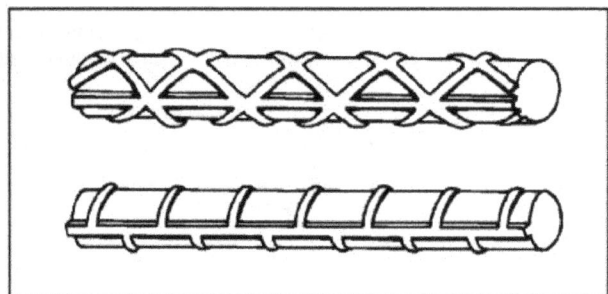

Figure 6-2. Deformed steel-reinforced bars

BENDING STRENGTH

6-6. A beam subjected to a bending moment deflects because its compression side shortens and its tension side lengthens. Therefore, the weak tension area shown in view 1 of figure 6-3 must be reinforced with steel as shown in view 2. (View 1 exaggerates possible cracking to show the beam condition if loaded sufficiently.) The concrete in the compression areas usually does not require reinforcement.

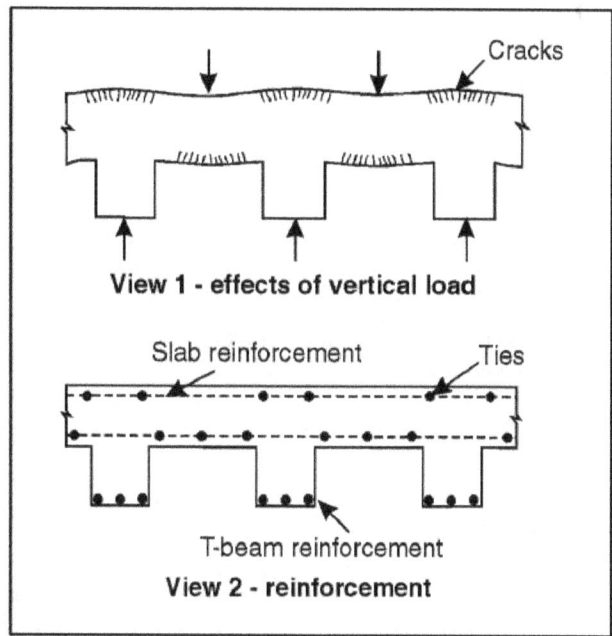

Figure 6-3. Concrete beams subjected to vertical load without and with steel reinforcement

CREEP

6-7. *Creep or plastic flow* is the tendency for loaded concrete members to deform after time. Creep occurs over the whole stress range, rapidly at first, then much more slowly, becoming small or negligible after a year or two. Some experts believe that the first few service loads remove the initial strains on well-designed, reinforced-concrete structures and if they are not overloaded, they will respond elastically. Because of creep, you cannot predict concrete deflection with any satisfactory degree of accuracy using the common deflection formulas. However, you cannot always trace failures to creep, because it usually disappears in well-proportioned structures before too much deflection occurs.

HOMOGENEOUS BEAMS

6-8. *Homogeneous* beams are nonreinforced beams composed of the same material throughout, such as steel or timber beams.

NEUTRAL AXIS

6-9. The neutral axis is the plane in a beam where the bending stress equals zero. This is shown as line N-A in figure 6-1.

REINFORCED-CONCRETE DESIGN

6-10. The design of a reinforced-concrete structure consists mainly of predicting both the position and direction of potential tension cracks in concrete and in preventing the cracks by locating sufficient reinforcing steel across their positions.

Chapter 6

SPECIFICATIONS

6-11. The practical experience of many structural engineers combined with comprehensive tests and investigations conducted at universities and elsewhere, provide the solutions to many common problems in reinforced-concrete design. For most practical designs, engineers refer to standard specifications.

BASIS OF DESIGN

6-12. The expression "to design a beam" means to determine the size of a beam and the materials required to construct a beam that can safely support specified loads under specific conditions of span and stress. Economy in the use of materials and efficiency in strength, spacing, and arrangements of reinforcing steel are the main factors that influence the design. This manual does not attempt to cover how to design reinforced-concrete structures or members because there are many authoritative texts available in this field.

SECTION II - STRUCTURAL MEMBERS

REINFORCEMENT

6-13. Reinforced-concrete structures consist of many types of reinforced structural members.

TYPES OF STRUCTURAL MEMBERS

6-14. Structural members include beams, columns, girders, walls, footings, and slabs. Each of these different structural members interacts with one another to a considerable degree, thus forming a monolithic whole.

BEAM REINFORCEMENT

6-15. Figure 6-4 shows four common shapes of reinforcing steel for beams. The purpose of both the straight and bent-up bar is to resist the bending tension in the bottom of a beam that is over the central portion of the span. A beam requires fewer bars near the ends of the span where the bending moment is small. This is where the inclined portion of a bent-up bar is placed to resist diagonal tension. The reinforcing bars of continuous beams continue across the intermediate supports to resist top tension in the support area. Two types of beam reinforcements are—

- Stirrups. Add stirrups or U-shaped bars when the bent-up bars cannot resist all the diagonal tension. Because of the tensile stress on the stirrups, they pass under the bottom bar and are inclined or perpendicular to it to prevent lateral slippage.
- Bolsters or high chairs. Devices called bolsters or high chairs (see figure 6-5) usually support the horizontal reinforcing bars and hold them in place during construction. The purpose of a high chair is to ensure that rebar does not rest on the earth or vapor barrier, but instead is suspended inside the concrete beam where it is more effective and protected from moisture. Stress-carrying reinforcing bars must be placed according to the ACI Code 318-83. Select high chairs with skids for use on vapor barrier or wet subbase. Select metal highchairs for two or more levels of rebar or when workers will be required to walk on the robber during placement.

Reinforced-Concrete Construction

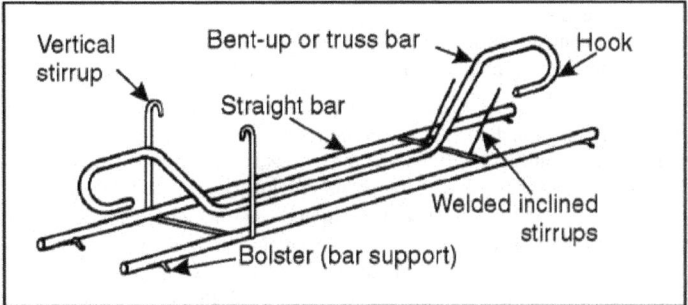

Figure 6-4. Typical shapes of reinforcing steel for beams

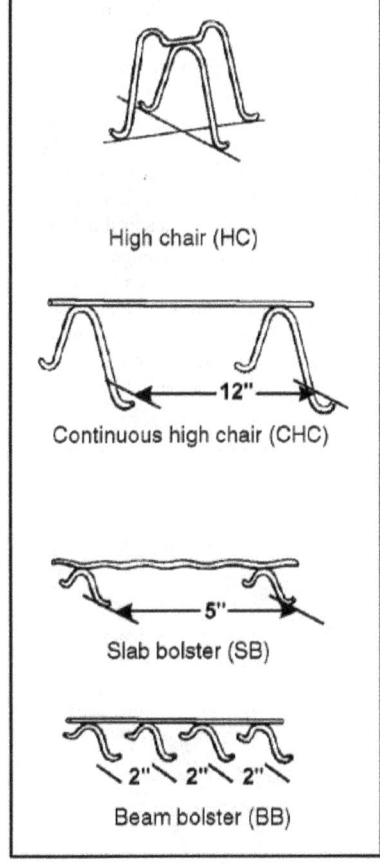

Figure 6-5. Supports for horizontal reinforcing steel

COLUMN REINFORCEMENT

6-16. A *column* is a slender vertical member that carries a superimposed load unless its height is less than three times its least lateral dimension. In that case, the vertical member is called a pier or pedestal.

Chapter 6

Allowable column loads and minimum column dimensions are governed by ACI Code 318-83. Concrete columns must always have steel reinforcement because they are subject to bending. Figure 6-6 shows two types of column reinforcement. Vertical reinforcement is the main type. Lateral reinforcement consists of individual ties or a continuous spiral that surrounds the column. These reinforcements are discussed below:

- Vertical reinforcement. A loaded-concrete column shortens vertically and expands laterally, see view 1 of figure 6-6. Four vertical reinforcing bars help to carry the direct axial load as the column shortens under load, while another four serves as intermediate vertical reinforcement. All eight bars are located around the column periphery to resist bending.
- Lateral reinforcement. This mainly provides intermediate lateral support for the vertical reinforcement. The vertical bars bulge outward in the direction of least opposition. They need lateral reinforcement to hold them securely—at close vertical intervals—and too restrain the expansion. A tied column is when lateral ties (see T_1 figure 6-6) restrain the expansion. A second system of ties (see T_2 in the cross section of view 1 in figure 6-6) confines the four intermediate vertical reinforcing bars to prevent the eight-bar group from tending to form a slightly circular configuration under load. Without the second system of ties, the square concrete shell would bulge and crack, and the column would fail. A round column has obvious advantages in this respect.
- Spiral columns. A *spiral column* is a round column that has a continuous spiral winding of lateral reinforcement (see view 2 of figure 6-6) encircling its core and vertical reinforcing bars to restrain expansion. A spiral column is generally considered to be stronger than a tied column. The spiral reinforcement is superior to the many imperfect anchorages that are at the ends of the individual lateral ties in a tied column. The pitch of the spiral reinforcement can be reduced to provide more effective lateral support. The design engineer specifies the pitch of the spiral, the tie size, and the number of vertical bars for a spiral column.
- Composite and combination columns. A *composite column* is a structural steel or cast-iron column thoroughly encased in concrete and reinforced with both vertical and spiral reinforcements. The cross-sectional area of the metal core cannot exceed 20 percent of the gross column area. A structural steel column that's encased in concrete at least 2 1/2 inches thick overall and reinforced with welded-wire fabric is called a *combination column*. Composite and combination columns are often used in constructing large buildings.

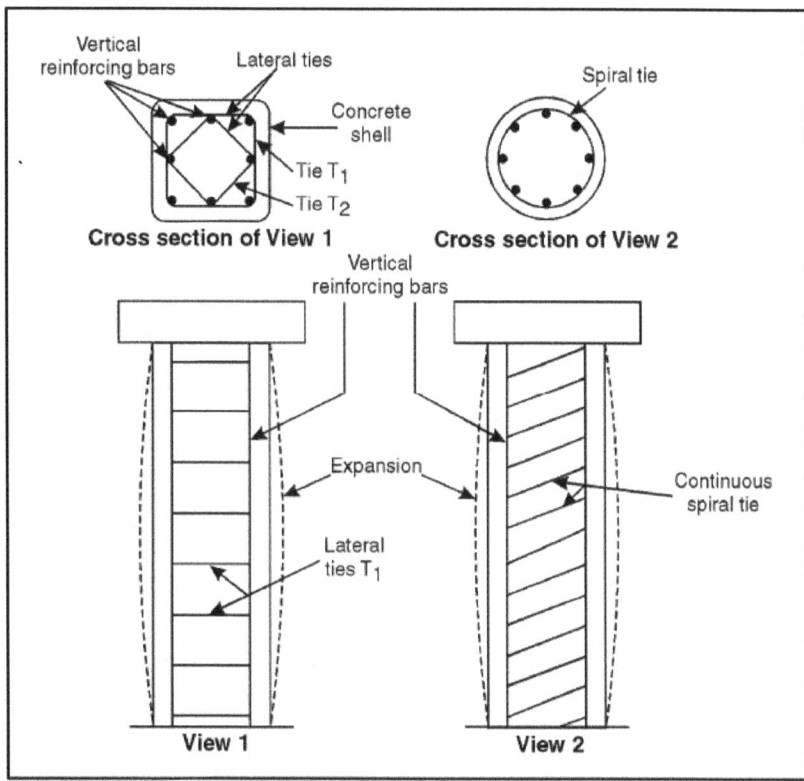

Figure 6-6. Reinforcing concrete columns

SLAB AND WALL REINFORCEMENT

6-17. Slabs and walls are reinforced both vertically and laterally. The vertical reinforcement counteracts applied loads. The lateral reinforcement in the form of wire mesh resists reduction in concrete volume due to drying, cooling and, in some cases, chemical reaction. Depending on how much disturbance adjacent construction creates, this movement tends to cause tension stress in the concrete. The amount of lateral reinforcement usually provided is approximately one-fifth of 1 percent of the cross-sectional area of the concrete, as required by the specifications.

SECTION III - REINFORCING STEEL

GRADES, DESIGNATIONS, AND METHODS

6-18. The main types of reinforcing steel are vertical and horizontal bars (see figure 6-4, page 6-5), wire for tying columns (see figure 6-6), and wire mesh for shrinkage and contraction reinforcement of slabs and walls. Bars are available in 11 sizes designated by numbers (see table 6-1, page 6-8) that range in size from 3/8 inch to about 2 1/4 inches in diameter. Reinforcing steel is specified in ASTM A615, A616, and A617. Minimum-yield strengths are: 40,000 psi, 50,000 psi, 60,000 psi, and 75,000 psi.

Table 6-1. Standard steel reinforcing bar

Bar Designation Number*	Unit Weight, in Pounds per Foot	Diameter, in Inches	Area, in Square Inches	Perimeter, in Inches
3	.376	0.375	0.11	1.178
4	.668	0.500	0.20	1.571
5	1.043	0.625	0.30	1.963
6	1.502	0.750	0.44	2.356
7	2.044	0.875	0.60	2.749
8	2.670	1.000	0.79	3.142
9	3.400	1.128	1.00	3.544
10	4.303	1.270	1.27	3.990
11	5.313	1.410	1.56	4.430
14	7.65	1.693	2.25	5.320
18	13.60	2.257	4.00	7.090

Note. The nominal dimensions of a deformed bar are equivalent to those of a plan round bar having the same weight per foot as the deformed bar.

*Bar numbers are based on the number of eighths of an inch included in the nominal diameter of the bars (example: a number 3 bar is 3/8 inch in diameter).

GRADES

6-19. The engineer specifies the sizes, amounts, and classifications of reinforcing steel to be used. Wire mesh is cold-drawn from hard-grade steel; but the steel in reinforcing bars is ductile so that you can generally cold-form hooks of relatively small radius on the job without breakage. However, the improved deformed bars (rebars) that conform to ASTM specifications are so superior in bonding value that hooking their ends does not add much strength. The latest ASTM specifications require permanent rolled-on identification because so many grades of steel reinforcing bars have the same deformation pattern. The two adopted grade-marking systems are shown in figure 6-7.

Reinforced-Concrete Construction

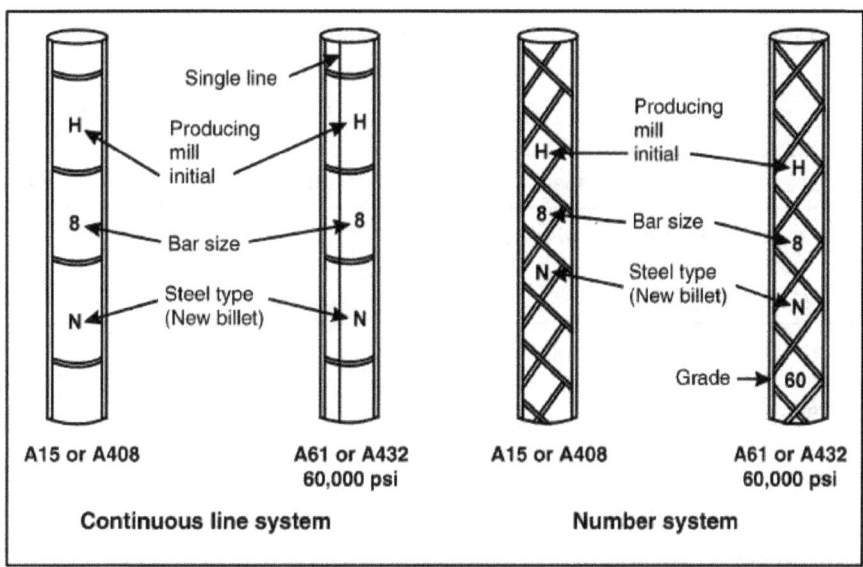

Figure 6-7. Reinforcing bar grade-marking systems

- Continuous Line System. The bars have a small, continuous, longitudinal grade line running between the main longitudinal ribs to identify 60,000 psi.
- Number System. A rolled-on grade number follows the producing mill, bar size, and steel type symbols. The number 40 identifies 40,000 psi, the number 50 identifies 50,000 psi, and the number 60 identifies 60,000 psi.

Bolsters or High Chairs

6-20. Bolsters or high chairs (see figure 6-5, page 6-5) are available in many heights to fit almost any requirement. They support the reinforcing steel and hold it a specified distance away from the exterior concrete shell poured around them, called protective concrete. Table 6-2, page 6-10, gives the minimum concrete cover that must be provided for different types of reinforcement under varying conditions. Cover must be provided for the bottom, sides, and tops of rebar.

Stirrups

6-21. Stirrups (see figure 6-4, page 6-5) help bent bars to resist diagonal tension, as well as to reinforce the beam web to keep cracks from spreading. At least one stirrup must cross every potential diagonal tension crack. Vertical stirrups pass underneath the bottom steel and are perpendicular to it to prevent lateral slippage. They are one of the most practical methods of web reinforcement because they are easy to arrange and set in the forms with the other bars. Vertical stirrups are anchored in different ways: by welding to or hooking tightly around longitudinal reinforcement or by embedding them sufficiently above a beam's midline so that they develop the required bond. Hooks should be fabricated to meet dimensional requirements stated in table 6-3, page 6-10. Welded stirrups are more efficient than vertical ones, because they can be welded at any angle parallel to the diagonal stress. However, practical considerations offset this advantage. Inclined stirrups must be welded to longitudinal reinforcements on-site to prevent slippage and displacement during concrete placement. This work is not only expensive, it is somewhat troublesome.

Chapter 6

Table 6-2. Minimum concrete cover requirements for steel reinforcement

Forms	Minimum Cover, in Inches
Concrete cast against and permanently exposed to the earth	3
Concrete exposed to the earth or weather: #6 through #18 bars #5, W31 or D31 wire and smaller	 2 1 1/2
Concrete not exposed to weather nor in contact with the earth: Slabs, walls, joist: #14 and #18 bars #11 bar and smaller Beams, columns: Primary reinforcement, ties, stirrups, spirals	 1 1/2 3/4 1 1/2
Concrete exposed to salt water	4
Adapted from ACI 318-83.	

Table 6-3. Recommended end hooks (all grades)

Bar Size Number	D*	180° Hooks		90° Hooks
		A or	J	A or G,
3	2 1/4"	5"	3"	6"
4	3"	6"	4"	8"
5	3 3/4"	7"	5"	10"
6	4 1/2"	8"	6"	12"
7	5 1/4"	10"	7"	14"
8	6"	11"	8"	16"
9	9 1/2"	13"	11 3/4"	19"
10	10 3/4"	15"	11 1/4"	22"
11	12"	17"	12 3/4"	24"
14	18 3/4"	23"	19 3/4"	31"
18	24"	30"	24 1/2"	41"
* D = Finished bend diameter.				

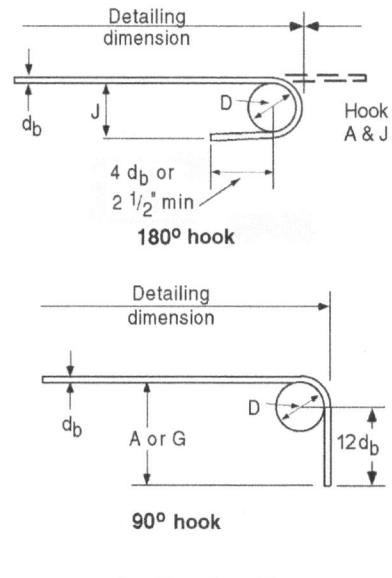

d_b = diameter of bar

SPLICES

6-22. Reinforcing bars are available only in certain lengths; splice them together for longer runs. A common way to splice bars is to lap them. If one bar is not long enough for the span, never butt reinforcing bars.

PRINCIPLES

6-23. Lapping the bars (see figure 6-8) allows bond stress to transfer the load from one bar to a second bar. Although you could hook the bars, it is not always practical nor even desirable to bend them. Engineering design dictates the actual length of the lap after considering the anticipated beam stresses, but the length is about 24 to 36 bar diameters, depending on bar size. Table 6-4, page 6-12, lists the recommended lap for particular bar designation numbers. Note that the minimum lap for plain bars is twice that for deformed bars. Except as shown on plans, do not splice steel reinforcement without the approval of an engineer, and never splice bars at the points of maximum bending. It is usually best to locate splices beyond the center of the beam. When possible, stagger the splices so that they all do not fall at the same point. Every bar requires at least two supports with bolsters spaced at about 5-foot intervals.

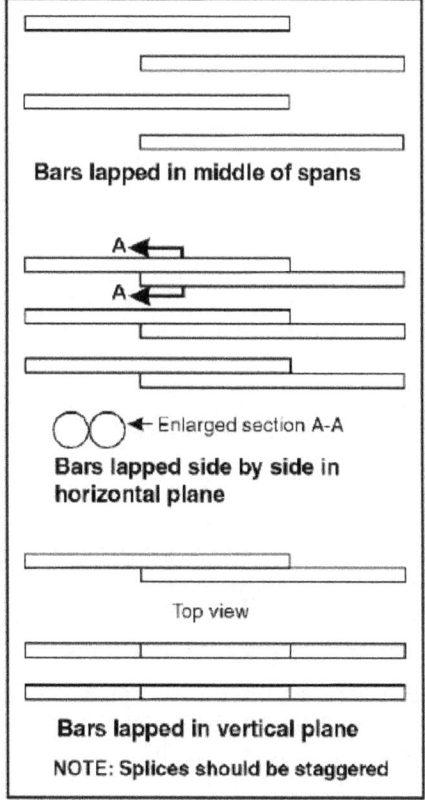

Figure 6-8. Method of splicing reinforcing bars

Table 6-4. Minimum splice overlap

Bar Size in Number	Overlap in Inches							
	f'c = 3,000		f'c = 4,000		f'c = 5,000		f'c = 6,000	
	Top Bars	All Others	Top Bars	All Others	Top Bars	All others	Top Bars	All Others
3	21	20	21	20	21	20	21	20
4	29	20	29	20	29	20	29	20
5	36	26	36	26	36	26	36	26
6	46	33	43	31	43	31	43	31
7	63	45	54	39	50	36	50	36
8	82	59	71	51	64	46	58	42
9	104	74	90	65	81	58	74	53
10	132	95	115	82	103	73	94	67
11	163	116	141	101	126	90	115	82

Notes:
1. This data is recommended for normal-weight concrete.
2. Top bars are defined as horizontal bars with 12 or more inches of fresh concrete placed below. This information is based on a class C splice (ACI 318-83).

Methods

6-24. The splicing method shown in view 1 of figure 6-8, page 6-11, is satisfactory when bar spacing is large. Do not use this method—

- In a beam.
- With similar members having several closely spaced bars.
- When the overlapped section interferes with proper bar covering.
- When form filling.

Lapping bars in a horizontal plane, shown in view 2 of figure 6-8 is the most practical arrangement, if the spacing provides enough clearance for the aggregate to pass. Both this method and the one shown in view 3 of figure 6-8 facilitate tying the bars to hold them in position during concreting. However, tying does not add much to splice strength and creates possible problems of air pockets and poor bond in the space under the bar overlap. Although lapping bars in a vertical plane, shown in view 3 of figure 6-8, permits better encasement, the top bars will not fit the stirrups properly, and the beam will have a smaller effective depth at one location than at another. In practice, these bars would probably be locked into the positions shown in view 2 of figure 6-8.

CARE

6-25. Care must be taken to prevent reinforcing steel from rusting too much during outdoor storage. Before storing outdoors, remove any objectionable coating from the steel, particularly heavy corrosion caused by previous outdoor storage. Other such coatings include oil, paint, grease, dried mud, and weak dried mortar. If the mortar is difficult to remove, leave it. A thin film of rust or mill scale is not seriously objectionable either; in fact, it can increase the bond between the steel and concrete. Remove all loose rust or scale by rubbing the steel with burlap or other similar means. Remember, anything that destroys the bond between the concrete and the steel can create serious problems by preventing the stress in the steel from performing its function properly. After cleaning, cover the steel to protect it from weather.

ON-SITE PREFABRICATION

6-26. Large numbers of reinforcing bars can be prefabricated on the job to the varied lengths and shapes shown on the drawings. You can usually cold-bend stirrups and column ties that are less than 1/2 inch in diameter, as well as steel bars no larger than 3/8 inch in diameter. Heating is not usually necessary, except for bars more than 1 1/8 inch in diameter. Make bends around pins whose diameters are not less than six times the bar diameter (see table 6-1, page 6-8), except for hooks and straight bars larger than 1 inch in diameter. For these bar diameters, the minimum bending pin diameter should be eight times the bar size. If possible, it is better to bend steel bars greater than 3/8 inch in diameter on a bar-bending machine; use a hickey (a lever for bending bars) or a bar-bending table like the one shown in figure 6-9, page 6-14. However, be aware that the bends these devices make are usually too sharp and weaken the bar. It is possible to improvise a hickey by attaching a 2- by 1 1/2- by 2-inch pipe tee section to the end of a 1 1/4-inch pipe lever that is 3 feet long, then sawing a section from one side of the tee.

PLACEMENT

6-27. All steel reinforcement must be accurately located in the forms and held firmly in place both before and during concreting. To do this, use built-in concrete blocks, metal bolsters and chairs, spacer bars, wires, or other devices that prevent steel displacement during construction, as well as retain it at the proper distance from the forms.

SUPPORT

6-28. Be sure to use enough supports and spacers to hold the steel firmly, even when subjected to construction loads. Never use rocks, wood blocks, or other unapproved supports. Wood blocks deteriorate, eventually creating a pathway for water to penetrate and degrade reinforcing rebar. Support horizontal bars at minimum intervals of 5 or 6 feet, and secure all bars to supports and other bars using tie wires not smaller than 18 gage. Twisted-tie ends should project away from an interior surface to avoid contact with the concrete surface.

SPACERS

6-29. Some specifications state that no metal be left in concrete within a certain distance of the surface. Generally, the minimum clearance between parallel bars in beams, footings, walls, and floor slabs is not less than 1 1/3 times the largest size aggregate particle in the concrete, but in no case less than 1 inch. To meet such specifications, make spacer and supporting blocks from mortar having the same consistency as the concrete, but no CA. Spacer blocks are usually 1 1/2 inch square or larger, varying in length as required. Cast tie wires into the blocks to secure them to the reinforcing bars. Do not remove this type of spacer block when placing the concrete.

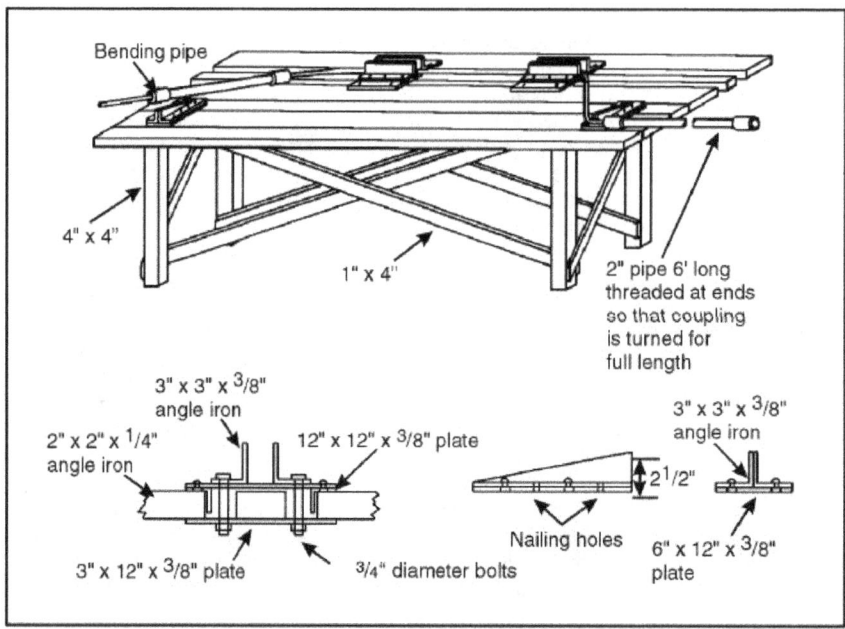

Figure 6-9. Bar bending table

COLUMNS

6-30. The minimum clearance between parallel bars in columns is not less than 1 1/2 times the bar diameter. First, tie the column steel together and position it as a unit. Then erect the column form around the unit and tie the reinforcing steel to the form at 5-foot intervals. Minimum clear space between reinforcing steel and concrete forms must also be checked with the MSA in the concrete mix design. The aggregate must not become lodged between the form and reinforcing steel, preventing passage by any further concrete.

FLOOR SLABS

6-31. Figure 6-10 shows a typical arrangement of reinforcing steel in a floor slab. The required thickness of the concrete protective cover determines the height of the bolster. Instead of the bolster, you can use spacer blocks made from sand-cement mortar as described above. Hold the bars firmly in place by tying the intersections together at frequent intervals with one turn of wire.

BEAMS

6-32. Figure 6-11 shows the position of bolsters and stirrups with the reinforcing bars in a reinforced-concrete beam. Note that the stirrups pass under the main reinforcing bars and are tied to them with one turn of wire.

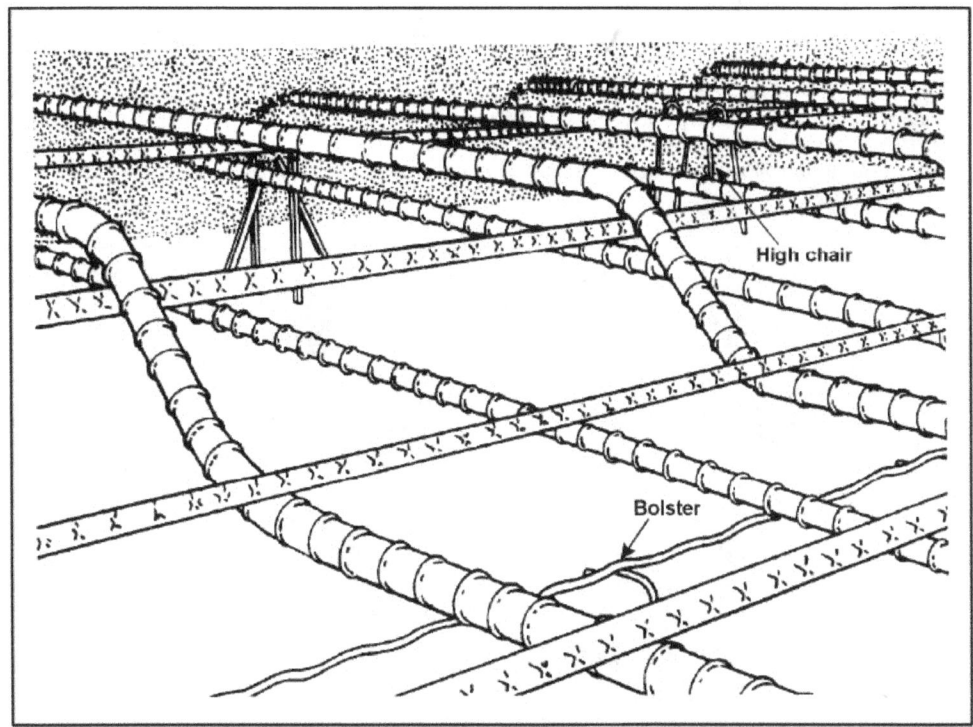

Figure 6-10. Reinforcing steel arrangement for a floor slab

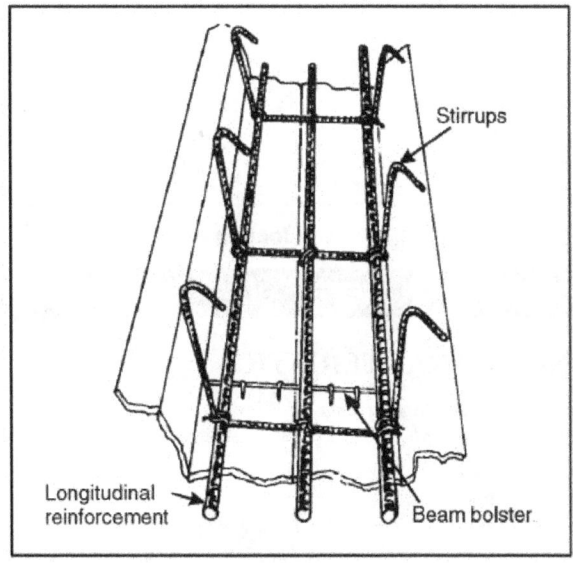

Figure 6-11. Beam reinforcing steel

Chapter 6

WALLS AND FOOTINGS

6-33. Unless you are using wire-mesh fabric, erect the reinforcing steel for concrete walls in place. Do not preassemble it as you do for columns. Use ties between the bars, from top to bottom, for high walls (see view 1 of figure 6-12). You can remove the wood blocks after you fill the forms to block level. Place reinforcing steel for footings as shown in view 3 of figure 6-12. Use spacer bars to support the steel above the subgrade to the proper distance. Welded-wire fabric (see view 2 of figure 6-12) is also used as limited reinforcement for concrete footings, walls, and slabs, but its primary use is to control crack widths due to temperature changes.

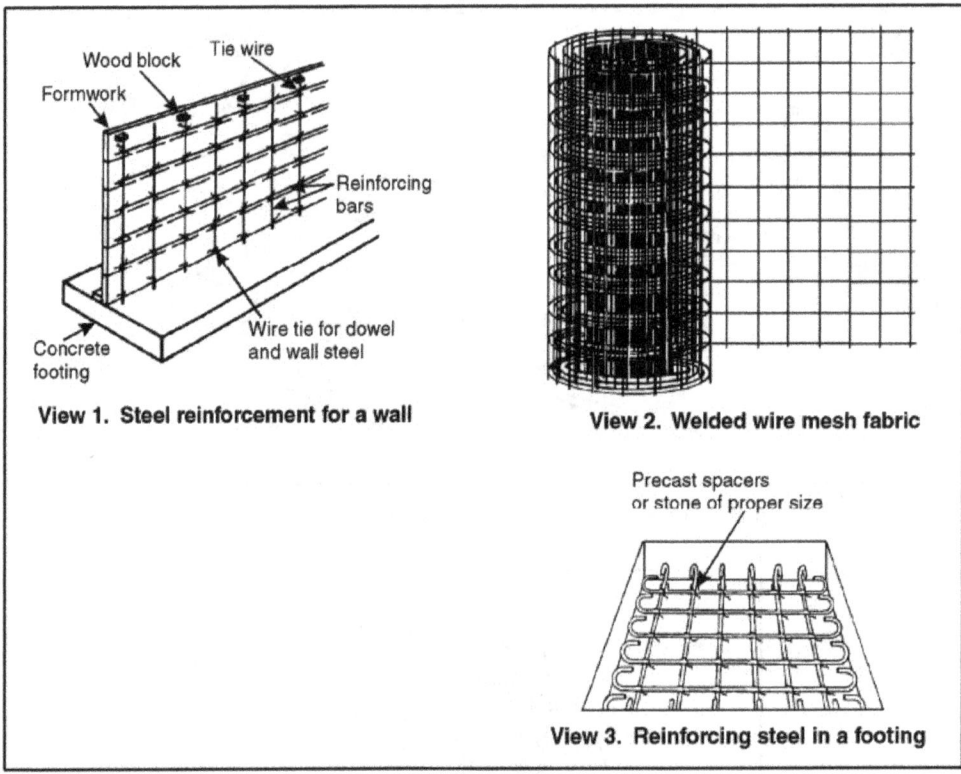

Figure 6-12. Wall and footing reinforcement

SECTION IV - PRECAST CONCRETE

DEFINITION AND CHARACTERISTICS

6-34. Nearly all prestressed concrete is precast. Prestressed concrete is subjected to compressive stresses before any external loads are applied. The compressive stresses counteract the stresses that occur when a unit is subjected to loads.

DEFINITION

6-35. *Precast concrete* is concrete cast in other than its final position. Unlike cast-in-place concrete, precast concrete requires field connections to tie the structure together. The connections can be a major design problem. Although precasting is best suited to a manufacturer's factory or yard, job-site precasting is common on large projects and in area's remote from a precasting source. Prestressed concrete is a combination of high-tensile steel and high-strength concrete. Prestressing is accomplished by stretching the high-strength steel wire, wire strands, or bars, either by pretensioning or posttensioning (stretched before or after pouring the concrete).

CHARACTERISTICS

6-36. Long-span, precast floor and roof units are usually prestressed with tongue-and-groove edges and special clips to anchor them to supporting members. Short members, 30 feet or less, are often made with ordinary steel reinforcement. Available types of precast floor and roof units include solid or ribbed slabs, hollow-core slabs, single and double tees, rectangular beams, L-shaped beams, inverted T-beams, and I-beams. Hollow-core slabs are usually available in normal or structural lightweight concrete. Units range from 16 to 96 inches wide and from 4 to 12 inches deep. Hollow-core slabs may have grouted shear keys to distribute loads to adjacent units over a slab width as great as one-half the span. Precast-reinforced slabs are also available for curtain walls in buildings having either steel or concrete frames. Such wall panels include plain panels, decorative panels, natural stone-faced panels, sandwich panels, solid panels, ribbed panels, load-bearing and nonload-bearing panels, and thin-section panels. Prestressing these panels makes it possible to handle and erect large units and thin sections without cracking. Two other convenient forms of precast concrete are tilt-up and lift-slab that save formwork expense since wall and floor surfaces are cast horizontally at a convenient level. In tilt-up, the wall is cast horizontally, then tilted up to its final, vertical position. Lift-slab involves precasting the floor and roof slabs of a multistory building horizontally, then jacking or lifting them one upon another to their final positions.

PRODUCT AVAILABILITY

6-37. Precast concrete tends to be a local business. Although there are many manufacturers, few deliver more than 200 to 300 miles from their facility. For this reason, there is a wide variation in the types and sizes of precast items available in different areas. Check those manufacturers who will deliver to the project area to find out what is available. Many firms do not maintain facilities for prestressing or special surface treatments that you may require. In general, precast concrete products fall roughly into five groups.

- Architectural cast stone and other type of ornamental concrete.
- Steps, paving flags, curbing, and so forth, that must be wear-resisting.
- Load-bearing structural members (beams, columns, flooring, roofing) and other standard units such as piles, highway girders, electric poles, masts, lintels, posts, pipes, water tanks, and troughs. Many of these units are prestressed.
- Roofing units and tiles that must be waterproof and weather-resistant.
- Concrete blocks, bricks, and slabs.

ADVANTAGES

6-38. Form costs are much less with precast concrete, because you do not have to support the forms on scaffolding; simply set the forms on the ground in a convenient position. A thin concrete wall is very difficult to construct if cast vertically, because the concrete must be poured high up in the narrow opening at the top of the form. But a thin wall is easily precasted flat on the ground. Thus, the large-side forms are eliminated, as well as the braces that hold a vertical form in place. In some types of precast construction, you lose no time waiting for concrete to gain strength at one level before you can place the next. Such delays are common in cast-in-place construction. In addition, permanent precast units can often be used as a working platform, thus eliminating the need for scaffolding. Precast units can be standardized and mass-produced. Great savings can result from repeated reuse of forms and assembly line production techniques. You can also maintain high quality control of precast products. Precast units poured on-site require minimal transportation, but any off-site precast concrete requires transportation, usually by truck. Precast concrete buildings are built so that all the structural concrete (except footings) is precast and put together piece by piece. The variety of shapes and designs available is limited only by the imagination and cost. Most are available in either reinforced or prestressed conditions.

Chapter 6

DISADVANTAGES

6-39. The decision as to whether concrete is cast-in-place, purchased precast, or precast on-site is made by the design engineer, architect or construction supervisor. However, precasting on-site is not recommended unless experienced precast personnel are available to supervise the work. In making the decision, the engineer must consider the following:

- Special requirements. Precast concrete requires large placing equipment such as cranes, derricks and a specialty crew to set the pieces. Precasting also takes plenty of space. Is enough space available on-site or should the material be precast off-site and transported to the job-site? If precasting is accomplished off-site, the necessary facilities and transportation costs must be added to the cost of the items precast.
- Manufacture. Should forms be made from steel, wood, fiberglass, or a combination of materials? Who will make the forms? How long will manufacturing take? How much will it cost? The cost of the forms must also be charged to the precast items.
- Supervision. Is a specialist available to supervise manufacture? Someone also has to coordinate the work and prepare the shop drawings. These costs must also be included, along with the following:
- Extra materials. Materials must be purchased for—
    - Reinforcement.
    - CA and FA.
    - Cement.
    - Water.
    - Anchors and inserts.
- Extra labor. Labor is required to—
    - Clean forms.
    - Apply oil or retarders to forms.
    - Place the reinforcement (including pretensioning, if required).
    - Mix and place concrete (including troweling, and so forth).
    - Cover the concrete and apply the curing method.
    - Uncover the concrete after initial curing.
    - Strip the concrete from the forms and stockpile it to finish curing.
    - Erect the precast concrete on the job.
    - Load and unload transport truck, if concrete is precast off-site.
- Extra equipment. Necessary equipment may include mixers, lift trucks, cranes, derricks, equipment to cure the concrete, and miscellaneous equipment such as hoes, shovels, wheelbarrows, hammers, and vibrators. Special equipment is required to prestress concrete. All equipment requires storage space.

DESIGN

6-40. The design of a structure composed of precast members requires both cleverness and engineering skill. There are inherent problems to resolve in this type of construction such as provisions for seating members and tying the structure together. When using precast members, the accuracy of the building layout is extremely important. The precast pieces must fit together like a jigsaw puzzle leaving little room for error. Because the purpose of this chapter is only familiarization with the end products and the materials and methods used to produce them, neither the engineering nor design of precast concrete is discussed.

PREFABRICATION

6-41. You can precast structural members either in central off-site or on-site prefabrication plants depending upon the product, the numbers required, and its use. On-site or temporary prefabrication plants are suitable for military operations. They are subject to prevailing weather and climate conditions because they have no roofing. Lay out the prefabrication yard to suit the type and the quantity of members you will precast. Bridge T-beams, reinforced-concrete arches, end walls, and concrete logs are precast structural members. The yard must be on firm, level ground that has ample working space and access routes. Figure 6-13 is a schematic layout of a prefabrication yard suitable for producing typical members. Table 6-5, page 6-20, recommends the personnel requirements. You can expect such a prefabrication facility to produce approximately 6,000 square feet of precast walls per day, although output will vary according to personal experience, equipment capabilities, and product requirements.

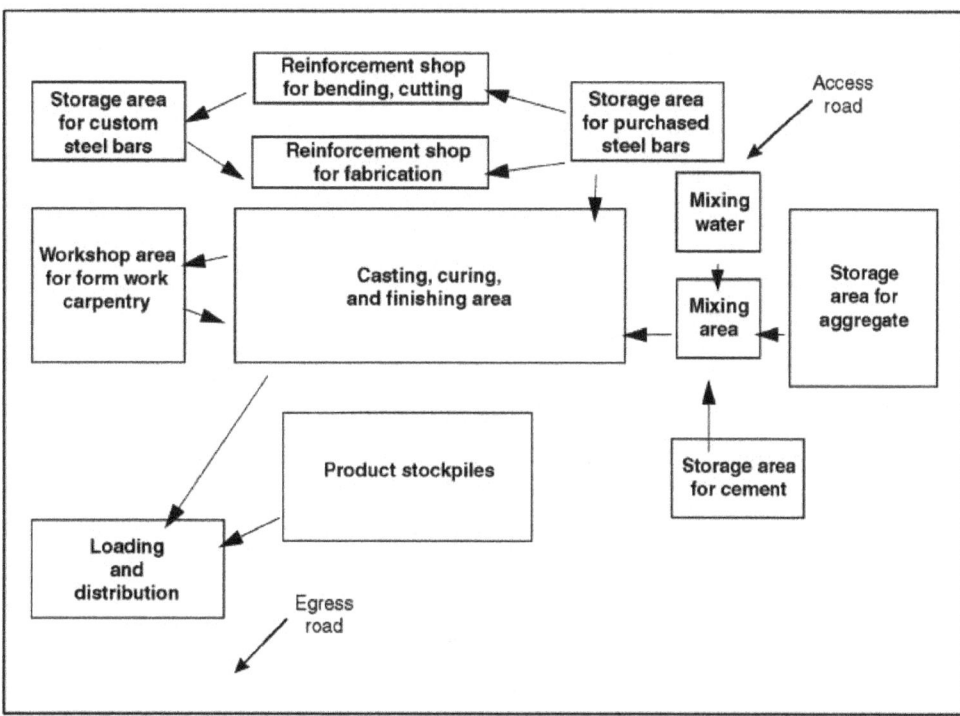

Figure 6-13. Schematic layout of on-site or temporary prefabrication yard for military operations

TRANSPORTATION

6-42. Precast members can withstand farther hauling and rougher treatment then either plain or reinforced members; all members require adequate support so that they do not undergo either strain or loading greater than the design loading. Various types of hauling and handling equipment are needed in precast concrete operations. You can transport heavy girders long distances on tractor trailers or with a tractor and dolly arrangement in which the girder acts as the tongue or tie between the tractor and the dolly. Smaller members can be transported (columns, piles, slabs) on flatbed trailers, but must be protected from too much bending stress caused by their own dead weight.

Table 6-5. Recommended precasting team personnel requirements

Area	Personnel
Reinforcement shops and steel storage	1 NCO 4 EMs
Workshop	2 EMs
Casting, curing and finishing Placing reinforcement Refurbishing forms	2 NCOs 13 EMs
Mixing Water, cement, and aggregate, Storage	1 NCO 5 EMs
Product stockpiles Loading and distribution	2 EMs

ERECTION

6-43. Erecting precast members is similar to erecting reinforcing steel and requires cranes or derricks of sufficient capacity. Use spreader bars to handle the elements at their correct pickup points. Sometimes two cranes or derricks are needed to lift very long or heavy members.

PART TWO
Masonry

The original idea of masonry was nothing more than placing stones in an orderly fashion normally by laying them in rows. As time continued, with improved quality and type of materials available, this orderly lying of stone has progressed into laying them with the use of masonry cements that bonds them together. Part two of this manual covers basic tools and equipment, properties, and mixing of mortar. Concrete masonry will include the characteristics of concrete blocks, construction procedures, and rubble stone masonry. The characteristics of brick, and brick laying methods are discussed in the final chapter.

Chapter 7
Basic Equipment and Components

Many of the longest-lasting structures in the world are constructed with masonry materials, such as stones, bricks, or blocks. Masons are highly skilled people who use specialized equipment to lay out and construct masonry walls and other building features. Along with erecting these structures, masons are responsible for the care of their tools and equipment. They are also responsible for the correct mixing proportions of the mortar and the safe erection of scaffolding.

SECTION I - MASON'S TOOLS AND EQUIPMENT

DEFINITION

7-1. Masonry originally meant the art of building a structure from stone. Today, masonry means to build a structure from any building materials, such as, concrete blocks, stones, bricks, clay tile products, gypsum blocks, and sometimes glass blocks, that consist of units held together with mortar; The characteristics of masonry work are determined by the properties of the masonry units and mortar, and the methods of bonding, reinforcing, anchoring, tying, and joining the units into a structure.

TOOLS

7-2. Figure 7-1, page 7-2, shows a set of typical basic mason's tools. Care of the tools are extremely necessary. Be sure to keep wheel-barrows, mortar boxes, and mortar tools clean because hardened mortar is difficult to remove. Clean all tools and equipment thoroughly at the end of each day or when the job is finished. A full set includes—

Chapter 7

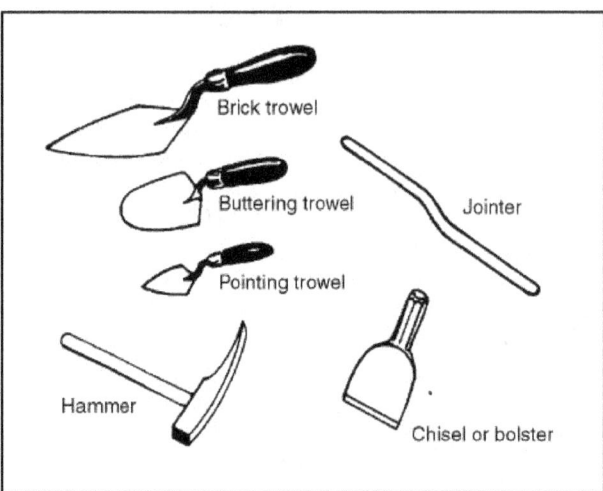

Figure 7-1. Basic mason's tools

- A trowel. A trowel is used to mix and pick up mortar from the board, to place mortar on the unit, to spread mortar, and to tap the unit down into the bed. A common trowel is usually triangular in shape ranging in size up to about 11 inches long and from 4 to 8 inches wide. Its length and weight depend on the mason. Generally, short, wide trowels are best because they do not put too much strain on the wrist. Trowels used to point and strike joints are smaller, ranging from 3 to 6 inches long and 2 to 3 inches wide.
- A chisel or bolster. A chisel is used to cut masonry units into parts. A typical chisel is 2 1/2 to 4 1/2 inches wide.
- A hammer. The mason's hammer has a square face on one end and a long chisel peen on the other. It weighs from 1 1/2 to 3 1/2 pounds. It is used to split and rough-break masonry units.
- A jointer. As its name implies, this tool is used to make various mortar joints. There are several different types of jointer—rounded, flat, or pointed, depending on the shape of the mortar joint you want.
- A square. The square that is shown in figure 7-2 is used to measure right angles and to lay out corners.
- A mason's level. The level is used to plumb and level walls. A level ranges from 42 to 48 inches in length and is made from either wood or metal. Figure 7-2 shows a level in both the horizontal and vertical positions. When you place it on the masonry horizontally and the bubble falls exactly in the middle of the center tube, the masonry is level. When you place the level against the masonry vertically and the bubbles fall exactly in the middle of the two end tubes, the masonry is plumb.
- A straightedge. A straightedge, shown in figure 7-2, can be any length up to 16 feet, from 1 1/8 inches to 1 1/2 inches thick, and the middle portion of the top edge from 6 to 10 inches wide. The middle portion of the top edge must be parallel to the bottom edge. Use a straightedge to extend a level to either plumb or level distances longer than the level length.
- Miscellaneous tools. Other mason's tools and equipment include shovels, mortar hoes, wheelbarrows, chalk lines, plumb bobs, and a 200-foot ball of good quality mason's line.

Basic Equipment and Components

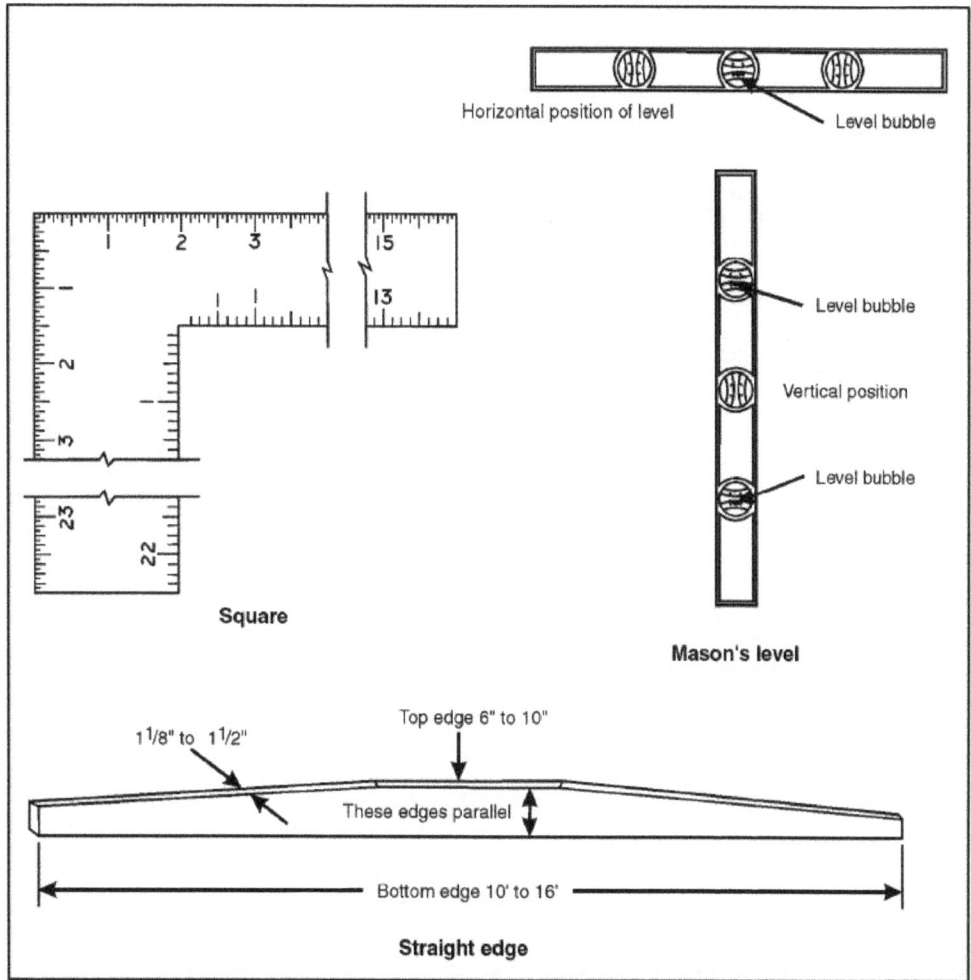

Figure 7-2. Square, mason's level, and straightedge

EQUIPMENT

7-3. Mortar is mixed by hand in a mortar box. It should be as watertight as possible. A mortar board (see figure 7-3, page 7-4) can range from 3 to 4 feet square. Wet down a mortar board thoroughly before placing any mortar on it to prevent the wood from drying it out and absorbing moisture from the mortar. Figure 7-3 shows the proper way to fill a mortar board. Note the mounds of mortar in the center of the board; this minimizes drying. After filling the mortar board, keep the mortar rounded up in the center of the board and the outer edges of the board clean. Any mortar spread in a thin layer dries out quickly, and lumps form in it. Be sure to maintain the proper mortar consistency at all times.

Chapter 7

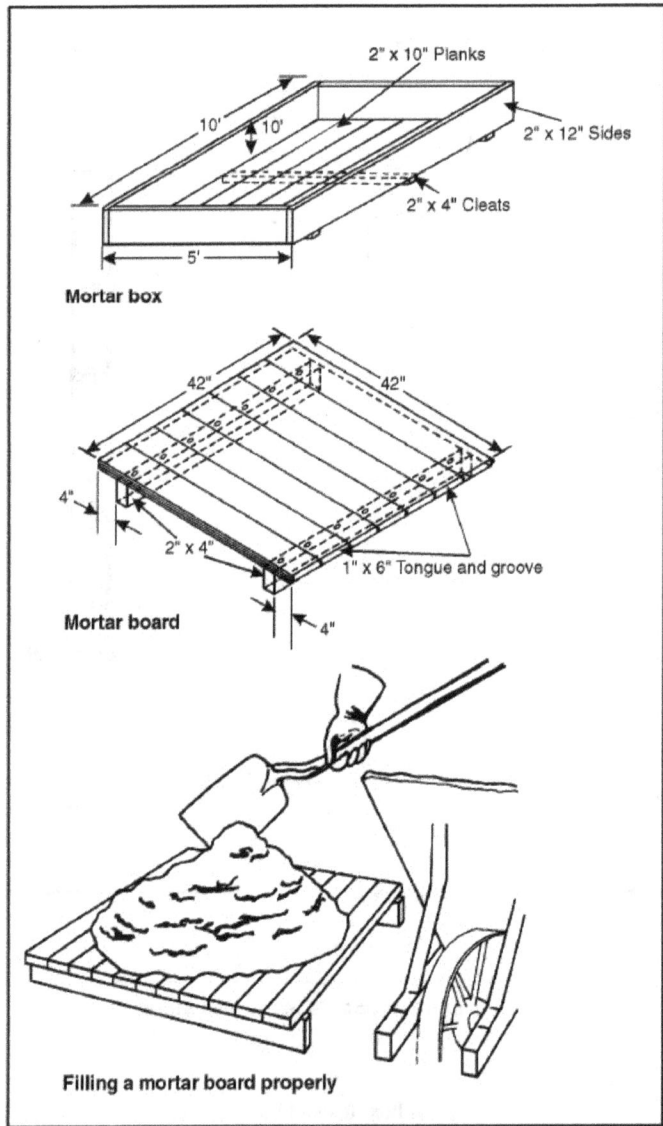

Figure 7-3. Mortar box and mortar board

SECTION II - MORTAR

DESIRABLE PROPERTIES

7-4. Good mortar is necessary for good workmanship and good masonry service because it must bond the masonry units into a strong well-knit structure.

Bond Considerations

7-5. The mortar that bonds concrete blocks, bricks, or clay tiles together will be the weakest part of the masonry unless you mix and apply it properly. When masonry leaks are encountered, they are usually through the mortar joints. The strength of masonry and its resistance to rain penetration depends largely on the strength of the bond between the masonry unit and the mortar. Various factors affect bond strength including the type and quantity of mortar, its workability or plasticity, its water retentivity, the surface texture of the mortar bed, and the quality of workmanship in laying the units. You can correct irregular brick dimensions and shape with a good mortar joint.

Plasticity

7-6. Mortar must be flexible enough to work with a trowel. You can obtain good plasticity or workability by—

- Using mortar having good water retentivity.
- Using the proper grade of sand and thoroughly mixing.
- Using less cementitious materials.

Mortar properties depend largely upon the type of sand the mortar contains. Clean, sharp sand produces excellent mortar; too much sand causes mortar to segregate, drop off the trowel, and weather poorly.

Water Retentivity

7-7. Mortar property resists rapid water loss to highly absorbent masonry units. Mortar must have water to develop the bond. If it does not contain enough water, the mortar will have poor plasticity and workability and the bond will be weak and spotty. Sometimes you must wet the brick to control water absorption before applying mortar, but never wet concrete masonry units.

Strength and Durability

7-8. The type of service that the masonry must give determines the mortar's strength and durability requirements. For example, walls subject to severe stresses or to severe weathering must be laid with more durable, stronger mortars than walls for ordinary service. Table 7-1, page 7-6, gives mortar mix proportions that provide adequate strength and durability for the conditions listed. You can convert the unit volume proportions to weight proportions by multiplying the unit volumes given by the weight per cubic foot of the materials. Those specifications are —

    Masonry cement ....Weight printed on bag
    Portland cement................................... 94 lb
    Hydrated lime...................................... 50 lb
    Mortar sand, damp and loose........... 85 lb

Types of Mortar

7-9. The following mortar types are proportioned on a volume basis:

- Type M mortar consist of one part portland cement, 1/4 part hydrated lime or lime putty and, 3 parts sand or 1 part portland cement, 1 part type II masonry cement, and 6 parts sand. Type M mortar is suitable for general use but is recommended specifically for below-grade masonry that contacts earth, such as foundations, retaining walls, and walks.

Chapter 7

Table 7-1. Recommended mortar mix proportions by unit volume

Type of Services	Cement	Hydrated Lime	Mortar Sand in Damp, Loose Conditions
Ordinary	1 unit masonry cement or 1 unit portland cement	1/2 to 1 1/4	2 1/4 to 3 4 1/2 to 6
Isolated piers subjected to extremely heavy loads, violent winds, earthquakes, or severe frost action	1 unit masonry cement* plus 1 unit portland cement or 1 unit portland cement	0 to 1/4	4 1/2 to 6 2 1/4 to 3
*ASTM specification C91 type II.			

- Type S mortar consist of one part portland cement, 1/2 part hydrated lime or lime putty, and 4 1/2 parts sand, or 1/2 part portland cement, 1 part type II masonry cement, and 4 1/2 parts sand. Type S mortar is also suitable for general use, but is recommended where high resistance to lateral forces is required.
- Type N mortar consist of one part portland cement, 1 part hydrated lime or lime putty, and 6 parts sand, or 1 part type II masonry cement and 3 parts sand. Type N mortar is suitable for general use in above-grade exposed masonry where high compressive and/or lateral strengths are not required.
- Type O mortar consist of one part portland cement, 2 parts hydrated lime or lime putty, and 9 parts sand, or 1 part type I or type II masonry cement and 3 parts sand. Type O mortar is recommended for load-bearing, solid-unit walls when the compressive stresses do not exceed 100 psi, and the masonry is not subject to freezing and thawing in the presence of a lot of moisture.

STORING MORTAR MATERIALS

7-10. Store all mortar materials, except sand and slaked quicklime, in a dry place. Sand and lime should be covered to prevent excessive losses or gains of surface moisture.

MIXING MORTAR

7-11. When blending or mixing mortar, always use the best consistency for the job.

MACHINE MIXING

7-12. Mix large quantities of mortar in a drum-type mixer, like a concrete mixer. Mix a minimum of 3 minutes. Place all dry ingredients in the mixer first, mix them for 1 minute before adding the water.

HAND MIXING

7-13. Mix small amounts of mortar by hand in a mortar box (see figure 7-3, page 7-4). Mix all ingredients thoroughly to obtain a uniform mixture. Mix all dry materials together first before adding water. Keep a steel drum of water close to the mortar box to use as the water supply. Use a second drum of water to store shovels and hoes when not in use.

MORTAR MIXING WITH LIME PUTTY

7-14. When machine mixing, measure the lime putty using a pail and place it into the skip on top of the sand. When hand mixing, add the sand to the lime putty. Wet pails before placing mortar in them and clean them immediately after emptying them.

WATER QUALITY

7-15. Mixing water for mortar must meet the same requirements as mixing water for concrete. Do not use water containing large amounts of dissolved salts, because the salts will cause efflorescence and weaken the mortar.

RETEMPERING MORTAR

7-16. The workability of any mortar that stiffens on the mortar board due to evaporation by remixing can be restored. Add water as necessary, but discard any mortar stiffened by initial setting. It is difficult to tell the cause of stiffening; a practical guide is to use mortar within 2 1/2 hours after the original mixing when the air temperature is 80°F or higher, and within 3 1/2 hours when the air temperature is below 80°F. Discard any mortar not used within these limits.

ANTIFREEZE MATERIALS

7-17. Do not use an admixture to lower the freezing point of mortar during winter construction. The quantity of antifreeze materials necessary to lower the freezing point of mortar to any appreciable degree is so large that it would seriously impair the mortar's strength and other desirable properties. Never use frozen mortar; freezing destroys its bonding ability.

ACCELERATORS

7-18. Make a trial mix to find the percentage of calcium chloride that gives the desired hardening rate. Do not add more than 2 percent calcium chloride, by weight of cement to mortar, to accelerate its hardening rate and increase early strength. Do not add more than 1 percent calcium chloride to masonry cements. Calcium chloride should not be used for steel-reinforced masonry. You can also accelerate hardening rate in mortars with high-early-strength portland cement.

REPAIRING AND TUCK-POINTING

7-19. Use the mortar mixes given in table 7-1 when repairing and tuck-pointing old masonry walls. Compact the joints thoroughly by tooling after the mortar partially stiffens.

SECTION III - SCAFFOLDING

CONSTRUCTION AND SAFETY

7-20. Extreme care should be taken in building scaffolds because lives depend on them. Use rough lumber for wood scaffolding. Never nail scaffolding in a temporary manner; always nail it securely. When you no longer need the planks at a lower level, remove them to avoid falling mortar splash.

TYPES OF SCAFFOLDING

7-21. A *scaffold* is a temporary, movable platform built with planks to support workers and materials. It allows bricklayers to work at heights not reachable when standing on the floor or ground. Scaffolds can be used in several functions and come in different sizes and heights.

Chapter 7

TRESTLE SCAFFOLD

7-22. Use a trestle scaffold shown in figure 7-4 when laying bricks from the inside of a wall. Erect the scaffold when the wall reaches a height of 4 or 5 feet. The height of the trestles should range from 4 to 4 1/2 feet. The planks should be made using 2 by 10s. Place the trestle at least 3 inches from the wall so that it does not press against the newly laid bricks and force them out of line. Build the wall to the next floor level working from the scaffold. When the rough flooring for the next floor is in place, repeat the procedure.

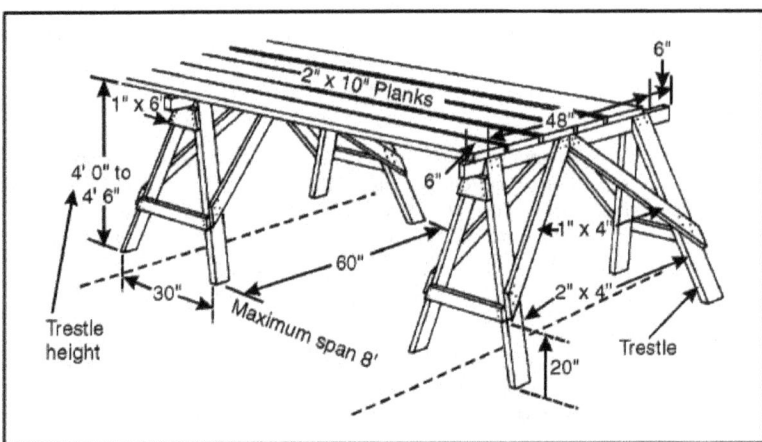

Figure 7-4. Trestle scaffold

FOOT SCAFFOLD

7-23. When reaching lower than a trestle scaffold permits, use a foot scaffold like the one shown in figure 7-5. Place 2 by 10 planks on blocks supported by the trestle scaffold. A foot scaffold should not exceed 18 inches in height.

Figure 7-5. Foot scaffold

Putlog Scaffold

7-24. A putlog scaffold (see figure 7-6) reaches from the ground to the height required. Its uprights are 4 by 4s supported on a 2 by 12 by 12 plank for bearing on the soil. Space the uprights on 8-foot centers and allow 4 1/2 feet of space between the wall and the uprights, as shown in figure 7-6. The putlog is 3-by 4-inch lumber that spans the gap between the wall and the ledger. One end of the putlog rests on top of the ledger and against the 4 by 4 uprights, while the other end fits into the wall (one brick is omitted to make an opening for it). Do not fasten the putlog to the ledger. Place five 2 by 12 planks on top of the putlog to form the scaffold platform. Do not nail the planks to the putlog. Two ways to use stays are—

- Tie the uprights to the wall with stays. You can either pass the stays through a window opening and fasten them to the inside structure or use spring stays as shown in figure 7-6. To make spring stays, omit one brick from the wall and insert the ends of two 2 by 6s in the opening. Then insert a brick between the 2 by 6s and force the brick toward the wall. Bring the other ends of the 2 by 6s together and nail them securely to the ledger.
- Use the putlog as a stay. You can also use the putlog as a stay by driving a wood wedge above the putlog into its hole in the wall. Then, nail the wedge to the putlog and nail the putlog to the ledger. Install longitudinal cross bracing as shown in figure 7-6.

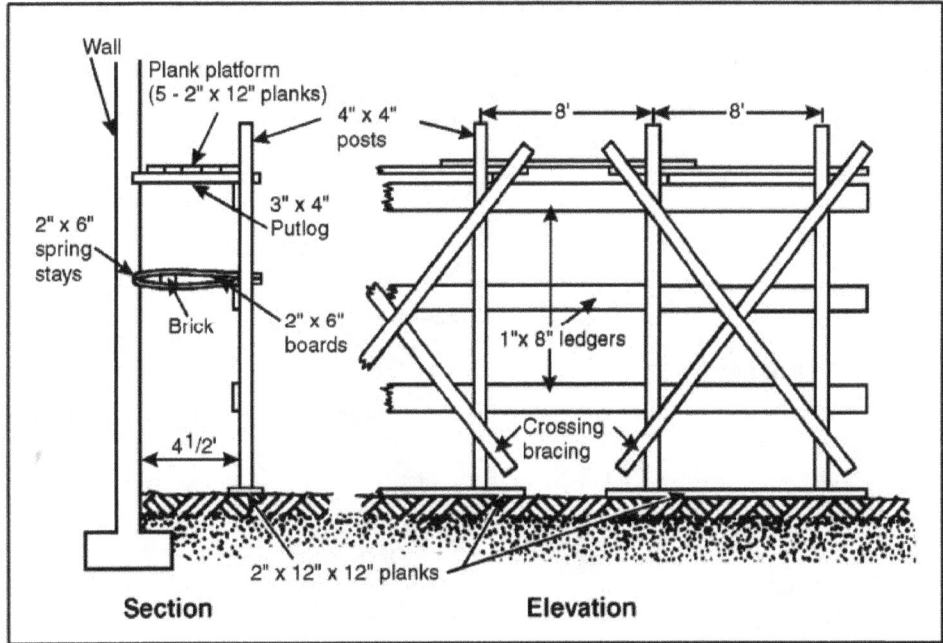

Figure 7-6. Putlog scaffold

Outrigger Scaffold

7-25. An outrigger scaffold (see figure 7-7, page 7-10) consists of a wood outrigger beam projecting from a window sill that supports 2 by 10 planks. Figure 7-7 shows how to brace a wood beam, but if you use a steel outrigger beam, fasten it to the structure's formwork using threaded U-bolts.

Chapter 7

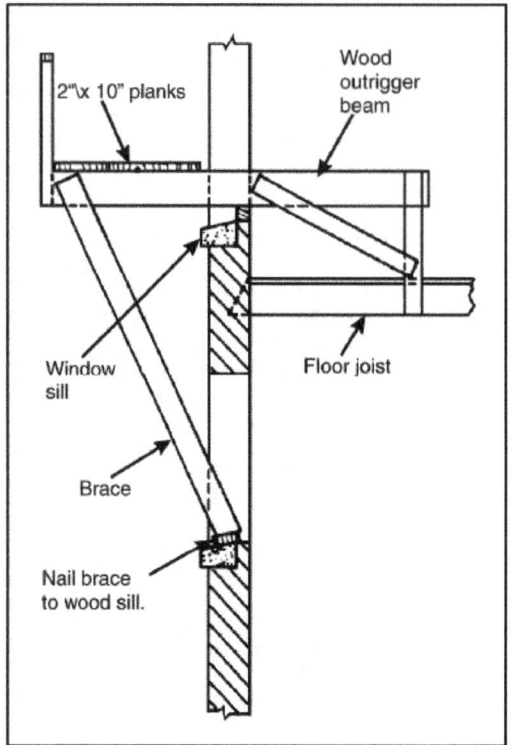

Figure 7-7. Outrigger scaffold

PREFABRICATED STEEL SCAFFOLD

7-26. If it is available, use prefabricated steel scaffolding (see figure 7-8) rather than building a scaffold.

Basic Equipment and Components

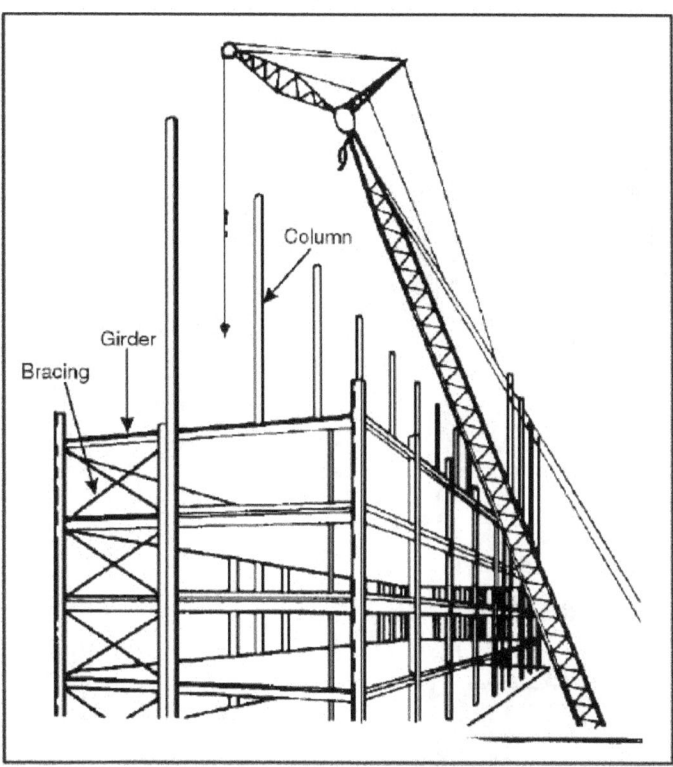

Figure 7-8. Prefabricated steel scaffold

MATERIALS TOWER

7-27. Use a steel material tower if construction details are available because it is easier to erect and generally safer. Otherwise, you can construct a wood tower to hoist materials to the working height, like the one shown in figure 7-9, page 7-12. Locate the tower where you can bring materials to it over the shortest haul, but far enough away from the structure to clear any external scaffolding. A clearance of 6 feet 8 inches is enough for scaffold platforms 5 feet wide. Construct the tower footing using two 2 by 12s, 2 feet long, placed under each 4 by 4 post. The height of the tower should extend at least 15 feet above the highest point where you need a landing. Then construct landings extending from the tower to the floors and scaffold platforms as needed. Use 2 by 10s or 2 by 12s for the landings.

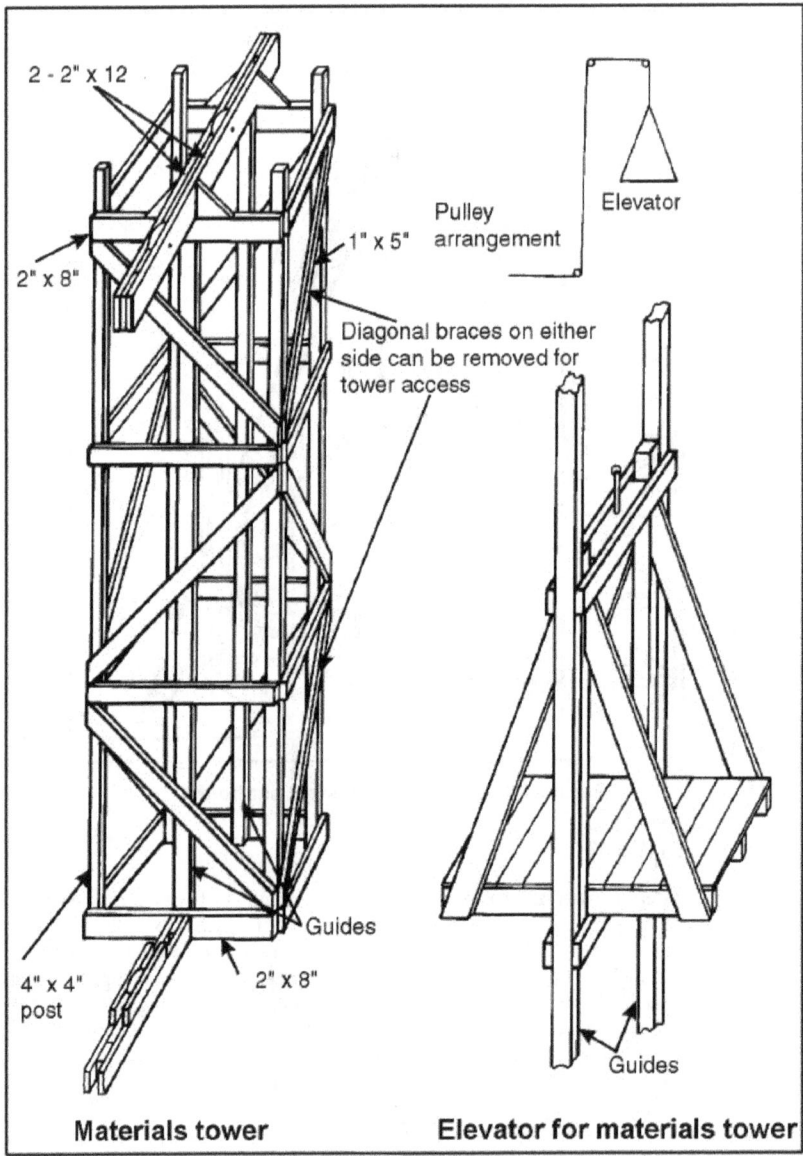

Figure 7-9. Materials tower and elevator

ELEVATOR

7-28. Figure 7-9 also shows the elevator, rope, and pulley arrangement that serves the materials tower. Note the guides at the base of the elevator that fit onto the guides running up from the base of the tower.

Chapter 8
Concrete Masonry

When portland cement, water, and suitable aggregates, such as sand, gravel, crushed stone, cinders, burned shale, or slag, are mixed and formed into individual pieces to be used in laying up walls and other structural details, the pieces thus formed are known as unit masonry, or units. However, most masons refer to them as concrete blocks. Concrete blocks will vary in size and shape as well as style. They are typically used for house foundations, decorative blocks for strong garden, or retaining walls. Rubble stone masonry is strong, durable, and offers an incomparable beauty and range of effect. Construction of concrete masonry is time consuming and requires highly skilled personnel.

SECTION I - CHARACTERISTICS OF CONCRETE BLOCK

NATURE AND PHYSICAL PROPERTIES

8-1. A concrete block is a masonry unit that either contains single or multiple hollows, or is solid. It is made from conventional cement mixes and various types of aggregate, including sand, gravel, crushed stone, air-cooled slag, coal cinders, expanded shale or clay, expanded slag, volcanic cinders (pozzolan), pumice, and scoria (refuse obtained from metal ore reduction and smelting). The term *concrete blocks* was formerly limited to only hollow masonry units made with such aggregates as sand, gravel, and crushed stone. But today the term covers all types of concrete blocks (both hollow and solid) made with any kind of aggregate. Concrete blocks are also available with applied glazed surfaces, various pierced designs, and a wide variety of surface textures. Although a concrete block is made in many sizes and shapes (see figure 8-1, page 8-2) and in both modular and nonmodular dimensions, its most common unit size is 7 5/8 by 7 5/8 by 15 5/8, known as 8- by 8- by 16-inch block nominal size. All concrete blocks must meet certain specifications covering size, type, weight, moisture content, compressive strength, and certain other special characteristics. Concrete masonry is an increasingly important type of construction due to technological developments in both the manufacture and the use of concrete blocks. Properly designed and constructed concrete masonry walls satisfy many building requirements, including fire prevention, safety, durability, economy, appearance, utility, comfort, and acoustics.

Chapter 8

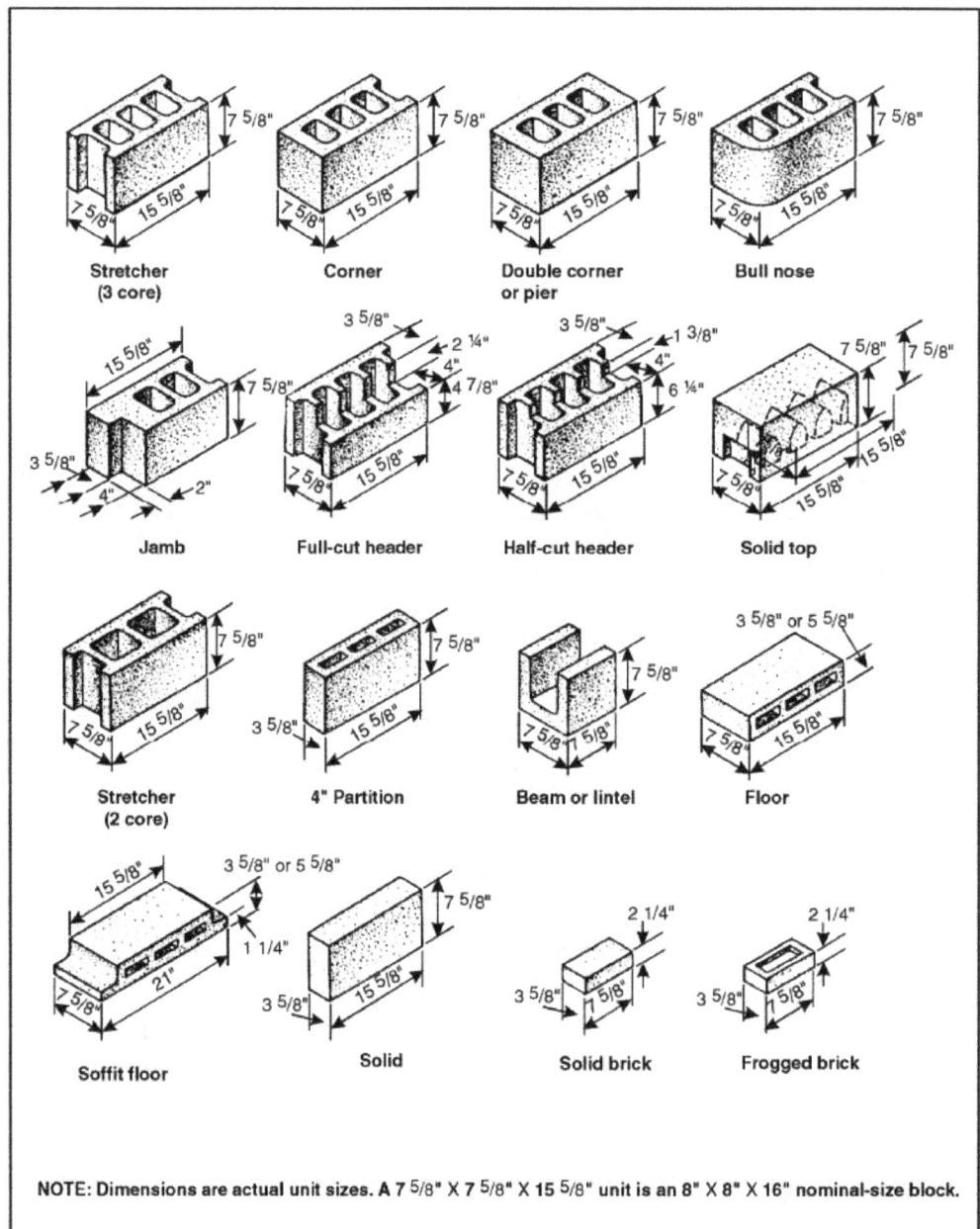

Figure 8-1. Typical unit sizes and shapes of concrete masonry units

Concrete Masonry

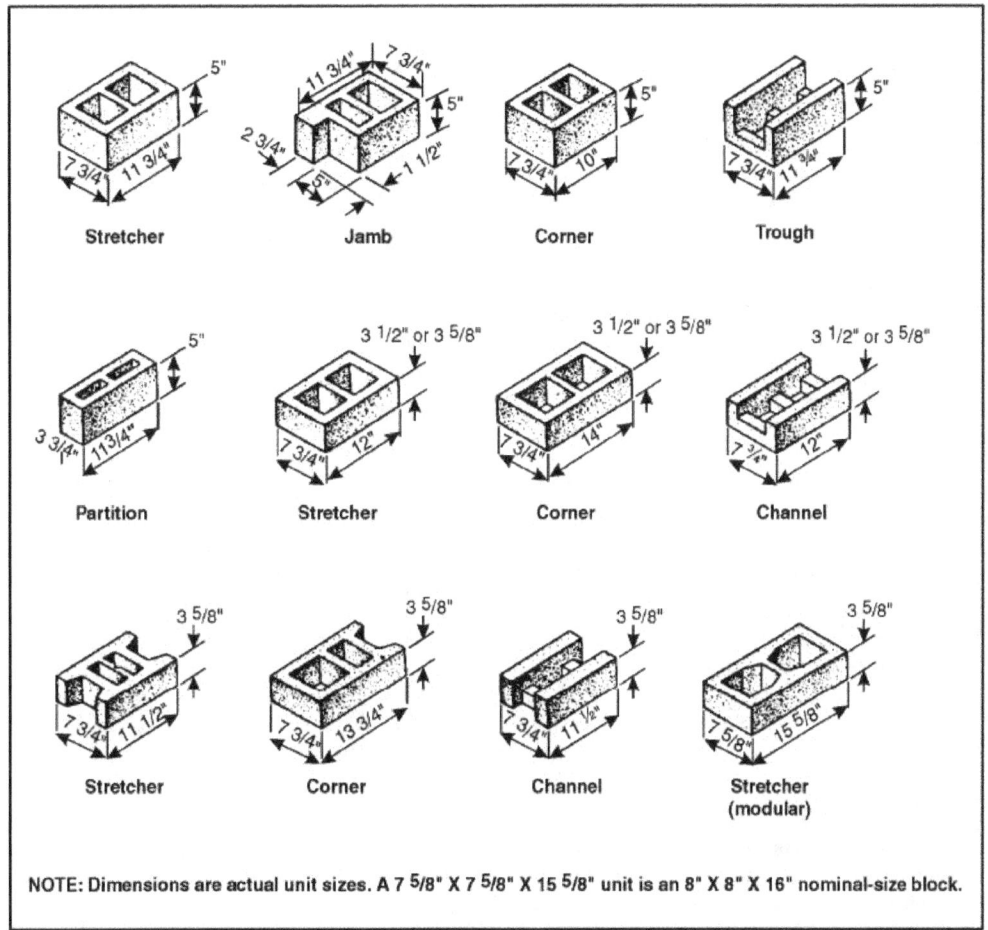

Figure 8-1. Typical unit sizes and shapes of concrete masonry units (continued)

CONCRETE BLOCK MASONRY UNIT

8-2. Concrete blocks are used in all types of masonry construction, such as—

- Exterior load-bearing walls (both below and above grade).
- Interior load-bearing walls.
- Fire walls and curtain walls.
- Partitions and panel walls.
- Backing for bricks, stones, and other facings.
- Fireproofing over structural members.
- Fire-safe walls around stairwells, elevators, and enclosures.
- Piers and columns.
- Retaining walls.
- Chimneys.
- Concrete floor units.

TYPES OF UNITS

8-3. The main types of concrete masonry units are—

- Hollow, load-bearing concrete block.
- Solid, load-bearing concrete block.
- Hollow, nonload-bearing concrete block.
- Concrete building tile.
- Concrete brick.

8-4. The load-bearing types of blocks have two grades. Grade N is for general use, such as exterior walls both above and below grade that may or may not be exposed to moisture penetration or weather, and for back-up and interior walls. Grade S is for above-grade exterior walls with weather-protective coating and for interior walls. The grades are further subdivided into two types: type I moisture-controlled units (for use in arid climates) N-I and S-I, and type II nonmoisture-controlled units N-II and S-II.

HEAVYWEIGHT AND LIGHTWEIGHT UNITS

8-5. The concrete masonry units made with either heavyweight or lightweight aggregates are referred to as such. A hollow, load-bearing concrete block is 8 by 8 by 16 inches with nominal-size weight from 40 to 50 pounds. These types of blocks are normally made with heavyweight aggregate such as sand, gravel, crushed stone, or air-cooled slag. The same type and nominal-size block weighs only from 25 to 35 pounds when made with coal cinders, expanded shale, clay, slag, volcanic cinders, or pumice. Your choice of masonry units depends on both availability and the requirements of the intended structure.

SOLID AND HOLLOW UNITS

8-6. ASTM specifications defines a solid concrete block as having a core area not more than 25 percent of the gross cross-sectional area. Most concrete bricks are solid and sometimes have a recessed surface like the frogged brick shown in figure 8-1, page 8-2. In contrast, a hollow concrete block has a core area greater than 25 percent of its gross cross-sectional area—generally 40 to 50 percent.

SIZES AND SHAPES

8-7. Concrete masonry units are available in many sizes and shapes to fit different construction needs. Both full- and half-length sizes are shown in figure 8-1. Because concrete block sizes usually refer to nominal dimensions, a unit actually measuring 7 5/8 by 7 5/8 by 15 5/8 inches is called an 8- by 8- by 16-inch block. When laid with 3/8-inch mortar joints, the unit will then occupy a space exactly 8 by 8 by 16 inches. Before designing a structure, contact local manufacturers for a schedule of their available unit sizes and shapes.

MAKING BLOCKS BY MACHINE

8-8. To precast concrete blocks, use a power-tamping machine available from several manufacturers. Tamp the concrete into the mold, then immediately strip off the mold. This way you can make many blocks rapidly using a single mold. The mix should be dry enough for the block to retain its shape.

MAKING BLOCKS BY HAND

8-9. To precast blocks by hand, pour concrete of fluid consistency into sets of iron molds, then strip off the molds when the concrete hardens. This procedure makes dense block with little labor; however, it requires a large number of molds.

MAKING WEATHER-EXPOSED BLOCKS

8-10. Make blocks subject to weathering with a concrete mix of at least six sacks of cement per cubic yards of mix. When using lightweight, porous aggregate, premix it with water for 2 minutes before adding the cement.

CURING CONCRETE BLOCKS

8-11. Steam is the best way to cure concrete blocks because it takes less time. Concrete blocks cured in wet steam at 125°F for 15 hours have 70 percent of their 28-day strength. If steam is not available, cure the blocks by protecting them from the sun and keeping them damp for 7 days.

SECTION II - CONSTRUCTION PROCEDURES

MODULAR COORDINATION AND PLANNING

8-12. *Modular coordination* is when the design of a building, its components, and the building-material units all conform to a dimensional standard based on a modular system. *Modular measure* is the system of dimensional standards for buildings and building components that permit field assembly without cutting. The basic unit is a 4-inch cube that allows a building to be laid around a continuous three-dimensional rectangular grid having 4-inch spacing. The modular system of coordinated drawings is based on a standard 4-inch grid placed on the width, length, and height of a building, as shown in figure 8-2, page 8-6.

Step 1. Use the grid on preprinted drawing paper for both small-scale plans and large-scale details. Do not use drawing paper with a scale less than 3/4 inch which equals 1 foot to show grid lines at 4-inch spacing.

Step 2. Select a larger planning module that is a multiple of 4 inches. For floor plans and elevations, for example, the module may be 2 feet 8 inches, 4 feet, 5 feet, 6 feet 4 inches, and so forth.

Chapter 8

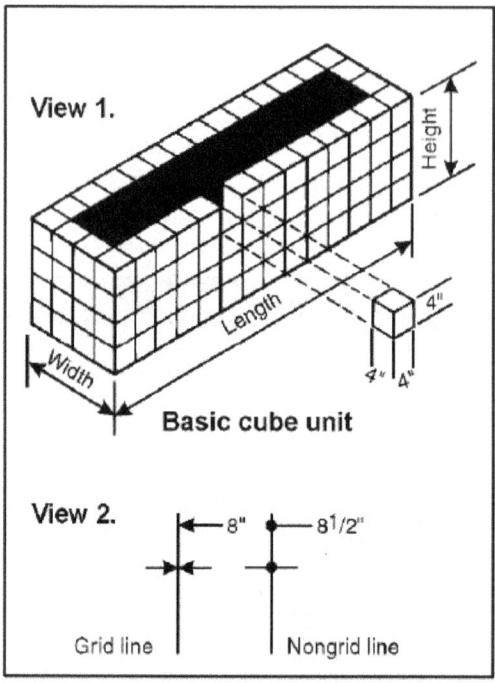

Figure 8-2. Elements of modular design

Step 3. Show materials actual size, or to scale, and locate them on or related to a grid line by reference dimensions. Dimensions falling on grid lines are shown by arrows; those not falling on grid lines, by dots (see figure 8-2).

- Maintain constant awareness of standard modular dimensions in planning, and preplan the work to make use of as many standard- sized building materials as possible. Such planning saves considerable labor, time, and materials.
- Make maximum use of full- and half-length units when laying out concrete masonry walls to minimize cutting and fitting units on the job.
- Plan the wall length and height, the width and height of openings, and wall areas between doors, windows, and corners to use full- and half-size units as shown in figure 8-3. Remember that window and door frames must have modular dimensions to fit modular full- and half-size masonry units.
- Keep all horizontal dimensions in multiples of nominal full-length masonry units. Both horizontal and vertical dimensions should be designed in multiples of 8 inches. Table 8-1 gives the nominal lengths of modular-concrete masonry walls in the number of stretchers, and table 8-2, page 8-9, gives the nominal heights of modular-concrete masonry walls in the number of courses.
- Plan the horizontal dimensions in multiples of 8 inches (half-length units) and the vertical dimensions in multiples of 4 inches, when using 8 by 4 by 16 blocks. If the wall thickness is either greater or less than the length of one half-length unit, use a special length unit at each corner in each course.

Concrete Masonry

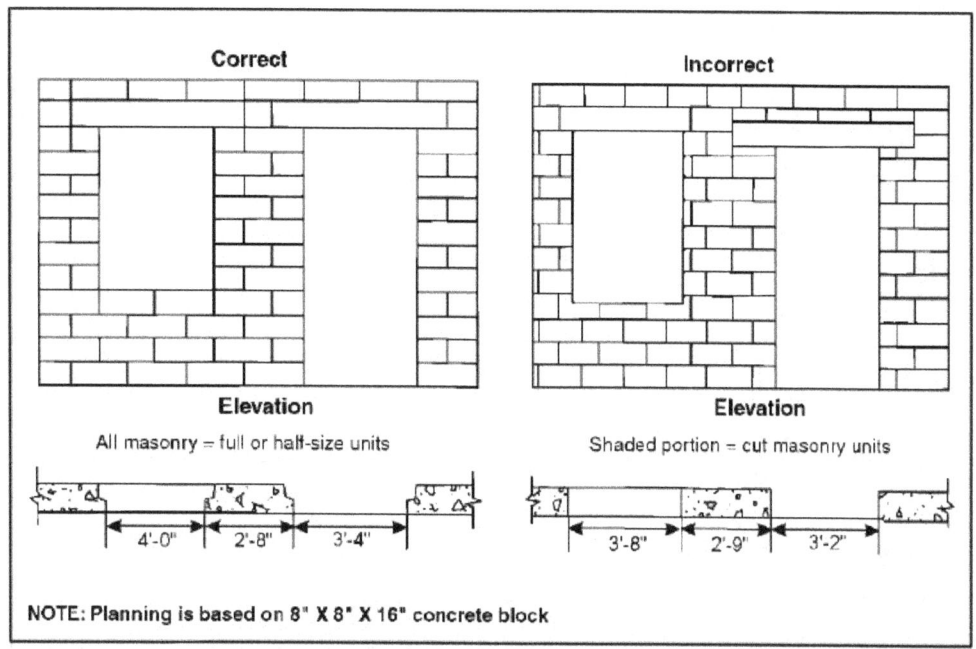

Figure 8-3. Planning concrete masonry wall openings

WALLS AND WALL FOOTINGS

8-13. Place masonry wall footings on firm, undisturbed soil having adequate load-bearing capacity to carry the design load and below frost penetration. Unless local requirements or codes specify otherwise, make the footings for small buildings twice as wide as the thickness of the walls they support. Table 8-3, page 8-9, gives both the unit weights and quantities for modular-concrete masonry walls. Footing thickness equals wall width (see figure 8-4, page 8-10).

Table 8-1. Nominal length of modular-concrete masonry walls in stretchers

Number of Stretchers	Nominal Length of Concrete Masonry Walls	
	Units 15 5/8" Long and Half Units 7 5/8" Long With 3/8" Thick Head Joints	Units 11 5/8" Long and Half Units 5 5/8" Long With 3/8" Thick Head Joints
1　　　　　　　　　　　1 1/2　　　　　　　　　　　2	1' 4"　　　　　　　　　　　2' 0"　　　　　　　　　　　2' 8"	1' 0"　　　　　　　　　　　1' 6"　　　　　　　　　　　2' 0"
2 1/2　　　　　　　　　　　3　　　　　　　　　　　3 1/2	3' 4"　　　　　　　　　　　4' 0"　　　　　　　　　　　4' 8"	2' 6"　　　　　　　　　　　3' 0"　　　　　　　　　　　3' 6"
4　　　　　　　　　　　4 1/2　　　　　　　　　　　5	5' 4"　　　　　　　　　　　6' 0"　　　　　　　　　　　6' 8"	4' 0"　　　　　　　　　　　4' 6"　　　　　　　　　　　5' 0"

Table 8-1. Nominal length of modular-concrete masonry walls in stretchers

Number of Stretchers	Nominal Length of Concrete Masonry Walls	
	Units 15 5/8" Long and Half Units 7 5/8" Long With 3/8" Thick Head Joints	Units 11 5/8" Long and Half Units 5 5/8" Long With 3/8" Thick Head Joints
5 1/2	7' 4"	5' 6"
6	8' 0"	6' 0"
6 1/2	8' 8"	6' 6"
7	9' 4"	7' 0"
7 1/2	10' 0"	7' 6"
8	10' 8"	8' 0"
8 1/2	11' 4"	8' 6"
9	12' 0"	9' 0"
9 1/2	12' 8"	9' 6"
10	13' 4"	10' 0"
10 1/2	14' 0"	10' 6"
11	14' 8"	11' 0"
11 1/2	15' 4"	11' 6"
12	16' 0"	12' 0"
12 1/2	16' 8"	12' 6"
13	17' 4"	13' 0"
13 1/2	18' 0"	13' 6"
14	18' 8"	14' 0"
14 1/2	19' 4"	14' 6"
15	20' 0"	15' 0"
20	26' 8"	20' 0"

Note. Actual wall length is measured from outside edge to outside edge of units and equals the nominal length minus 3/8" (one mortar joint).

Concrete Masonry

Table 8-2. Nominal height of modular-concrete masonry walls in courses

Number of Courses	Nominal Height of Concrete Masonry Walls	
	Units 7 5/8" High and 3/8" Thick Bed Joints	Units 3 5/8" High and 3/8" Thick Bed Joints
1	8"	4"
2	1' 4"	8"
3	2' 0"	1' 0"
4	2' 8"	1' 4"
5	3' 4"	1' 8"
6	4' 0"	2' 0"
7	4' 8"	2' 4"
8	5' 4"	2' 8"
9	6' 0"	3' 0"
10	6' 8"	3' 4"
15	10' 0"	5' 0"
20	13' 4"	6' 8"
25	16' 8"	8' 4"
30	20' 0"	10' 0"
35	23' 4"	11' 8"
40	26' 8"	13' 4"
45	30' 0"	15' 0"
50	33' 4"	16' 8"

Note. For concrete masonry units 7 5/8" and 3 5/8" in height laid with 3/8" mortar joints. Height is measured from center to center of mortar joints.

Table 8-3. Unit weight and quantities for modular concrete masonry walls

Actual Unit Size (Width x Height x Length), in Inches	Nominal Wall Thickness, in Inches	For 100 Sq Ft of Walls			
		Number of Units	Average Weight of Finished Wall		Mortar,*in Cu Ft
			Heavyweight Aggregate, in Lb*	Lightweight Aggregate, in Lb*	
3 5/8 x 3 5/8 x 15 5/8	4	225	3,050	2,150	9.5	6
5 5/8 x 3 5/8 x 15 5/8	6	225	4,550	3,050	9.5	6
7 5/8 x 3 5/8 x 15 5/8	8	225	5.700	3,700	9.5	6
3 5/8 x 7 5/8 x 15 5/8	4	112.5	2,850	2,050	6.0	7
5 5/8 x 7 5/8 x 15 5/8	6	112.5	4,350	2,950	6.0	7
7 5/8 x 7 5/8 x 15 5/8	8	112.5	5,500	3,600	6.0	7
11 5/8 x 7/5/8 x 15 5/8	12	112.5	7,950	4,900	6.0	7

Table 8-3. Unit weight and quantities for modular concrete masonry walls

Notes.
1. Table is based on 3/8-inch mortar joints.
2. Actual weight of 100 sq ft of wall can be computed by formula W (n) + 150(M).
where—
W = actual weight of a single unit.
n = number of units for 100 sq ft of wall.
M = cu ft of mortar for 100 sq ft of wall.
*Actual weight within ± 7% of average weight. With face-shell mortar bedding. Mortar quantities should include 10% allowance for waste.

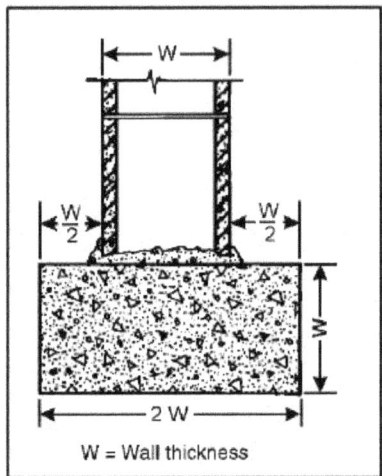

Figure 8-4. Dimensions of masonry wall footings

SUBSURFACE DRAINAGE

8-14. If you expect the groundwater level during a wet season to reach the basement floor elevation, place a line of drain tile along the exterior side of the footings. The tile line should fall at least 1/2 inch in 12 feet and drain to a suitable outlet. Place pieces of roofing felt over the joints to keep out sediment during backfilling. Cover the tile line to a depth of 12 inches with a permeable fill of coarse gravel or crushed stone ranging from 1 to 1 1/2 inches in size. When the first floor is in place, fill the balance of the trench with earth from the excavation.

BASEMENT WALLS

8-15. Always give exterior concrete masonry basement walls two 1/4-inch thick coats of parging, using either portland cement mortar (1:2 1/2 mix by volume) or joint mortar.

- **Step 1.** In hot, dry weather dampen the wall surface very lightly with a fog water spray before applying the first parging coat.
- **Step 2.** Roughen the first coat when it is partly hardens, to provide a bond for the second parging coat.
- **Step 3.** Wait for the first coat to harden for 24 hours, then dampen it lightly just before applying the second coat. Keep the second coat damp for at least 48 hours following application.

Step 4. For below-grade parged surfaces in very wet soils, use two continuous coatings of bituminous mastic brushed over a suitable priming coat. Make sure that the parging is dry before you apply the primer and that the primer is dry when you apply the bituminous mastic. Do not backfill against concrete masonry walls until the first floor is in place.

FLOOR AND ROOF SUPPORT

8-16. Use solid masonry courses to support floor beams or floor slabs to help distribute the loads over the walls as well as provide a termite barrier. Use either solid masonry units or fill the cores of hollow block with concrete or mortar. If using blocks that are filled with mortar, lay strips of expanded metal lath in the bed joint underneath to support the fill.

WEATHERTIGHT CONCRETE MASONRY WALLS

8-17. Good workmanship is a very important factor in building weathertight walls.

Step 1. Lay each masonry unit plumb and true.

Step 2. Fill both horizontal and vertical joints completely; compact them by tooling when the mortar partly stiffens.

Step 3. Add flashing at vertical joints in copings and caps, at the joints between the roofs and walls, and below cornices and other members that project beyond the wall face.

Step 4. Shed water away from the wall surface by providing drips for chimney caps, sills, and other projecting ledges. Make sure that drains and gutters are large enough so that overflowing water does not run down masonry surfaces.

FIRST COURSE

8-18. The first step in building a concrete masonry wall is to locate the corners of the structure. Then check the layout by placing the first course blocks without mortar (see figure 8-5, page 8-12).

Step 1. Use a chalked snapline to mark the footing and align the blocks accurately.

Step 2. Replace the loose blocks with a full bed of mortar, spreading and furrowing it with a trowel to ensure plenty of mortar under the bottom edges of the first course.

Step 3. Use care to position and align the corner block first.

Step 4. Lay the remaining first course blocks with the thicker end up to provide a larger mortar-bedding area.

Step 5. Apply mortar to the block ends for the vertical joints by placing several blocks on end and buttering them all in one operation. Make the joints 3/8 inch thick.

Step 6. Place each block in its final position and push it down vertically into the mortar bed and against the previously laid block to obtain a well-filled vertical mortar joint.

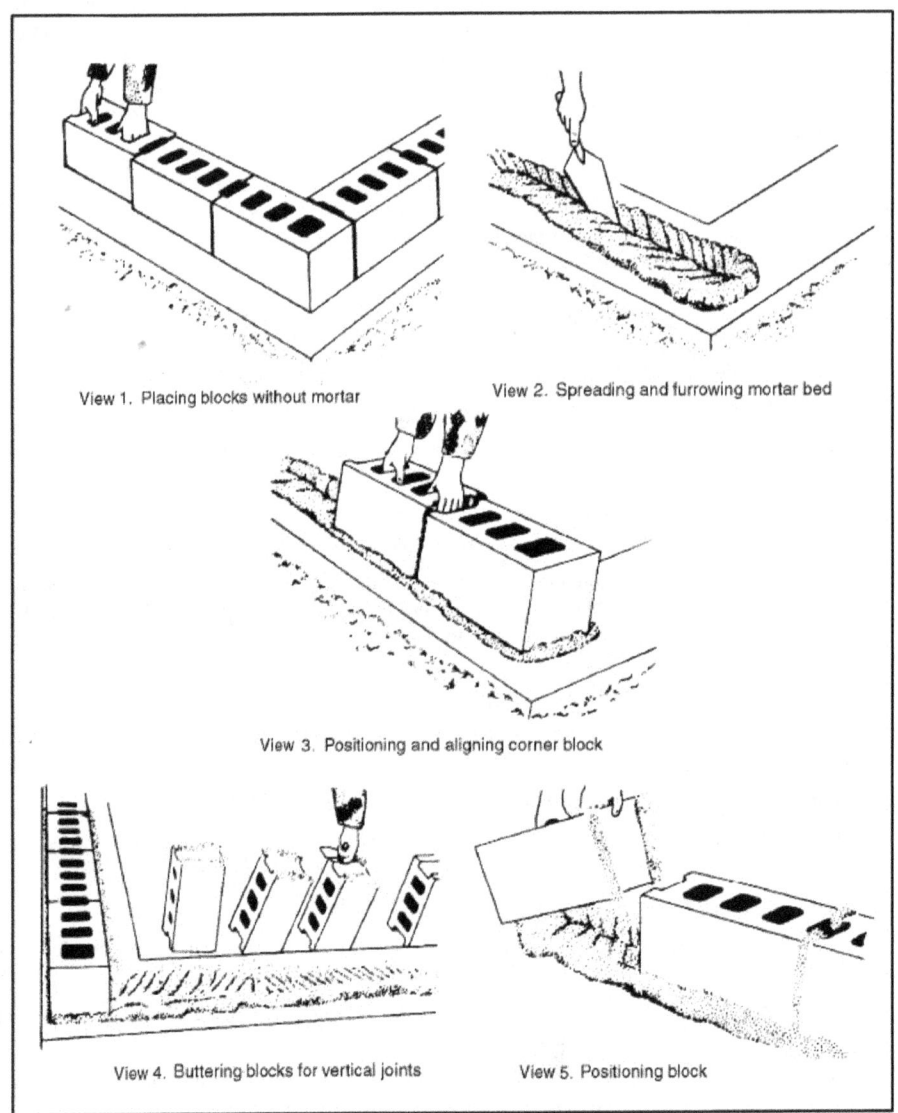

Figure 8-5. Laying first course of blocks for a wall

Step 7. Use a mason's level after laying three or four blocks as a straightedge to check correct block alignment (see figure 8-6).

Step 8. Use the level to bring the blocks to proper grade and make them plumb by tapping with a trowel handle as shown in figure 8-6.

Concrete Masonry

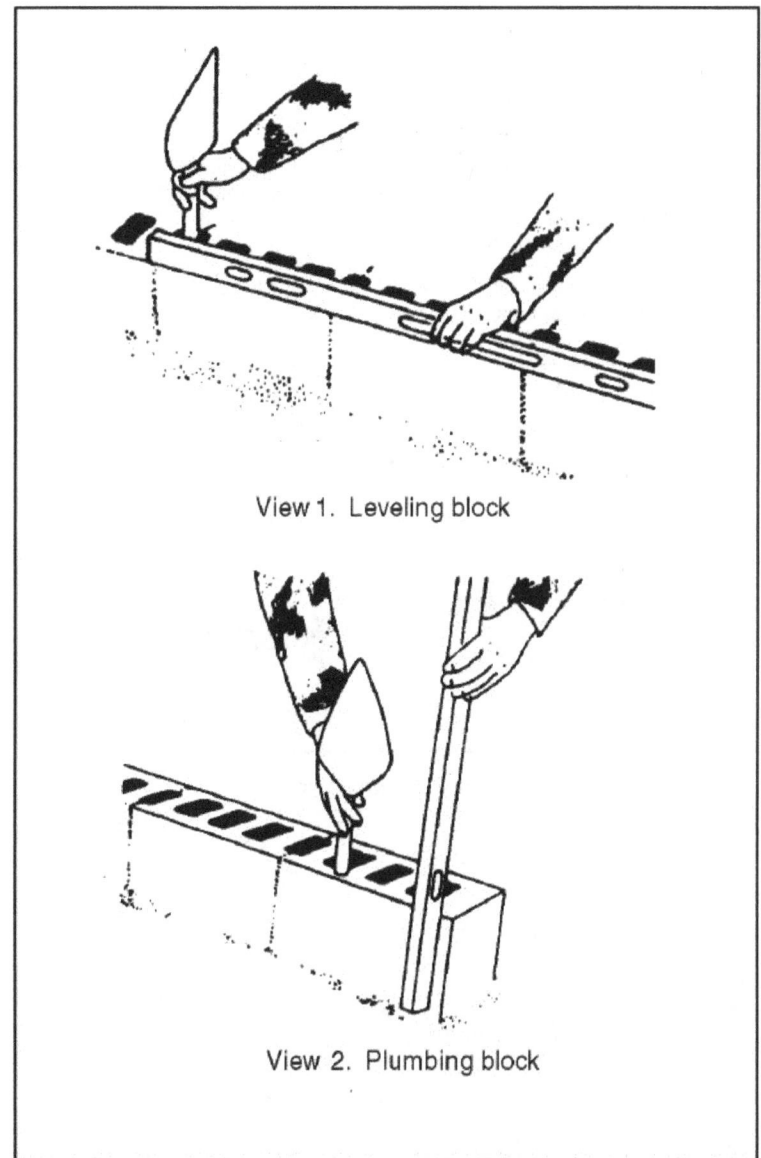

Figure 8-6. Leveling and plumbing first course of blocks for a wall

Step 9. Lay out the first course of concrete masonry very carefully, making sure that it is properly aligned, level, and plumb. This ensures that succeeding courses and the final wall are both straight and true.

LAYING UP THE CORNERS

8-19. After laying the first course, build up the corners of the wall next, usually four or five courses high.

Step 1. Move back each course one-half block.

Chapter 8

Step 2. Apply mortar only to the tops of the blocks of the horizontal joints already laid.

Step 3. Apply mortar to the vertical joints either to the ends of the new block or the end of the block previously laid, or both, to ensure well-filled joints (see figure 8-7).

Step 4. Lay each course at the corner, check it with a level for alignment for leveling and for ensuring that it is plumb (see figure 8-8).

Step 5. Use care to check each block with a level or straightedge to make sure that all the block faces are in the same plane to ensure true, straight walls. A story or course pole, which is a board with markings 8 inch apart as shown in figure 8-9, page 8-16, helps to accurately determine the top of each masonry course.

Step 6. Check the horizontal block spacing by placing a level diagonally across the corners of the blocks as shown in figure 8-10, page 8-16.

Figure 8-7. Vertical joints

Concrete Masonry

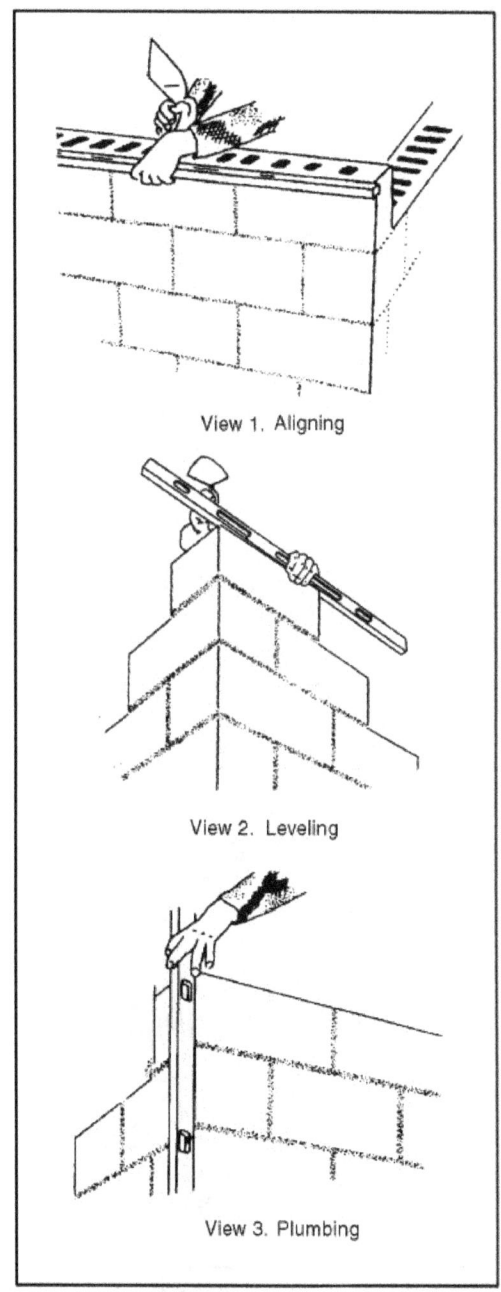

Figure 8-8. Checking each course at the corner

Chapter 8

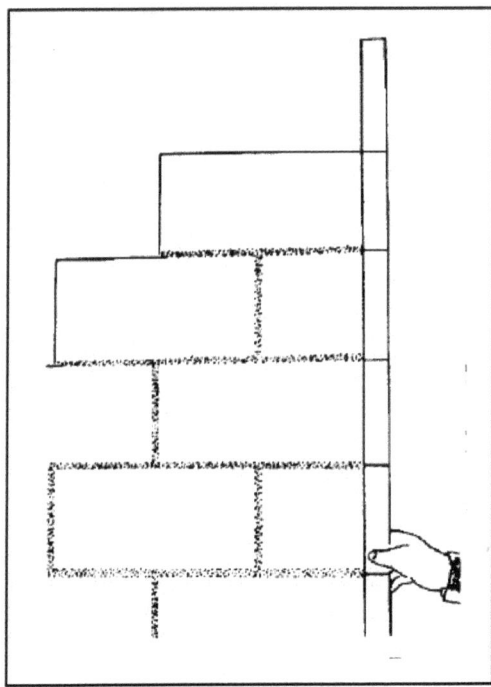

Figure 8-9. Using a story or course pole

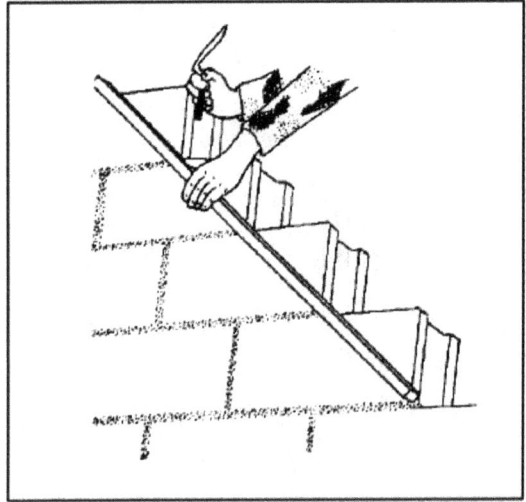

Figure 8-10. Checking horizontal block spacing

Laying Blocks Between Corners

8-20. When filling in the wall between the corners, follow the procedures below.

   Step 1. Stretch a mason's line along the exterior block edges from corner to corner for each course.

   Step 2. Lay the top outside edge of each new block to this line (see figure 8-11). How you grip a block before laying it is important.

   - First, tip it slightly toward you so that you can see the edge of the course below.
   - Place the lower edge of the new block directly on the edges of the blocks comprising the course below, as shown in figure 8-11.
   - Next, make all final position adjustments while the mortar is soft and plastic, because any adjustments you make after the mortar stiffens will break the mortar bond and allow water to penetrate.
   - Finally, level each block and align it to the mason's line by tapping it lightly with a trowel handle.

Figure 8-11. Filling in the wall between corners

Closure Block

8-21. Before installing the closure block, butter both edges of the opening and all four vertical edges of the closure block with mortar. Then lower the closure block carefully into place as shown in figure 8-12, page 8-18. If any mortar falls out leaving an open joint, remove the block and repeat the procedure.

Chapter 8

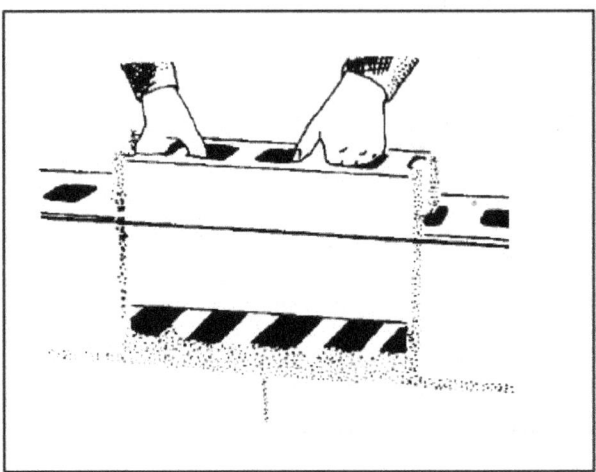

Figure 8-12. Installing a closure block

MORTAR JOINTS

8-22. To ensure a good bond, do not spread mortar too far ahead of actually laying blocks or it will stiffen and lose its plasticity. The recommended width of mortar joints for concrete masonry units is approximately 3/8 inch thick which—when properly made—helps to produce a weathertight, neat, and durable concrete masonry wall. As you lay each block, cut off excess mortar extruding from the joints using a trowel (see figure 8-13) and throw it back on the mortar board to rework into the fresh mortar. Do not, however, rework any dead mortar from the scaffold or floor.

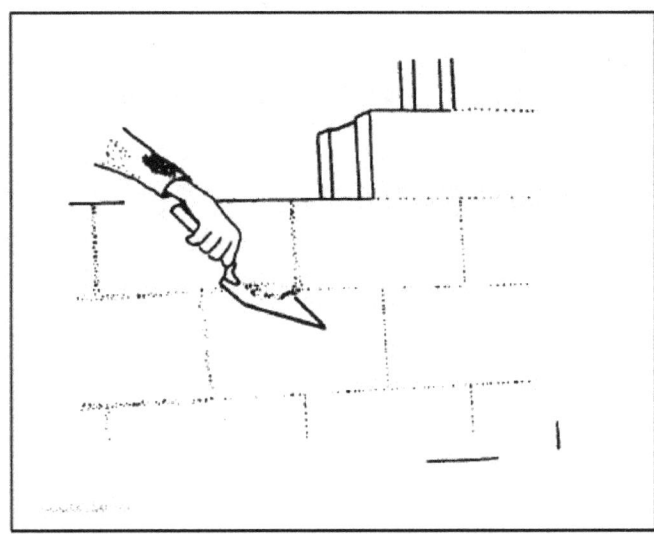

Figure 8-13. Cutting off excess mortar from the joints

Concrete Masonry

TOOLING

8-23. Weathertight joints and the neat appearance of concrete masonry walls depend on proper tooling. After laying a section of the wall, tool the mortar joints when the mortar becomes thumbprint hard. Tooling compacts the mortar and forces it tightly against the masonry on each side of the joint. Use either concave or V-shaped tooling on all joints (see figure 8-14). Tool vertical joints first, followed by striking the horizontal joints with a long jointer (see figure 8-15, page 8-20). Trim off mortar burrs from the tooling flush with the wall face using a trowel or soft bristle brush, or by rubbing with a burlap bag.

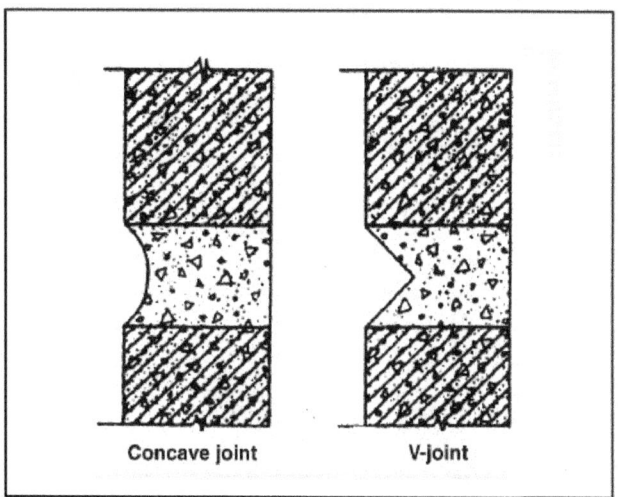

Figure 8-14. Tooled mortar joints for weathertight exterior walls

ANCHOR BOLTS

8-24. You must prepare in advance for installing wood plates on top of hollow concrete masonry walls with anchor bolts. To do this, place pieces of metal lath in the second horizontal mortar joint from the top of the wall under the cores that will contain the bolts. Use anchor bolts 1/2 inch in diameter and 18 inches long, spacing them up to a maximum of 4 feet apart. When you complete the top course, insert the bolts into the cores of the top two courses, and fill the cores with concrete or mortar. The metal lath underneath holds the concrete or mortar filling in place. The threaded end of the bolt should extend above the top of the wall as shown in figure 8-16, page 8-21.

Chapter 8

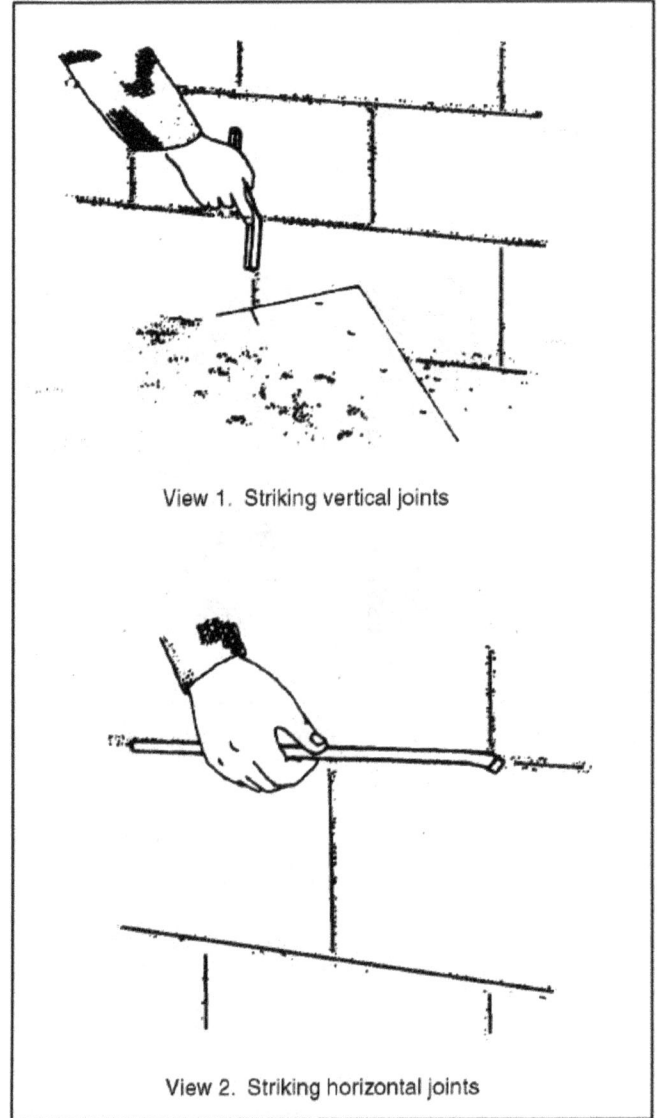

Figure 8-15. Tooling mortar joints

CONTROL JOINTS

8-25. *Control Joints* are continuous vertical joints that permit the masonry wall to move slightly under unusual stresses without cracking. A combination of full- and half-length blocks form the continuous vertical joint as shown in view 1 of figure 8-17, page 8-22. Lay up control joints in mortar just as any other joint, but if they are exposed to either the weather or to view, caulk them as well. After the mortar is stiff, rake it out to a depth of about ¾ inch to make a recess for the caulking compound as shown in view 2 of figure 8-17.

Concrete Masonry

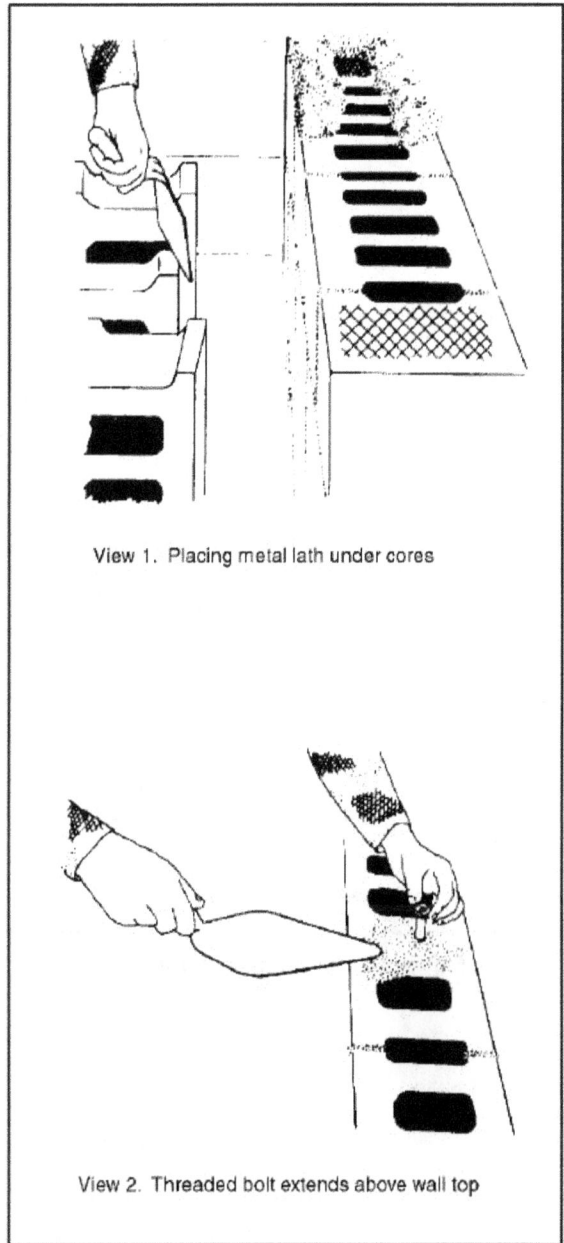

Figure 8-16. Installing anchor bolts for wood plates

8-26. Use a thin, flat caulking trowel to force the compound into the joint. You can make a second type of control joint by inserting building paper or roofing felt into the block end cores extending the full height of the joint (see figure 8-18, page 8-23). Cut the paper or felt to convenient lengths, but wide enough to extend across the joint. The paper or felt material prevents the mortar from bonding on that side of the joint. Use control joint blocks, if available (see figure 8-18).

Chapter 8

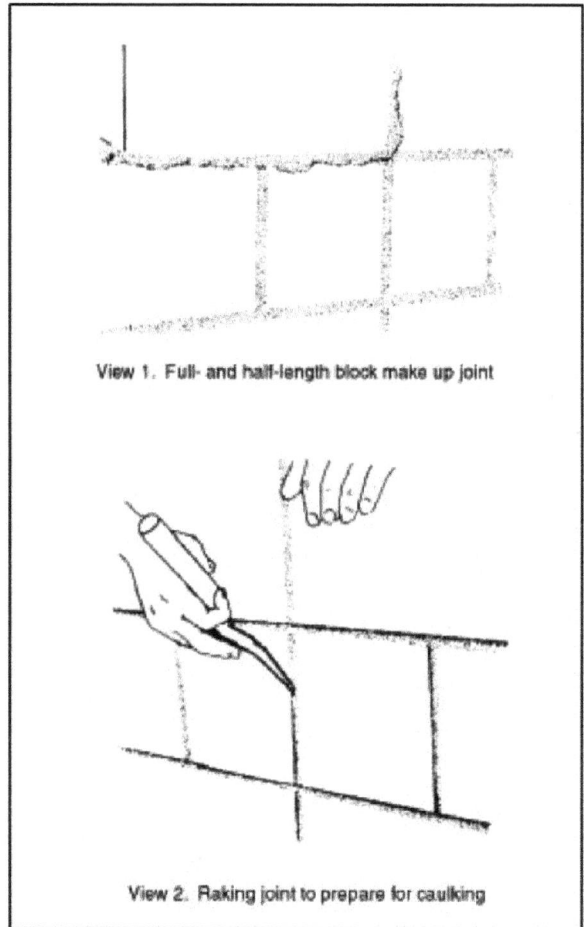

Figure 8-17. Making a control joint

Concrete Masonry

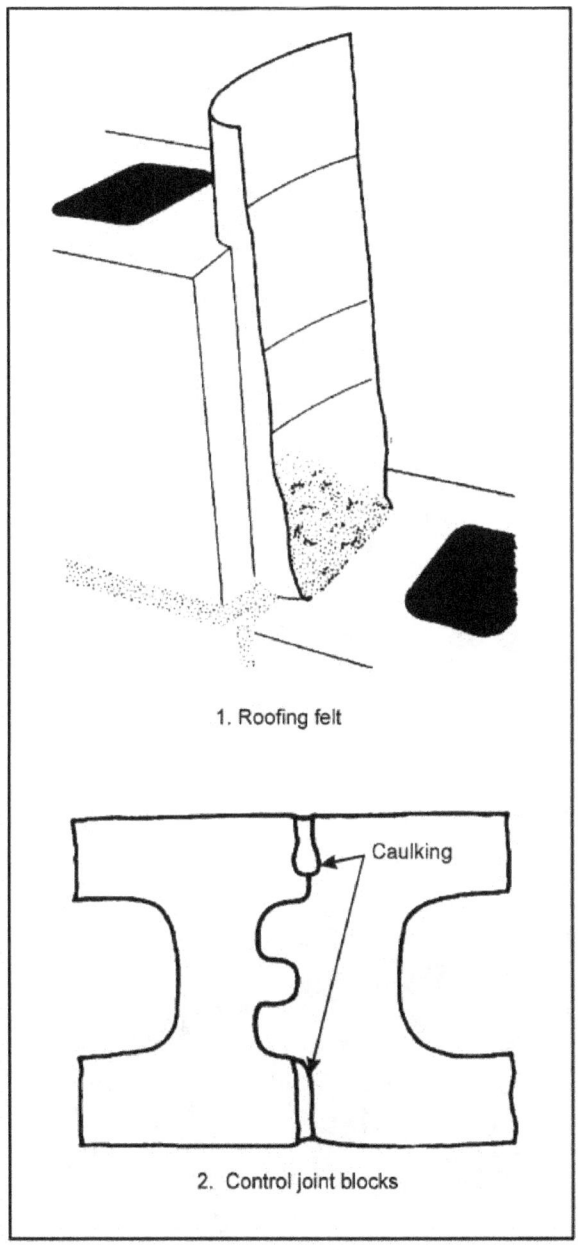

Figure 8-18. Control joints made using roofing felt or control joint blocks

Chapter 8

INTERSECTING WALLS

8-27. The two types of intersecting walls are bearing and nonbearing.

BEARING WALLS

8-28. Do not join intersecting concrete block-bearing walls with a masonry bond, except at the corners. Instead, terminate one wall at the face of the second wall with a control joint. Tie the intersecting walls together with Z-shaped metal-tie bars 1/4 by 1 1/4 by 28 inches in size, having a 2-inch right angle bend on each end (see figure 8-19). Space the tie bars no more than 4 feet apart vertically, and place pieces of metal lath under the block cores that will contain the tie-bar ends (see figure 8-16, page 8-21). Embed the right angle bends in the cores by filling them with mortar or concrete (see figure 8-19).

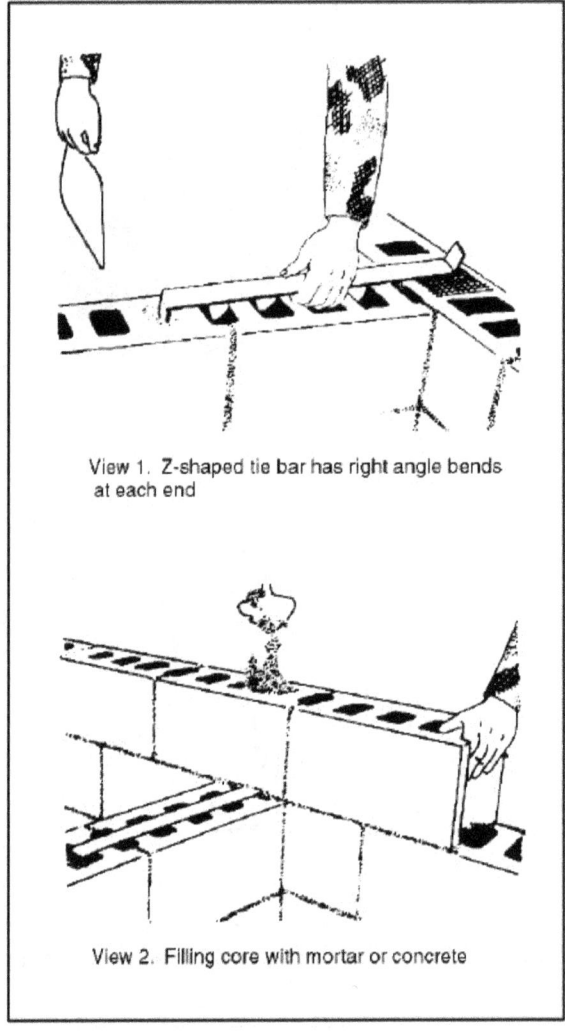

Figure 8-19. Tying intersecting bearing walls

NONBEARING WALLS

8-29. To join intersecting nonbearing block walls, terminate one wall at the face of the second with a control joint. Place strips of metal lath or 1/4-inch mesh galvanized hardware cloth across the joint between the two walls in alternate courses. Insert one half of the metal strips into one wall as you build it; tie the other halves into mortar joints as you lay the second wall (see figure 8-20, page 8-26).

LINTELS

8-30. Modular door and window openings usually require lintels to support the blocks over the openings. Use precast concrete lintels that contain an offset on the underside to fit the modular openings or use steel-lintel angles that you install with an offset on the underside (see figure 8-21, page 8-27) to fit modular openings. In either case, place a noncorroding metal plate under the lintel ends at the control joints to allow the lintel to slip and the control joints to function properly. Apply a full bed of mortar over the metal plate to uniformly distribute the lintel load.

SILLS

8-31. Install precast concrete sills following wall construction (see figure 8-22, page 8-28). Fill the joints tightly at the ends of the sills with mortar or a caulking compound.

PATCHING AND CLEANING BLOCK WALLS

8-32. When laying concrete masonry walls, be very careful not to smear mortar into the block surfaces, because you cannot remove hardened, embedded mortar smears, even with an acid wash; also paint will not cover them. Allow any droppings to dry and harden. You can then chip off most of the mortar with a small piece of broken concrete block (see figure 8-23, page 8-28), or with a trowel. A final brushing of the spot will remove practically all of the mortar. Always patch mortar joints and fill holes made by nails or line pins with fresh mortar.

8-33. The mason is responsible for laying out the job to do the work properly. Masons must make sure that the walls are plumb and that courses are level. They are also responsible for the quality of all the detail work such as cutting and fitting masonry units, making joints, and installing anchor bolts and ties in intersecting walls.

8-34. The mason's helper mixes mortar, keeps it tempered, and supplies concrete blocks and mortar to the mason as needed. Helpers aid the mason in laying out the job and sometimes lay out blocks ahead on an adjacent course to expedite the mason's work.

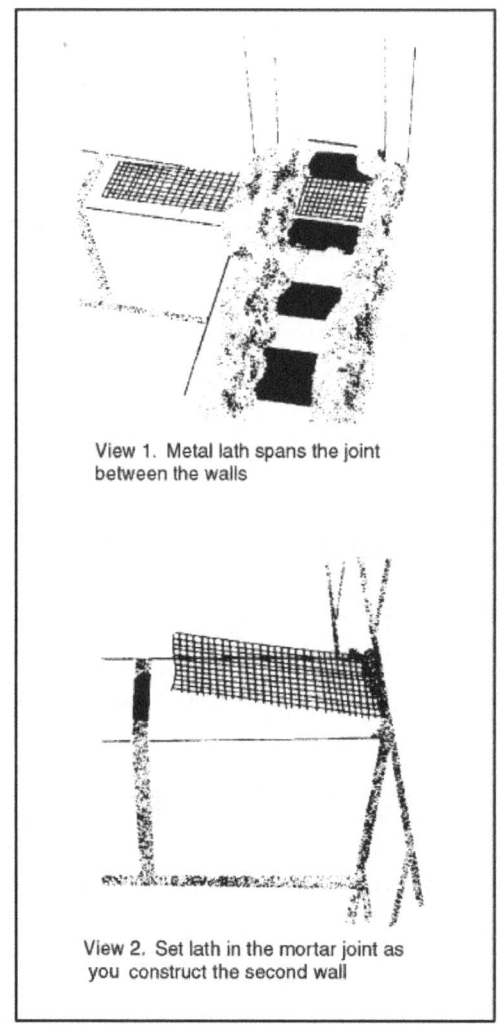

Figure 8-20. Tying intersecting nonbearing walls

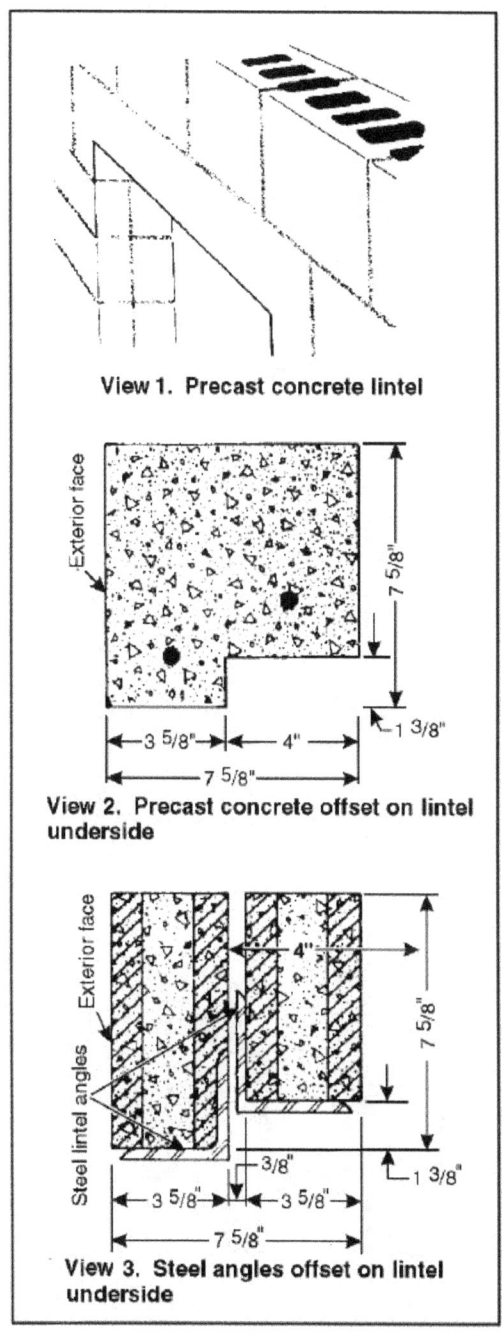

Figure 8-21. Installing precast concrete lintels without and with steel angles

Chapter 8

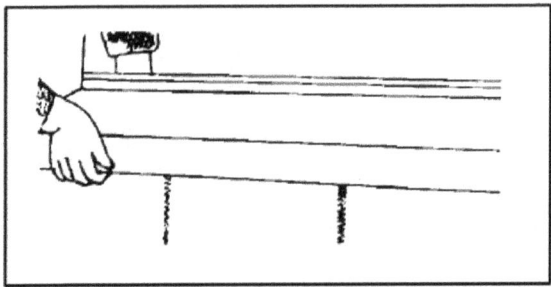

Figure 8-22. Installing precast concrete sills

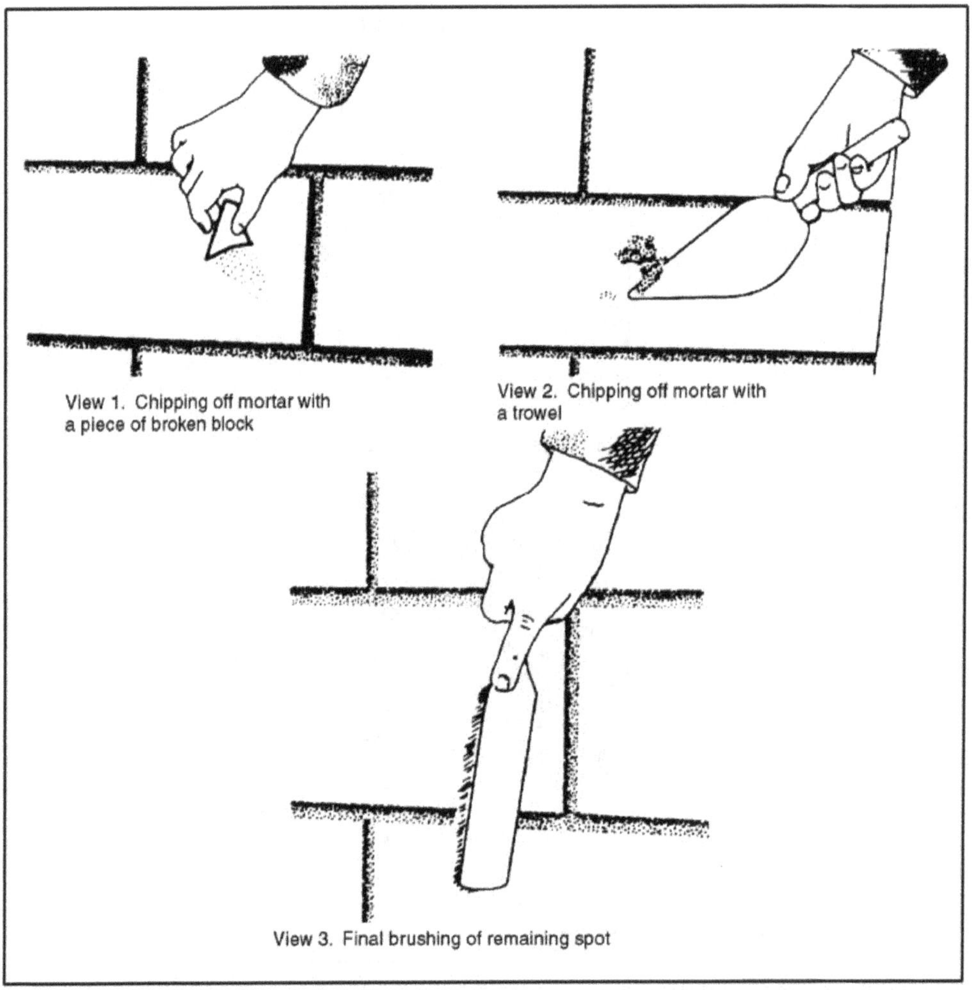

Figure 8-23. Cleaning mortar droppings from a concrete block wall

SECTION III - RUBBLE

RUBBLE STONE MASONRY

8-35. You can use rubble stone masonry (see figure 8-24) for walls—both above and below ground—and for bridge abutments, particularly when form lumber or masonry units are not available. You can lay up rubble masonry with or without mortar, but for strength and stability use mortar. There are two types of rubble-stone masonry: random and coursed.

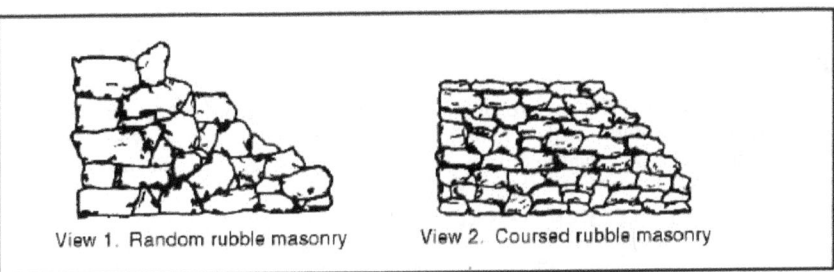

Figure 8-24. Rubble stone masonry

RANDOM RUBBLE MASONRY

8-36. This type is the crudest of all types of stonework. It does not require laying the stone in courses (see figure 8-24), but each layer must contain bonding stones that extend through the wall (as shown in figure 8-25) to tie the walls together. Make the bed joints horizontal for stability; the build or head joints can run in any direction.

COURSED RUBBLE MASONRY

8-37. This type contains roughly squared stones laid in nearly continuous horizontal bed joints as shown in figure 8-24.

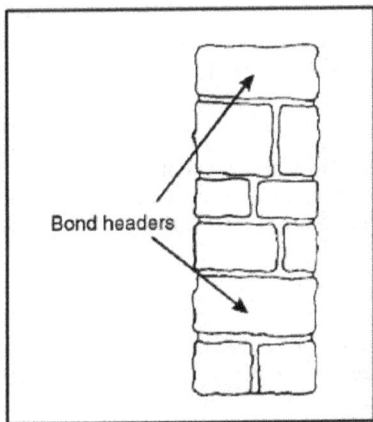

Figure 8-25. Bonding stones extend through a rubble stone masonry wall

Chapter 8

RANDOM RUBBLE MASONRY MATERIALS

8-38. The two main random rubble masonry materials are stones and mortar. Some of the more common suitable stones are limestone, sandstone, granite, and slate.

STONES

8-39. Use stones that are strong, durable, and cheap for random rubble masonry. Durability and strength depend on the stone's chemical composition and physical structure. Use unsquared stones or fieldstones from nearby ledges or quarries. No stones should be larger than what two persons can handle easily. The larger the variety of sizes you select, the less mortar you need.

MORTAR

8-40. Table 7-1, page 7-6, gives the proportions of the portland-cement-lime mortar mixture to use with random rubble masonry. Mortar made with ordinary portland cement stains most types of stone. To prevent staining, substitute nonstaining white portland cement. Lime usually does not stain the stone.

LAYING RUBBLE STONE MASONRY

8-41. The quality of workmanship affects the economy, durability, and strength of a rubble stone masonry wall more than any other factor. Lay out the wall—

- By eye. If the wall does not have to be exactly plumb and true to line, lay it out by eye without using a level and line. This requires frequent sighting.
- By line. If the wall must be exactly plumb and true to line, erect wood corner posts to serve as corner leads, and lay the stone with a line. Remember that some parts of the stone will extend farther away from the line than other parts. Do not try to lay the stone in level courses.

RULES FOR LAYING

8-42. Lay each stone on its broadest face. If appearance is important, place the larger stones in the lower courses. Lay stones of increasingly smaller sizes as you build to the top of the wall.

- Moistening. Moisten porous stones before placing them in mortar to prevent water absorption from the mortar, thereby weakening the bond.
- Packing and filling. Pack adjoining stones as tightly as practicable, completely filling any spaces between them with smaller stones and mortar.
- Removing. If removing a stone after placing it on the mortar bed, lift it clear and reset it.

FOOTINGS

8-43. Because a footing is always larger than the wall itself, use the largest stones in the footing to give it greater strength and lessen the risk of unequal settlement. Select footing stones as long as the footing is wide, if possible. Lay them in a mortar bed about 2 inches deep, and fill all the spaces between them with smaller stones and mortar.

BED JOINTS

8-44. Bed-joint thickness varies with the stone you use. Spread enough mortar on top of the lower course stone to completely fill the space between it and the stone you are placing. Take care not to spread mortar too far ahead of the stonelaying.

HEAD JOINTS OR BUILDS

8-45. Form the head joints before the bed joint mortar sets up. After laying three or four stones, make the head joints by slushing the small spaces with mortar and filling the larger spaces with both small stones and mortar.

BONDING STONES

8-46. Be sure to use one bonding stone for every 6 to 10 square feet of wall. Bonding stones pass all the way through the wall as shown in figure 8-25, page 8-29. Offset each head joint from adjacent head joints above and below it as much as possible (see figure 8-25) to bond the wall together and make it stronger.

This page intentionally left blank.

Chapter 9

Brick and Tile Masonry

For at least 5000 years, bricks have been popular for their strength, durability, and beauty. Although traditionally made of natural clay that is heated in a kiln at high temperatures, many bricks today are made from compressed concrete and come in a wide variety of shapes and colors. Bricks can be used for a variety of structures including house walls, garden walls, retaining walls, chimneys and fireplaces, and also pavers for driveways, walkways, and patios.

SECTION I - CHARACTERISTICS OF BRICK

PHYSICAL PROPERTIES AND CLASSIFICATION

9-1. Structural clay products includes brick, hollow tile of all types, and architectural terra cotta, but exclude thin wall tile, sewer pipe, flue linings, and drain tile.

BRICK MASONRY UNITS

9-2. Bricks are small masonry units that are either solid or cored but not more than 25 percent. They are kiln-fired (baked) from various clay and shale mixtures. The chemical and physical characteristics of the ingredients change considerably and combine with the kiln temperature to produce the brick in a variety of colors and hardnesses. The clay or shale pits in some regions yield a product that is simply ground, moistened, formed, and baked into durable brick. In other regions, the clay or shale from several pits must be mixed to produce durable brick. Bricks are small enough to place with one hand. Uniform units can be laid in courses with mortar joints to form walls of almost unlimited length and height.

BRICK SIZES AND WEIGHT

9-3. Standard United States (US) bricks are 2 1/4 by 3 3/4 by 8 inches actual size. They may have three core holes or ten core holes. Modular US bricks are (2 2/3 by 4 by 8 inches) nominal size, normally having three core holes. English bricks are 3 by 4 1/2 by 9 inches, Roman bricks are 1 1/2 by 4 by 12 inches, and Norman bricks are 2 3/4 by 4 by 12 inches nominal size. Actual brick dimensions are smaller, usually by an amount equal to mortar joint width. Brick weighs from 100 to 150 pounds per cubic foot, depending on its ingredients and firing duration. Well-burned brick is heavier than underburned brick.

CUT SHAPES

9-4. Sometimes you must cut a brick into various shapes to fill in spaces at corners and other locations where a full brick does not fit. Figure 9-1, page 9-2, shows the more common cut shapes: half or bat, three-quarter closure, quarter closure, king closure, queen closure, and split.

Chapter 9

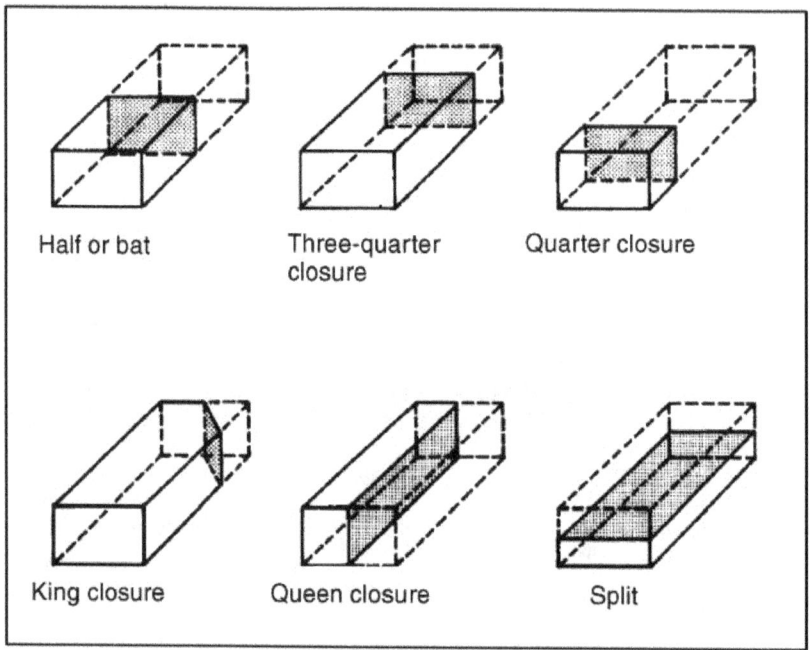

Figure 9-1. Common cut brick shapes

Surface Names

9-5. The five surfaces of a brick are called face, side, cull, end, and beds as shown in figure 9-2.

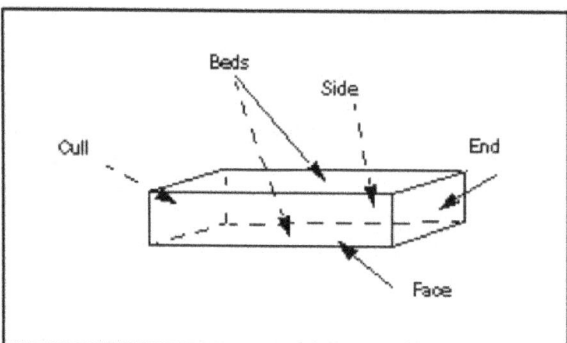

Figure 9-2. Names of brick surfaces

Brick Classification

9-6. The three general types of structural brick-masonry units are solid, hollow, and architectural terra cotta. All three can serve a structural function, a decorative function, or a combination of both. The three types differ in their formation and composition, and are specific in their use. Bricks commonly used in construction are—

- Building bricks. Also called common, hard, or kiln-run bricks, these bricks are made from ordinary clays or shales and fired in kilns. They have no special scoring, markings, surface texture, or color. Building bricks are generally used as the backing courses in either solid or cavity brick walls because the harder and more durable kinds are preferred.

- Face bricks. These are better quality and have better durability and appearance than building bricks because they are used in exposed wall faces. The most common face brick colors are various shades of brown, red, gray, yellow, and white.
- Clinker bricks. These bricks are oven-burnt in the kiln. They are usually rough, hard, durable, and sometimes irregular in shape.
- Pressed bricks. These bricks are made by the dry-press process rather than by kiln-firing. They have regular smooth faces, sharp edges, and perfectly square corners. Ordinarily, they are used as face bricks.
- Glazed bricks. These have one surface coated with a white or other color of ceramic glazing. The glazing forms when mineral ingredients fuse together in a glass-like coating during burning. Glazed brick is particularly suited to walls or partitions in hospitals, dairies, laboratories, and other structures requiring sanitary conditions and easy cleaning.
- Fire bricks. These are made from a special type of fire clay to withstand the high temperatures of fireplaces, boilers, and similar constructions without cracking or decomposing. Fire brick is generally larger than other structural brick, and often is hand-molded.
- Cored bricks. These bricks have ten holes (two rows of five holes each) extending through their beds to reduce weight. Walls built from all cored bricks are not much different in strength than walls built from all solid bricks, and both have about the same resistance to moisture penetration. Whether cored or solid, use the more easily available brick that meets building requirements.
- European bricks. Their strength and durability (particularly English and Dutch bricks) are about the same as US clay bricks.
- Sand-lime bricks. These bricks are made from a lean mixture of slaked lime and fine sand containing a lot of silica. They are molded under mechanical pressure and hardened under steam pressure. These are used extensively in Germany.

STRENGTH OF BRICK MASONRY

9-7. The strength of a single brick masonry unit varies widely, depending on its ingredients and manufacturing method. The main factors governing the strength of brick masonry are—

- Brick strength.
- Mortar strength and elasticity.
- Bricklayer workmanship.
- Brick uniformity.
- Bricklaying method used.

RANGES

9-8. Bricks can have an ultimate compressive strength as low as 1,600 psi, whereas some well-burned bricks have compressive strengths exceeding 15,000 psi.

PORTLAND-CEMENT-LIME MORTAR

9-9. Brick masonry laid with portland-cement-line mortar is stronger than an individual brick unit because this mortar is normally stronger than the brick. The load-carrying capacity of a wall or column made with plain lime mortar is much less than half that made with portland-cement-lime mortar. The compressive working strength of a brick wall or column laid with cement-lime mortar normally ranges from 500 to 600 psi.

Chapter 9

DRY BRICK

9-10. In order for mortar to bond to brick, sufficient water must be present to completely hydrate the portland cement in the mortar. Bricks sometimes have high absorption and, if not corrected, will suck the water out of the mortar preventing complete hydration. A field test to determine if the brick has absorptive qualities is as follows: Using a medicine dropper, place 20 drops of water in a 1-inch circle (about the size of a quarter). If the brick absorbs all the water in less than 1 1/2 minutes, then it will suck the water out of the mortar when laid. To correct this condition, thoroughly wet the bricks and allow time for the surfaces to air--dry before placing.

WEATHER RESISTANCE

9-11. A brick's resistance to weathering depends almost entirely upon its resistance to water penetration, because freeze-thaw action is almost the only type of weathering that affects it.

9-12. A brick wall made with superior workmanship will resist rain water penetration during a storm lasting as long as 24 hours accompanied by a 50 to 60 mile per hour wind.

9-13. Two important factors in preventing water penetration are tooled-mortar joints and caulking around windows and door frames. Mortar joints that bond tightly to the brick resist moisture penetration better than joints with loose bonds. Slushing or grouting the joints after laying the brick does not fill the joint completely. Fill the joints between the brick solidly, especially in the face tier. Tool the joint to a concave surface before the mortar sets up. When tooling, use enough force to press the mortar tightly against the brick on both sides of the joint. Although good bricklaying workmanship does not permit water penetration, it provides some means of removing moisture that does penetrate the masonry, such as properly designed flashing or the use of cavity walls.

FIRE RESISTANCE

9-14. Table 9-1 gives the hours of fire resistance for various thicknesses of brick walls determined by tests conducted on brick walls laid with portland-cement-lime mortar. The ASTM standard method for conducting fire tests was used.

Table 9-1. Fire resistance of brick load-bearing walls laid with Portland-cement-lime mortar

Normal Wall Thickness, in Inches	Types of Wall	Material	Ultimate Fire-Resistance Period. Incombustible Members Framed into Wall or not Framed in Members		
			No Plaster, in Hours	Plaster on One Side, in Hours	Plaster on Two Sides, in Hours
4	Solid	Clay or shale	1 1/4	1 3/4	2 1/2
8	Solid	Clay or shale	5	6	7
12	Solid	Clay or shale	10	10	12
8	Hollow rowlock	Clay or shale	2 1/2	3	4
12	Hollow rowlock	Clay or shale	5	6	7
9 to 10	Cavity	Clay or shale	5	6	7
4	Solid	Sand lime	1 3/4	2 1/2	3
8	Solid	Sand lime	7	8	9
12	Solid	Sand lime	10	10	12
Note. Not less than 1/2 inch of 1-3 sanded gypsum plaster is required to develop these rating.					

ABRASION RESISTANCE

9-15. A brick's resistance to abrasion depends largely upon its compressive strength, which is determined by how well it was fired. Well-burned brick has excellent wearing qualities.

INSULATING QUALITIES OF BRICK MASONRY

9-16. A brick masonry wall expands and contracts with temperature change. However, because the wall itself takes up a lot of the expansion and contraction, the amount of movement calculated theoretically does not actually occur. Therefore, walls up to 200 feet long do not need expansion joints, but longer walls require one expansion joint for every 200 feet.

HEAT

9-17. Solid-brick masonry walls provide very little insulation from heat and cold. A cavity wall or a brick wall backed with hollow clay tile gives much better insulating value.

SOUND

9-18. Brick walls are massive and provide good sound insulation. Generally, the heavier the wall, the better its sound-insulating value. However, the sound insulation provided by a wall more than 12 inches thick is not much greater than a wall 10 to 12 inches thick. Dividing a wall into two or more layers, such as a cavity wall, increases its resistance to sound transmission from one side of the wall to the other. Brick walls poorly absorb sound originating within the walls and reflect much of the sound back into the structure. However, impact sounds, such as a hammer striking the wall, travel a long way along the wall.

SECTION II - BRICKLAYING METHODS

FUNDAMENTALS

9-19. Good bricklaying procedures depend on good workmanship and efficiency. Efficiency means to do the work with the fewest possible motions. Each motion should have a purpose and accomplish a particular result. After learning the fundamentals, study your own work to eliminate unnecessary motions, thereby achieving maximum efficiency. Organize your work to ensure a continual supply of brick and mortar. Plan the scaffolding before the work begins, and build it so that it interferes as little as possible with other workers. Paragraphs 7-2 and 7-3 describe mason's tools and equipment, which are generally the same as, or similar to, those used in bricklaying.

BRICK MASONRY TERMS

9-20. You need to know the specific terms that describe the position of masonry units and mortar joints in a wall (see figure 9-3). These terms include —

- Course. One of several continuous, horizontal layers (or rows) of masonry units bonded together.
- Wythe. Each continuous, vertical section of a wall, one masonry unit thick, such as the thickness of masonry separating flues in a chimney. Sometimes called a tier.
- Stretcher. A masonry unit laid flat on its bed along the length of a wall with its face parallel to the face of the wall.
- Header. A masonry unit laid flat on its bed across the width of a wall with its face perpendicular to the face of the wall. Generally used to bond two wythes.
- Rowlock. A header laid on its face or edge across the width of a wall.
- Bull stretcher. A rowlock brick laid with its bed parallel to the face of the wall.
- Bull header. A rowlock brick laid with its bed perpendicular to the face of the wall.
- Soldier. A brick laid on its end with its face perpendicular to the face of the wall.

Brick and Tile Masonry

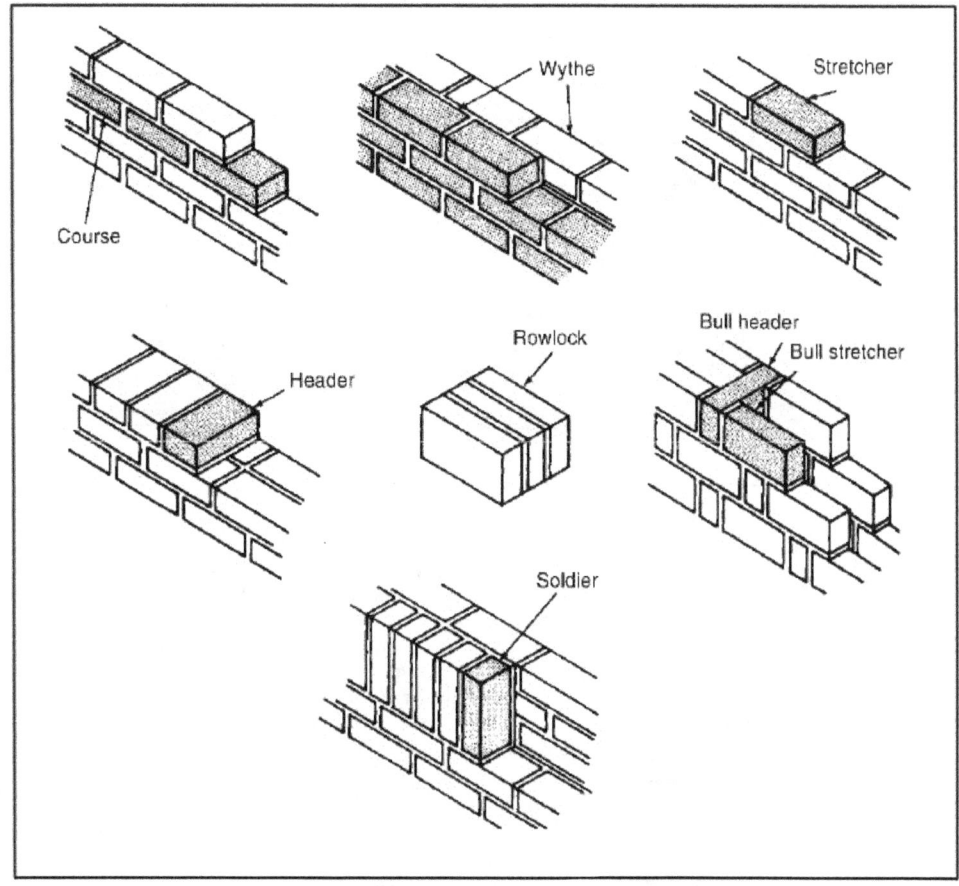

Figure 9-3. Masonry units and mortar joints

TYPES OF BONDS

9-21. The term bond as used in masonry has three different meanings: structural bond, mortar bond, or pattern bond. Metal ties are also used as bonds.

- Structural bond. This means how the individual masonry units interlock or tie together into a single structural unit. You can achieve structural bonding of brick and tile walls in one of three ways:
  - Overlapping (interlocking) the masonry units.
  - Embedding metal ties in connecting joints.
  - Using grout to adhere adjacent wythes of masonry.
- Mortar bond. This is adhesion of the mortar joint to the masonry units or to the reinforcing steel.
- Pattern Bond. This is pattern formed by the masonry units and mortar joints on the face of a wall. The pattern may result from the structural bond or may be purely decorative and unrelated to the structural bond. Figure 9-4, page 9-8, shows the six basic bond patterns in common use today: running bond, common or American bond, Flemish bond, English bond, stack bond, and English cross or Dutch bond.
  - Running bond. This is the simplest of the six bonds, consisting of all stretchers. The bond has no headers, therefore metal ties usually form the structural bond. The running bond is

used largely in cavity wall construction, brick veneer walls, and facing tile walls made with extra wide stretcher tile.

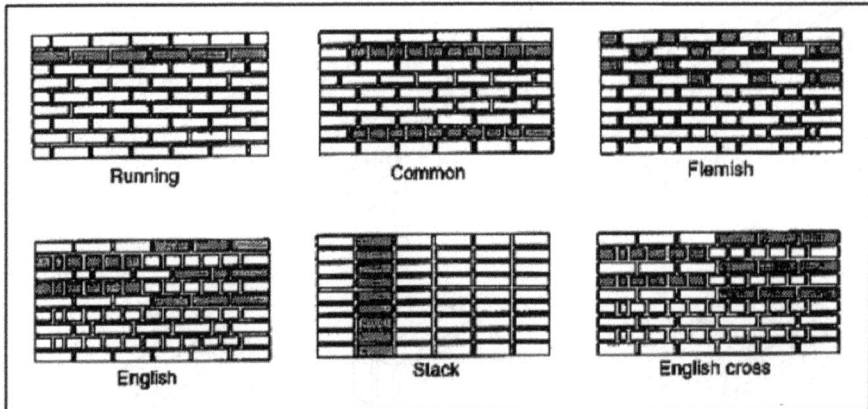

Figure 9-4. Types of masonry bond

- Common or American bond. This is variation of the running bond having a course of full-length headers at regular intervals that provide the structural bond as well as patterns. Header courses usually appear at every fifth, sixth, or seventh course, depending on the structural bonding requirements or the common bond that will vary with a Flemish header course. In laying out any bond pattern, be sure to start the corners correctly. In a common bond, use a three-quarter closure at the corner of each header course.
- Flemish bond. Each course consists of alternating headers and stretchers. The headers in every other course center over and under the stretchers in the courses in between. The joints between stretchers in all stretcher courses align vertically. When headers are not required for structural bonding, use bricks called blind headers. Start the corners two different ways; in the Dutch corner, a three-quarter closure starts each course, and with the English corner, a 2-inch or quarter closure starts the course.
- English bond. This pattern consists of alternating courses of headers and stretchers. The headers center over and under the stretchers. The joints between stretchers in all stretcher courses do not align vertically. Use blind headers in courses that are not structural bonding courses.
- Stack bond. This is purely a pattern bond, with no overlapping units and all vertical joints aligning. You must use dimensional-accurate or prematched units to achieve good vertical joint alignment. You can vary the pattern with combinations and modifications of the basic patterns shown in figure 9-4. This pattern usually bonds to the backing with rigid steel ties, or 8-inch-thick stretcher units when available. In large wall areas or for load-bearing construction, insert steel-pencil rods into the horizontal mortar joints as reinforcement.
- English cross or Dutch bond. A variation of the English bond, the English cross or Dutch bond differs only in that the joints between the stretchers in the stretcher courses align vertically. These joints center on the headers in the courses above and below.
- Metal ties. When a wall bond has no header courses, use metal ties to bond the exterior wall brick to the backing courses. Figure 9-5 shows three typical metal ties.

Brick and Tile Masonry

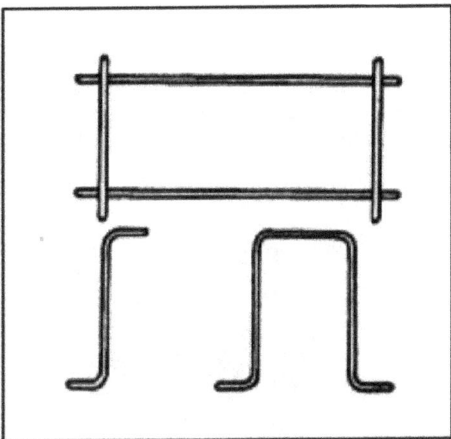

Figure 9-5. Metal ties

FLASHING

9-22. Install flashing at any spot where moisture is likely to enter a brick masonry structure. Flashing diverts the moisture back outside. Always install flashing under horizontal masonry surfaces such as sills and copings and at intersections between masonry walls and horizontal surfaces. This also includes roof and parapet or a roof and chimney, above openings such as doors and windows, and frequently at floor lines, depending on the type of construction. The flashing should extend through the exterior wall face and then turn downward against the wall face to form a drop. Provide weep holes at intervals of 18 to 24 inches to drain water that accumulates on the flashing to the outside. Weep holes are even more important when appearance requires the flashing to stop behind the wall face instead of extending through the wall. This type of concealed flashing with tooled mortar joints often retains water in the wall for long periods and, by concentrating the moisture at one spot it does more harm than good.

MAKING AND POINTING MORTAR JOINTS

9-23. Pointing is filling exposed joints with mortar immediately after laying a wall. You can also fill holes and correct defective mortar joints by pointing, using a pointing trowel.

MORTAR JOINT

9-24. There is no rule governing the thickness of a brick masonry mortar joint. Irregularly shaped bricks may require mortar joints up to 1/2 inch thick to compensate for the irregularities. However, mortar joints 1/4 inch thick are the strongest. Use this thickness whenever the bricks are regular enough in shape to permit it.

9-25. A slush joint is made simply by depositing the mortar on top of the head joints allowing it to run down between the bricks to form a joint. You cannot make solid joints this way. Even if you fill the space between the bricks completely, there is no way you can compact the mortar against the brick faces, and a poor bond will result.

PICKING UP AND SPREADING MORTAR

9-26. Figure 9-6, page 9-10, shows the correct way to hold a trowel firmly in the grip with your thumb resting on top of the handle, not encircling it. If you are right-handed, pick up mortar from the outside of the mortar-board pile with the left edge of your trowel (see figure 9-7, page 9-11). You can pick up enough mortar to spread one to five bricks depending on the wall space and your skills. A pickup for one brick forms only a small pile along the left edge of the trowel, however, a pickup for five bricks is a full load for a large trowel as shown in view 2 of figure 9-7.

9-27. If you are right-handed, spreading the mortar working from left to right along the wall. Holding the left edge of the trowel directly over the centerline of the previous course, tilt the trowel slightly and move it to the right (see view 3 of figure 9-7) dropping an equal amount of mortar on each brick until the course is completed or the trowel is empty. Return any leftover mortar to the trowel mortarboard.

Figure 9-6. Correct way to hold trowel

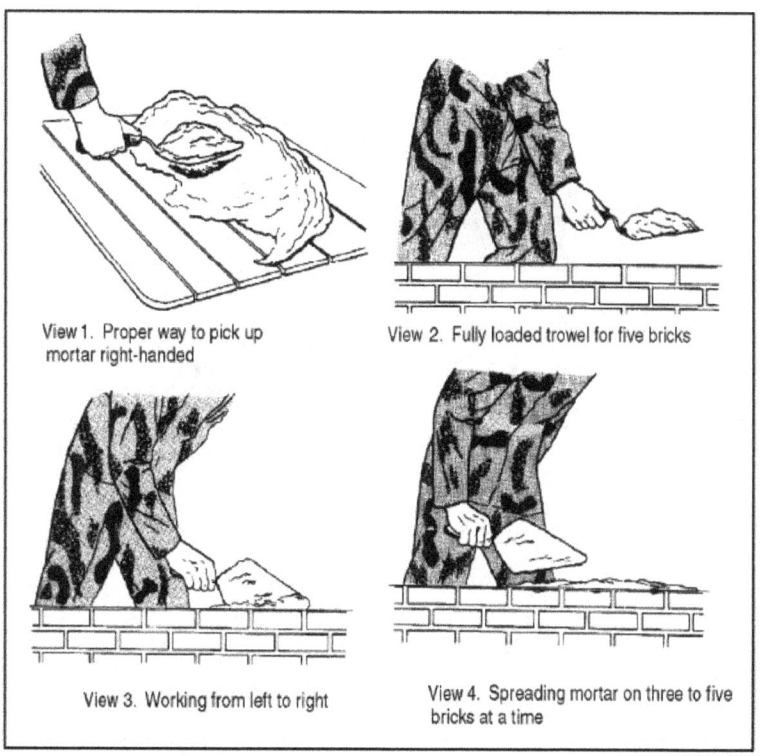

Figure 9-7. Picking up and spreading mortar

MAKING BED AND HEAD JOINT

9-28. Do not spread the mortar for a bed joint too far ahead of laying (the length of 4 or 5 bricks is best). Mortar spread out too far ahead dries out before the bricks bedded in it and causes a poor bond as shown in figure 9-8. The mortar must be soft and plastic so that the brick beds in it easily. Spread the mortar about 1 inch thick, and then make a shallow furrow in it (see figure 9-9, page 9-12). A furrow that is too deep leaves a gap between the mortar and the bedded brick, which will reduce the wall's resistance to water penetration.

Chapter 9

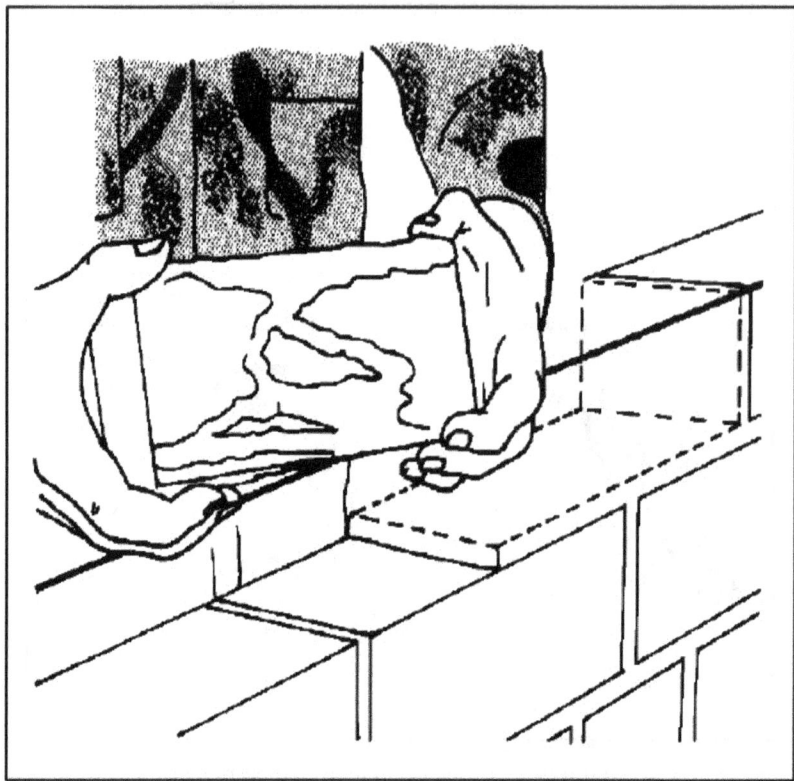

Figure 9-8. A poorly bonded brick

9-29. Cut off any mortar projecting beyond the wall line with the edge of the trowel (see figure 9-9). Use a smooth, even stroke. Retain enough mortar on the trowel to butter the left end of the first brick you will lay in the fresh mortar and throw the rest back on the mortar board.

9-30. Placing your thumb on one side of the brick and your fingers on the other, pick up the first brick to be laid (see figure 9-10, page 9-14). Apply as much mortar as will stick to the end of the brick and then push it into place, squeezing out excess mortar at the head joint and at the sides as shown in figure 9-11, page 9-15. Make sure that the mortar completely fills the head joint. After bedding the brick, cut off the excess mortar and use it to start the next end joint. Throw any surplus mortar on the back of the mortar board for retampering if necessary.

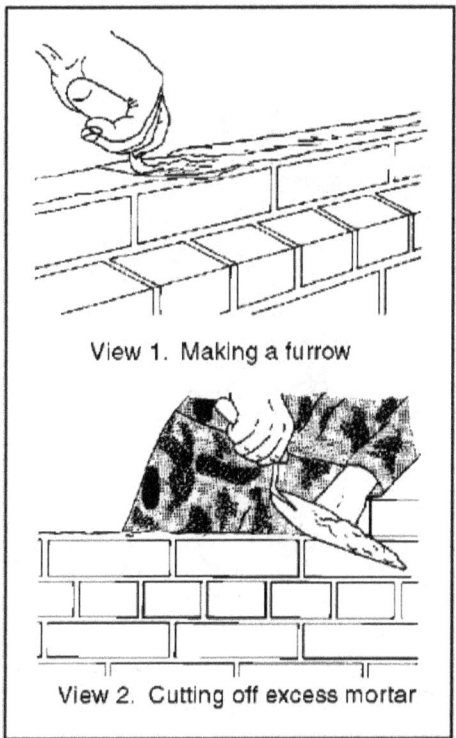

Figure 9-9. Making a bed joint in a stretcher course

Chapter 9

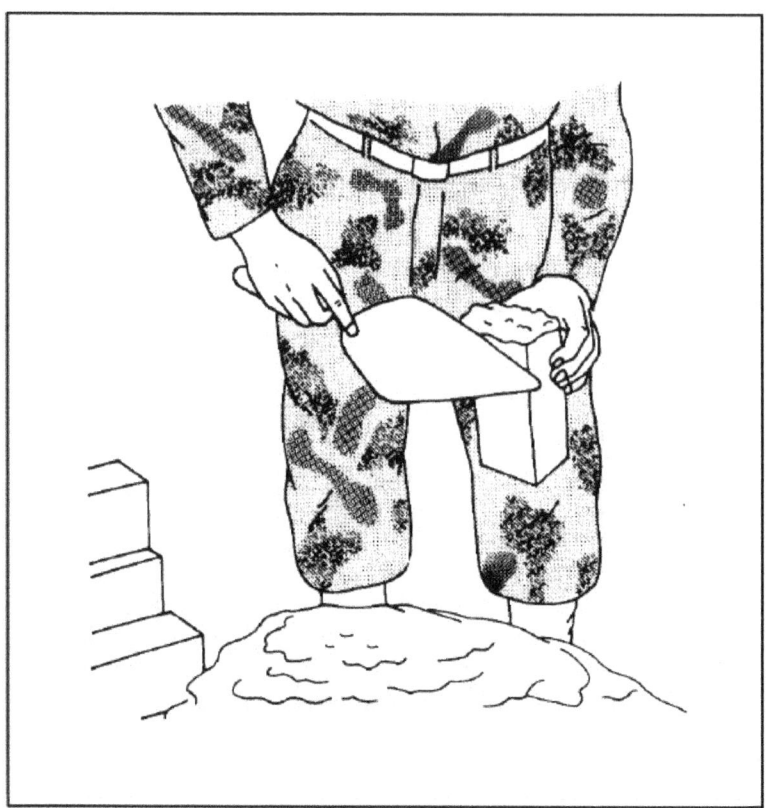

Figure 9-10. Proper way to hold a brick when buttering the end

Figure 9-11. Making a head joint in a stretcher course

INSERTING A BRICK IN A WALL

9-31. Figure 9-12, page 9-16, shows how to insert a brick in a space left in a wall. First, spread a thick bed of mortar (see figure 9-12), and then shove the brick into it (see view 2 of figure 9-12) until mortar squeezes out of the four joints (see view 3 of figure 9-12). In doing so you are assured that the joints are full of mortar at every point.

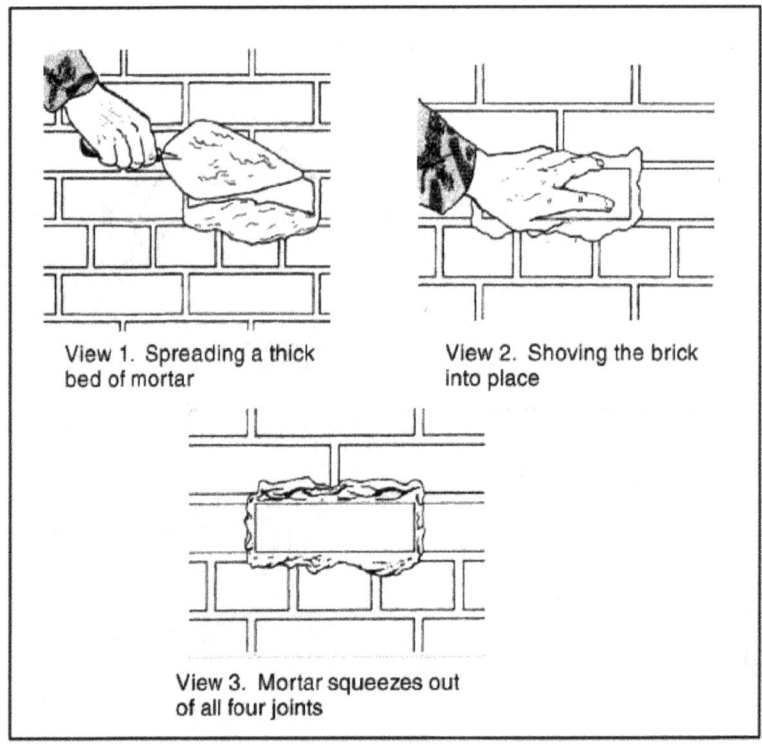

Figure 9-12. Inserting a brick in a wall

MAKING CROSS JOINTS AND JOINTS CLOSURE

9-32. Spread the bed-joint mortar several brick widths in advance. Then spread mortar over the face of the header brick before placing it in the wall (see view 1 of figure 9-13). Next shove the brick into place, squeezing out mortar at the top of the joint. Finally cut off the excess mortar as shown in view 2 of figure 9-13.

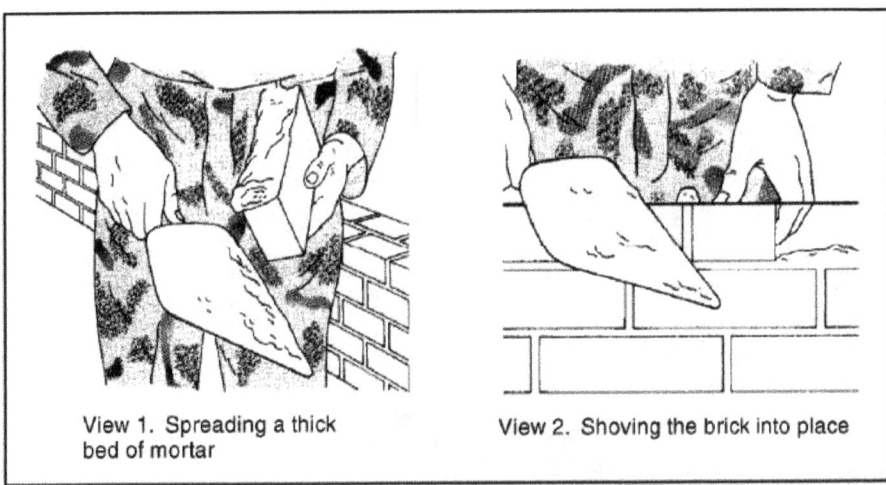

Figure 9-13. Making a cross joint in a header course

9-33. Figure 9-14, page 9-18, shows how to lay a closure brick in a header course. Spread about 1 inch of mortar on the sides of the brick already in place (see view 1 of figure 9-14), as well as on both sides of the closure brick (see view 2 of figure 9-14). Then lay the closure brick carefully into position, without disturbing the brick already laid (see view 3 of figure 9-14).

9-34. To make a closure joint in stretcher courses, first spread plenty of mortar on the ends of the brick already in place (see view 1 of figure 9-15, page 9-19), as well as both ends of the closure brick (see view 2 of figure 9-15). Then carefully lay the closure brick without disturbing the brick already in place (see view 3 of figure 9-15). If you do disturb any adjacent bricks, you must remove and relay them. Otherwise, cracks will form between the brick and mortar, allowing moisture to penetrate the wall.

Chapter 9

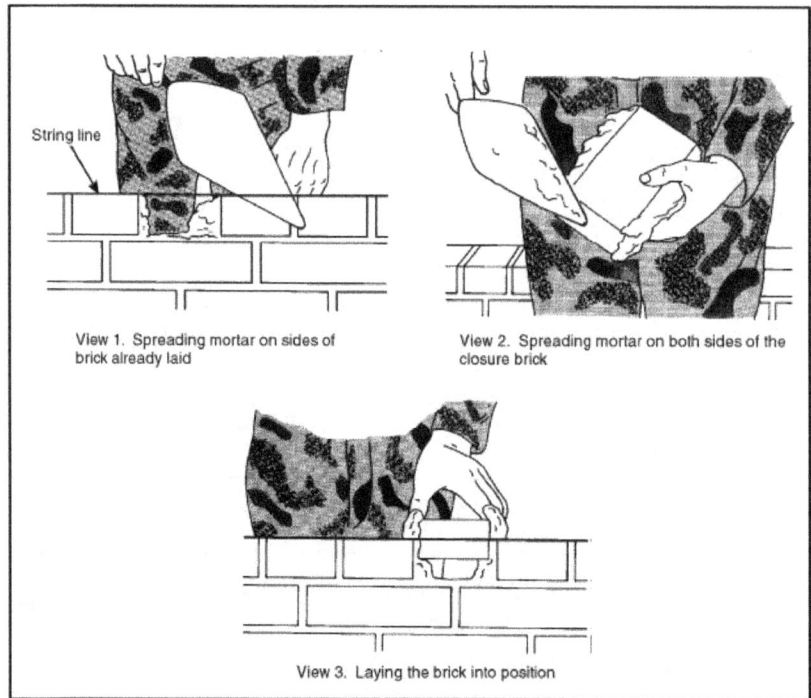

Figure 9-14. Making a closure joint in a header course

CUTTING BRICK

9-35. Bricks will either be cut with a bolster or a brick set, using a brick hammer.

- Using a bolster or brick set. When you must cut a brick to exact line, use a bolster (see figure 9-16) or brick set. The straight side of the tool's cutting edge should face both the part of the brick to be saved, and the bricklayer. One mason's hammer blow should break the brick. For an extremely hard brick, first cut it roughly using the brick hammer head, but leave enough brick to cut accurately with the brick set.
- Using a brick hammer. Use a brick hammer for normal cutting work, such as making the closure bricks and bats around wall openings or completing corners. Hold the brick firmly while cutting it. First cut a line around the brick using light blows. Hitting a sharp blow to one side of the cutting line should split the brick at the cutting line (see view 1 of figure 9-17, page 9-20). Trim rough spots using the hammer blade as shown in view 2 of figure 9-17.

Brick and Tile Masonry

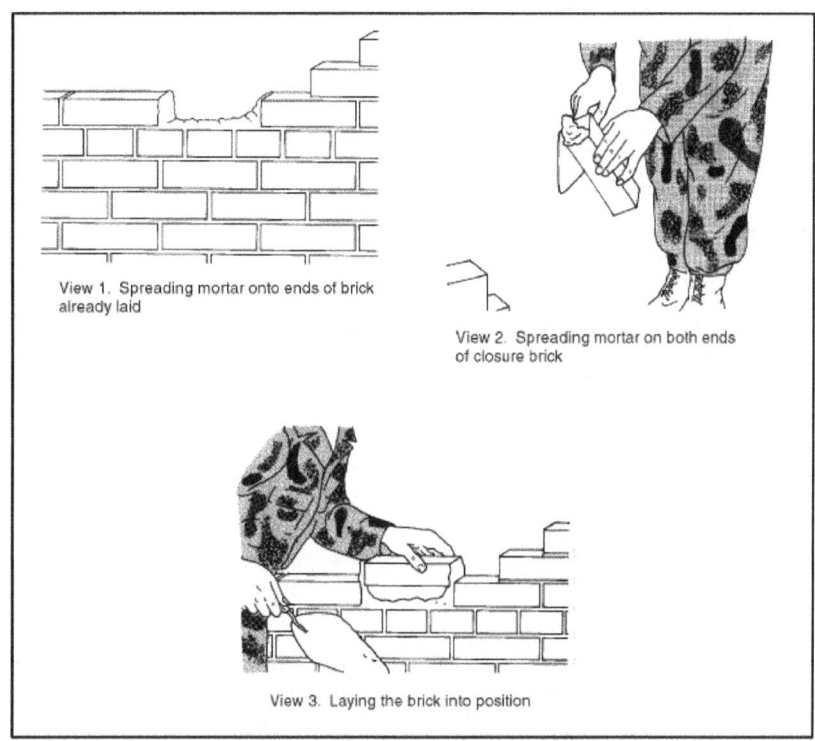

Figure 9-15. Making a closure joint in a stretcher course

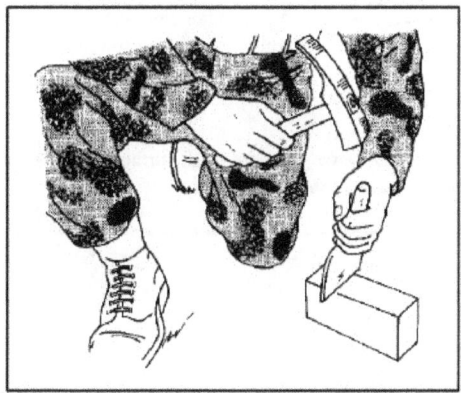

Figure 9-16. Cutting brick with a bolster

Chapter 9

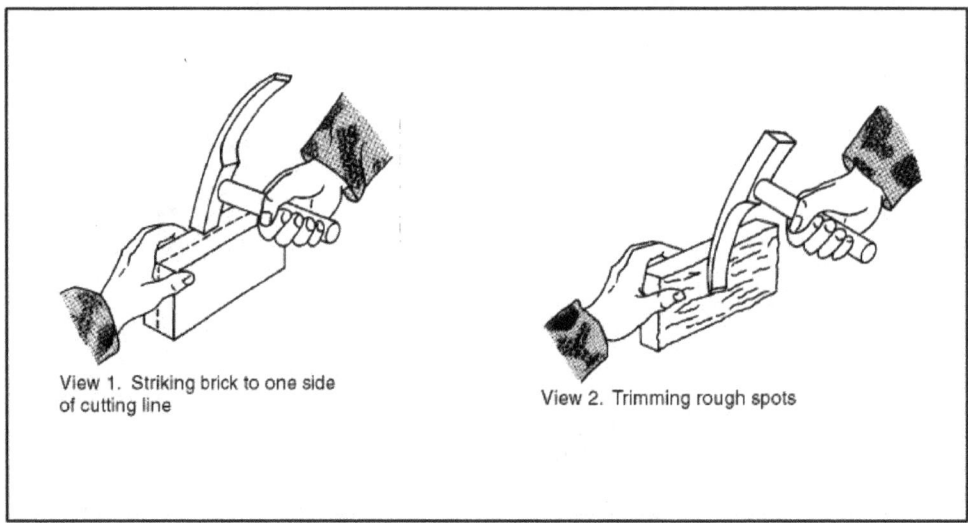

Figure 9-17. Cutting a brick with a hammer

FINISHING JOINTS

9-36. Purpose. The exterior surfaces of mortar joints are finished to make brick masonry more waterproof and give it a better appearance. If joints are simply cut flush with the brick and not finished, shallow cracks develop immediately between the brick and the mortar. Finishing or tooling the joint using the jointer shown in figure 7-1, page 7-2, prevents such cracks. Always finish a mortar joint before the mortar hardens too much. Figure 9-18 shows several types of joint finishes, the more important of which are discussed below.

- Concave joint. It is very weather tight. After removing the excess mortar with a trowel, make this joint using a jointer that is slightly larger than the joint. Use force against the tool to press the mortar tight against the brick on both sides of the mortar joint.
- Flush joint. It is made by holding the trowel almost parallel to the face of the wall while drawing its point along the joint.
- Weather joint. It sheds water from a wall surface more easily. To make it, simply push downward on the mortar with the top edge of the trowel.

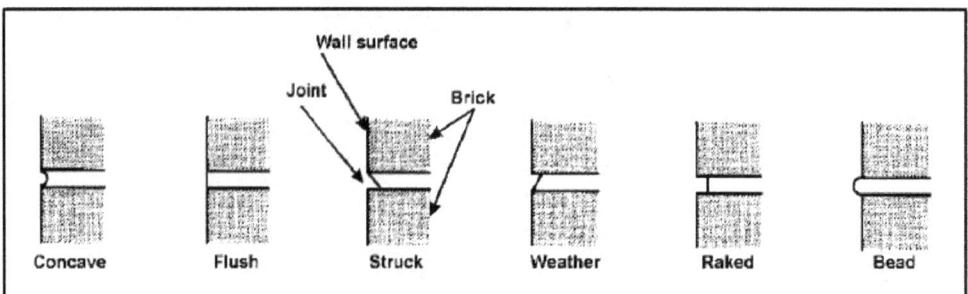

Figure 9-18. Joint finishes

SECTION III - BRICK CONSTRUCTION

BRICKLAYER'S DUTIES

9-37. The bricklayer actually lays the brick and is responsible for laying out the job so that the finished masonry has the proper quality and appearance. In wall construction, the bricklayer must make sure that the walls are plumb and the courses are level.

BRICKTENDER'S DUTIES

9-38. The bricktender mixes mortar and carries brick, mortar, and supplies material to the bricklayer, as needed. The bricktender fills the mortar board and places it in a convenient spot for the bricklayer. He assists in layout and rapid backup bricklaying by laying out bricks in a line on an adjacent course so that the bricklayer only has to move each brick a few inches to lay them. He also wets brick during warm weather. There are four reasons why bricks must be wet just before laying them:

- A better bond is created between the brick and the mortar.
- Dust and dirt are washed from the brick surfaces because mortar adheres better to a clean brick.
- Mortar spreads more evenly under a wet brick surface.
- Dry brick absorbs water from the mortar rapidly, particularly portland-cement mortar. To harden properly, cement requires sufficient moisture to complete the hydration process. Therefore, if the brick absorbs too much water from the mortar, the cement will not harden properly.

LAYING FOOTINGS

9-39. A qualified engineer must determine actual footing width and thickness for high walls and walls that will carry a heavy load. A footing must rest below the frost line to prevent foundation heaving and settlement.

WALL FOOTINGS

9-40. A wall requires a footing when the supporting soil cannot withstand the wall load without a further means of load redistribution. The footing must be wider than the wall thickness as shown in figure 9-19, page 9-22. For an ordinary one-story building having an 8-inch-thick wall, a footing 16 inches wide and approximately 8 inches thick is usually large enough. Although brick masonry footings are satisfactory, footings are normally made from concrete leveled on top to receive the brick or stone foundation walls. After preparing the subgrade, place a mortar bed about 1 inch thick on the subgrade to compensate for all irregularities. Lay the first course of the foundation on the mortar bed followed by succeeding courses (see figure 9-19).

Chapter 9

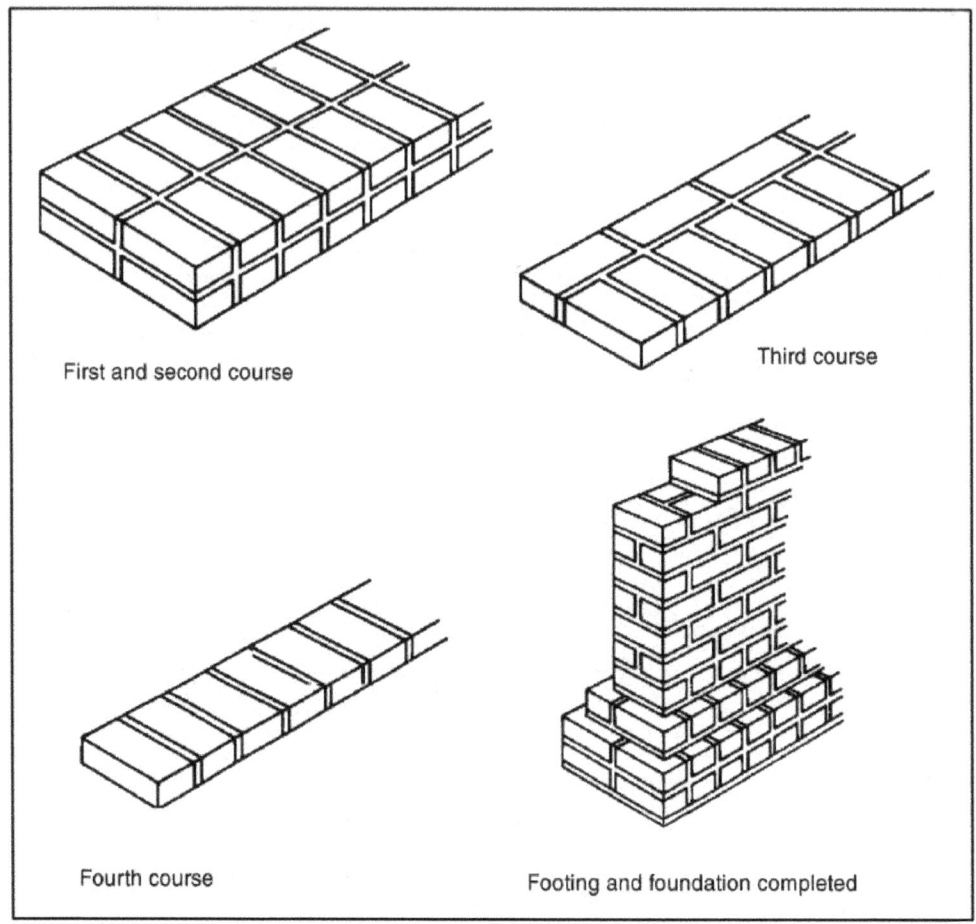

Figure 9-19. Laying a wall footing

COLUMN FOOTINGS

9-41. Figure 9-20 shows a footing for a 12 by 16-inch brick column. This footing requires the same construction method as the wall footing.

Brick and Tile Masonry

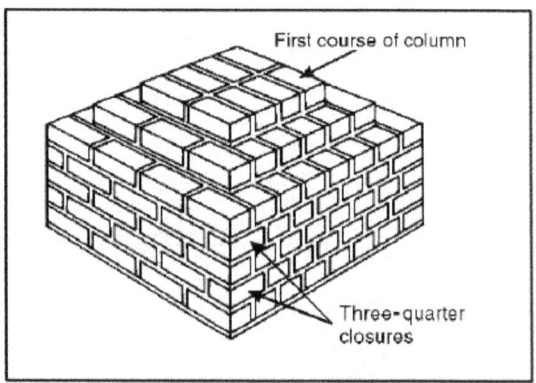

Figure 9-20. Column footing

LAYING AN 8-INCH COMMON-BOND BRICK WALL

9-42. You can build both solid and hollow walls from brick masonry. The solid 8- and 12-inch walls in common bond are the most familiar ones in the US.

LAYING OUT THE WALL

9-43. To build a wall of a given length, adjust the width of the head joints so that a particular number of bricks or a particular number plus a half-brick will equal the given length. To do this use the following steps:

   Step 1. Lay out the bricks for the foundation without mortar first, as shown in figure 9-21.

   Step 2. Space them equally. The distance between them will equal the thickness of the head joints. Tables 9-2 through 9-4, pages 9-24 through 9-26, gives the number of courses in a wall of a given height using standard brick and different joint widths.

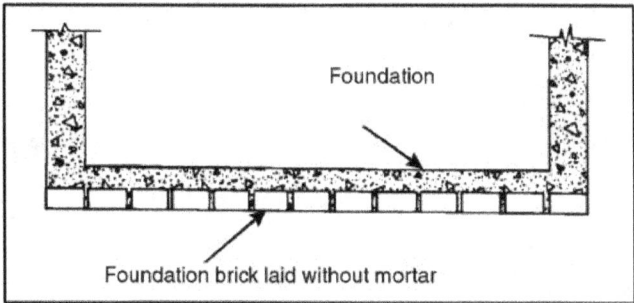

Figure 9-21. Determining number of bricks in one course and head joint widths

Table 9-2. Height of course using 2 1/4-inch brick, 3/8 inch joint

Course	Height	Course	Height	Course	Height	Course	Height	Course	Height
1	0' 2 5/8"	21	4' 7 1/8"	41	8' 11 5/8"	61	13' 4 1/8"	81	17' 8 5/8"
2	0' 5 1/4"	22	4' 9 3/4"	42	9' 2 1/4"	62	13' 6 3/4"	82	17' 11 1/4"
3	0' 7 7/8"	23	5' 0 3/8"	43	9' 4 7/8"	63	13' 9 3/8"	83	18' 1 7/8"
4	0' 10 1/2"	24	5' 3"	44	9' 7 1/2"	64	14' 0"	84	18' 4 1/2"
5	1' 1 1/8"	25	5' 5 5/8"	45	9' 10 1/8"	65	14' 2 5/8"	85	18' 7 1/8"
6	1' 3 3/4"	26	5' 8 1/4"	46	10' 0 3/4"	66	14' 5 1/4"	86	18' 9 3/4"
7	1' 6 3/8"	27	5' 10 7/8"	47	10' 3 3/8"	67	14' 7 7/8"	87	19' 0 3/8"
8	1' 9"	28	6' 1 1/2"	48	10' 6"	68	14' 10 1/2"	88	19' 3"
9	1' 11 5/8"	29	6' 4 1/8"	49	10' 8 5/8"	69	15' 1 1/8"	89	19' 5 5/8"
10	2' 2 1/4"	30	6' 6 3/4"	50	10' 11 1/4"	70	15' 3 3/4"	90	19' 8 1/4"
11	2' 4 7/8"	31	6' 9 3/8"	51	11' 1 7/8"	71	15' 6 3/8"	91	19' 10 7/8"
12	2' 7 1/2"	32	7' 0"	52	11' 4 1/2"	72	15' 9"	92	20' 1 1/2"
13	2' 10 1/8"	33	7' 2 5/8"	53	11' 7 1/8"	73	15' 11 5/8"	93	20' 4 1/8"
14	3' 0 3/4"	34	7' 5 1/4"	54	11' 9 3/4"	74	16' 2 1/4"	94	20' 6 3/4"
15	3' 3 3/8"	35	7' 7 7/8"	55	12' 0 3/8"	75	16' 4 7/8"	95	20' 9 3/8"
16	3' 6"	36	7' 10 1/2"	56	12' 3"	76	16' 7 1/2"	96	21' 0"
17	3' 8 5/8"	37	8' 1 1/8"	57	12' 5 5/8"	77	16' 7 1/2"	97	21' 2 5/8"
18	3' 11 1/4"	38	8' 3 3/4"	58	12' 8 1/4"	78	17' 0 3/4"	98	21' 5 1/4"
19	4' 1 7/8"	39	8' 6 3/8"	59	12' 10 7/8"	79	17' 3 3/8"	99	21' 7 7/8"
20	4' 4 1/2"	40	8' 9"	60	13' 1 1/2"	80	17' 6"	100	21' 10 1/2"

LAYING CORNER LEADS

9-44. The following steps should be used in laying the corner leads:

Step 1. Erect the wall corners first. This is called laying the leads.

Step 2. Use the leads as a guide when laying the rest of the wall.

Step 3. Lay the face tier.

Step 4. Build the corner leads up about six or seven courses, or to the height of the next header course.

Header course

9-45. Normally, the first course is a header course. Step 1 of figure 9-22, page 9-27, shows how to start laying a corner lead.

Step 1. Lay a 1-inch mortar bed on the foundation.

Table 9-3. Height of course using 2 1/4-inch brick, 1/2-inch joint

Course	Height	Course	Height	Course	Height	Course	Height	Course	Height
1	0' 2 3/4"	21	4' 9 3/4"	41	9' 4 3/4"	61	13' 11 3/4"	81	18' 6 3/4"
2	0' 5 1/2"	22	5' 0 1/2"	42	9' 7 1/2"	62	14' 2 1/2"	82	18' 9 1/2"
3	0' 8 1/4"	23	5' 3 1/4"	43	9' 10 1/4"	63	14' 5 1/4"	83	19' 0 1/4"
4	0' 11"	24	5' 6"	44	10' 1"	64	14' 8"	84	19' 3"
5	1' 1 1/4"	25	5' 8 3/4"	45	10' 3 3/4"	65	14' 10 3/4"	85	19' 5 3/4"
6	1' 4 1/2"	26	5' 11 1/2"	46	10' 6 1/2"	66	15' 1 1/2"	86	19' 8 1/2"
7	1' 7 1/4"	27	6' 2 1/4"	47	10' 9 1/4"	67	15' 4 1/4"	87	19' 11 1/4"
8	1' 10"	28	6' 5"	48	11' 0"	68	15' 7"	88	20' 2"
9	2' 0 3/4"	29	6' 7 3/4"	49	11' 2 3/4"	69	15' 9 3/4"	89	20' 4 3/4"
10	2' 3 1/2"	30	6' 10 1/2"	50	11' 5 1/2"	70	16' 0 1/2"	90	20' 7 1/2"
11	2' 6 1/4"	31	7' 1 1/4"	51	11' 8 1/4"	71	16' 3 1/4"	91	20' 10 1/4"
12	2' 9"	32	7' 4"	52	11' 11"	72	16' 6"	92	21' 1"
13	2' 11 3/4"	33	7' 6 3/4"	53	12' 1 3/4"	73	16' 8 3/4"	93	21' 3 3/4"
14	3' 2 1/2"	34	7' 9 1/2"	54	12' 4 1/2"	74	16' 11 1/2"	94	21' 6 1/2"
15	3' 5 1/4"	35	8' 0 1/4"	55	12' 7 1/4"	75	17' 2 1/4"	95	21' 9 1/4"
16	3' 8"	36	8' 3"	56	12' 10"	76	17' 5"	96	22' 0"
17	3' 10 3/4"	37	8' 5 3/4"	57	13' 0 3/4"	77	17' 7 3/4"	97	22' 2 3/4"
18	4' 1 1/2"	38	8' 8 1/2"	58	13' 3 1/2"	78	17' 10 1/2"	98	22' 5 1/2"
19	4' 4 1/4"	39	8' 11 1/4"	59	13' 6 1/4"	79	18' 1 1/4"	99	22' 8 1/4"
20	4' 7"	40	9' 2"	60	13' 9"	80	18' 4"	100	22' 11"

Step 2. Cut two three-quarter closures and press one into the mortar bed until it makes a bed joint 1/2 inch thick (see a in step 2 of figure 9-22).

Step 3. Spread mortar on the end of the second three-quarter closure and form a 1-inch thick head joint as described in paragraph 9-28 (see b in step 2 of figure 9-22).

Step 4. Cut off the mortar that squeezes out of the joints.

Step 5. Lay a mason's level in the two positions shown in step 2 of figure 9-22 and check the levels of the two three-quarter closures. The exterior edges of both closures must be flush with the exterior face of the foundation.

Step 6. Spread mortar on one bed of a whole brick (see c in step 3 of figure 9-22) and lay it as shown.

Table 9-4. Height of course using 2 1/4-inch brick, 5/8-inch joint

Course	Height	Course	Height	Course	Height	Course	Height	Course	Height
1	0' 2 7/8"	21	5' 0 3/8"	41	9' 9 7/8"	61	14' 7 3/8"	81	19' 4 7/8"
2	0' 5 3/4"	22	5' 3 1/4"	42	10' 0 3/4"	62	14' 10 1/4"	82	19' 7 3/4"
3	0' 8 5/8"	23	5' 6 1/8"	43	10' 3 5/8"	63	15' 1 1/8"	83	19' 10 5/8"
4	0' 11 1/2"	24	5' 9"	44	10' 6 1/2"	64	15' 4"	84	20' 1 1/2"
5	1' 2 3/8"	25	5' 11 7/8"	45	10' 9 3/8"	65	15' 6 7/8"	85	20' 4 3/8"
6	1' 5 1/4"	26	6' 2 3/4"	46	11' 0 1/4"	66	15' 9 3/4"	86	20' 7 1/4"
7	1' 8 1/8"	27	6' 5 5/8"	47	11' 3 1/8"	67	16' 0 5/8"	87	20' 10 1/8"
8	1' 11"	28	6' 8 1/2"	48	11' 6"	68	16' 3 1/2"	88	21' 1"
9	2' 1 7/8"	29	6' 11 3/8"	49	11' 8 7/8"	69	16' 6 3/8"	89	21' 3 7/8"
10	2' 4 3/4"	30	7' 2 1/4"	50	11' 11 3/4"	70	16' 9 1/4"	90	21' 6 3/4"
11	2' 7 5/8"	31	7' 5 1/8"	51	12' 2 5/8"	71	17' 0 1/8"	91	21' 9 5/8"
12	2' 10 1/2"	32	7' 8"	52	12' 5 1/2"	72	17' 3"	92	22' 0 1/2"
13	3' 1 3/8"	33	7' 10 7/8"	53	12' 8 3/8"	73	17' 5 7/8"	93	22' 3 3/8"
14	3' 4 1/4"	34	8' 1 3/4"	54	12' 11 1/4"	74	17' 8 3/4"	94	22' 6 1/4"
15	3' 7 1/8"	35	8' 4 5/8"	55	13' 2 1/8"	75	17' 11 5/8"	95	22' 9 1/8"
16	3' 10"	36	8' 7 1/2"	56	13' 5"	76	18' 2 1/2"	96	23' 0"
17	4' 0 7/8"	37	8' 10 3/8"	57	13' 7 1/8"	77	18' 5 3/8"	97	23' 2 7/8"
18	4' 3 3/4"	38	9' 1 1/4"	58	13' 10 3/4"	78	18' 8 1/4"	98	23' 5 3/4"
19	4' 6 5/8"	39	9' 4 1/8"	59	14' 1 5/8"	79	18' 11 1/8"	99	23' 8 5/8"
20	4' 9 1/2"	40	9' 7"	60	14' 4 1/2"	80	19' 2"	100	23' 11 1/2"

Step 7. Check its level using the mason's level in the positions shown in step 3 of figure 9-22. The end of this brick must also be flush with the exterior face of the foundation.

Step 8. After laying this brick in the proper position, cut the quarter closures e and f, and lay them as described in paragraph 9-33 for laying closure bricks.

Step 9. Remove all excess mortar and check the tops of the quarter closures to make sure that they are flush with bricks a and b.

Step 10. Spread mortar on brick g (see step 4 of figure 9-22), shove it into position as shown, and remove any excess mortar.

Step 11. Lay bricks h, i, j, and k the same way. Check their levels by placing the mason's level in the positions shown in step 4 of figure 9-22. All brick ends must be flush with the foundation surface.

Step 12. Lay bricks l, m, n, o, and p in the same manner (see step 5 of figure 9-22). You must lay 12 header bricks in the first course of the corner lead—six bricks on each side of the three-quarter closures a and b.

Brick and Tile Masonry

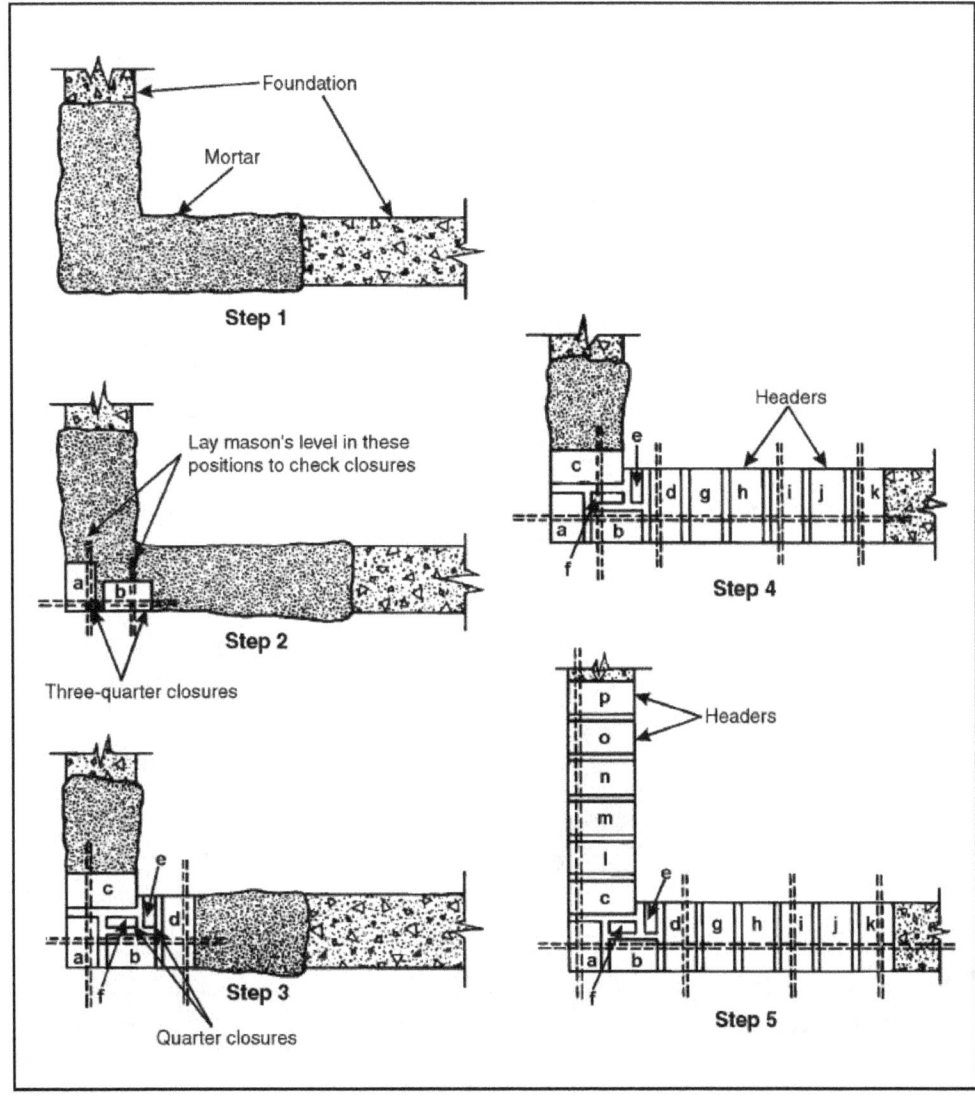

Figure 9-22. Laying first course of corner lead for 8-inch common-bond brick wall

Stretcher course

9-46. Lay the second course of the corner lead (a stretcher course) as shown in steps 1 and 2 of figure 9-23, page 9-28.

    Step 1. Spread a 1-inch mortar bed over the first course and make a shallow furrow in it.

    Step 2. Push brick (a) (see step 2 of figure 9-23) into the mortar bed until it makes a joint 1/2 inch thick.

    Step 3. Spread mortar on the end of the brick and shove it into place. Remove the excess mortar and check the joints for thickness.

Chapter 9

Step 4. Lay bricks c, d, e, f, and g the same way. Check them by placing the mason's level in the position shown in step 2 of figure 9-23 to make sure they are level.

Step 5. Plumb the corners in several places by placing the mason's level in the vertical position as shown in figure 9-24. As step 3 of figure 9-23 shows, the second course requires seven bricks.

Step 6. Lay the remaining bricks in the corner lead as you did the bricks in the second course.

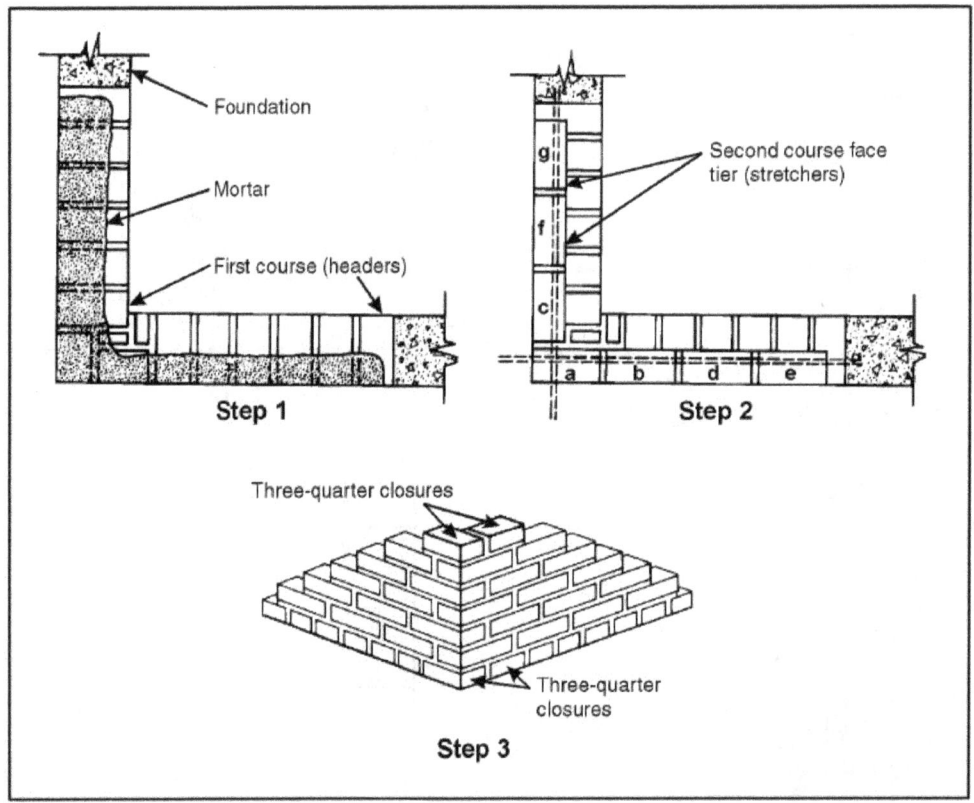

Figure 9-23. Laying second course of corner lead for 8-inch common-bond brick wall

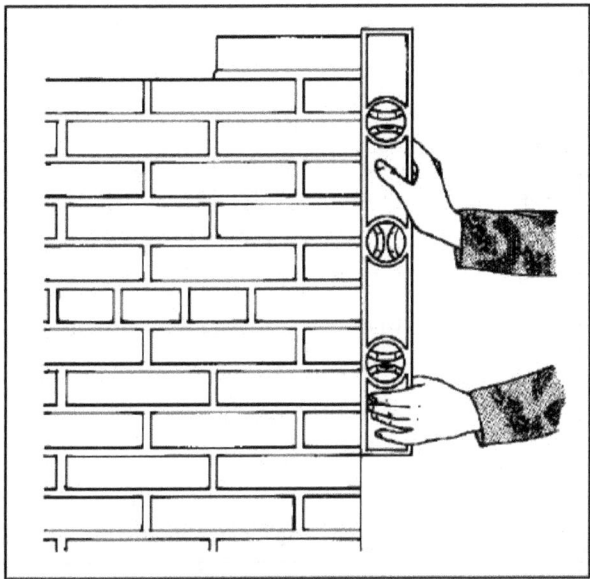

Figure 9-24. Plumbing a corner

Leveling

9-47. It is not good practice to move brick after it is laid in the mortar. Take great care to place the bricks accurately the first time. Be sure to finish or tool the joints before the mortar sets.

- Step 1. Lay the portion of the wall between the leads using the leads as a guide.
- Step 2. Check the level of the lead courses continuously.
- Step 3. Plumb the lead after laying the first few courses. If the masonry is not plumb, move the bricks either in or out until the lead is exactly plumb.

Opposite Corner Leads

9-48. Build the opposite corner lead the same way. Make sure that the tops of corresponding courses are the same level in each lead. For example, the top of the second course in one corner lead must be the same height above the foundation as the second course in the opposite corner. Mark a long 2-by 2-inch pole with the correct course heights above the foundation, and then use it to check the course height in the corner leads as you build them.

LAYING THE FACE TIER BETWEEN THE CORNER LEADS

9-49. Use a line, as shown in figure 9-25, page 9-30, to lay the face tier of brick for the wall between the leads.

- Step 1. Drive nails into the top of the cross joints.
- Step 2. Attach the line to the nail in the left-hand lead, pull it taut, and attach it to the nail in the right-hand lead. Position the line 1/16 inch outside the wall face, level with the top of the brick. It is better to use a tool called a line pin that resembles a triangular-shaped nail to attach the line at the right-hand or pull end. The line pin prevents the taut line from unwinding.

Chapter 9

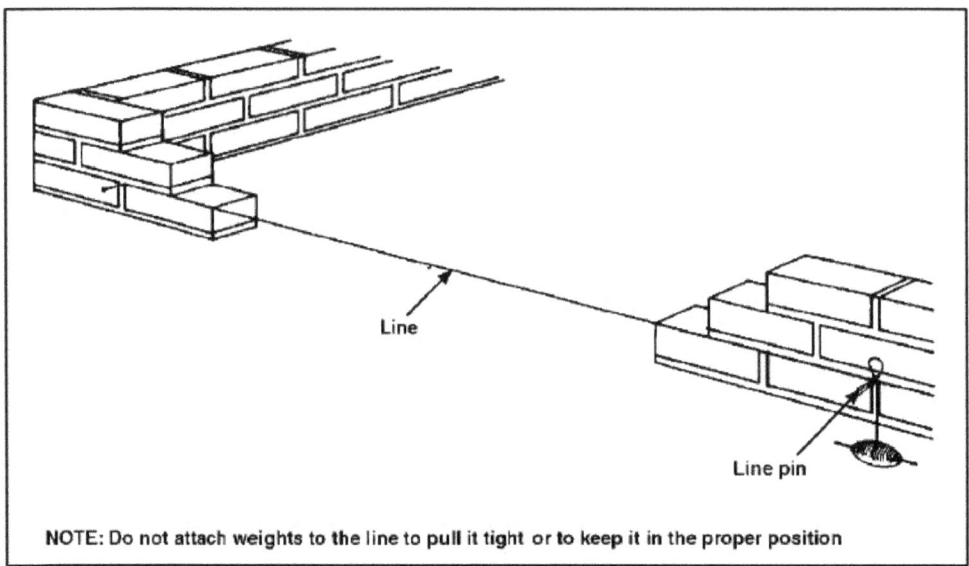

Figure 9-25. Using a line to lay face tier of brick between corner lead

Step 3. Lay the first or header course in between the two corner leads as described in paragraph 9-46, when the line is in place.

Step 4. Push the brick into position with its top edge 1/16 inch behind the line. Be sure not to crowd the line. If the corner leads are built accurately, the entire wall will be level and plumb. You need not use a mason's level continually when laying the wall between the leads, but check it occasionally at several points.

Step 5. Move the line to the top of the next mortar joint for the second or stretcher course.

Step 6. Lay the stretcher course as described in paragraph 9-46, and finish the face joints before the mortar hardens.

Step 7. Lay the face tier of the wall between the leads up to, but not including, the second header course (normally five stretcher courses). Then lay the backup tier.

Laying the Backup Tier Between the Corner Leads

9-50. Lay the backup brick for the corner leads first, as shown in figure 9-26, followed by the remaining brick. For an 8-inch wall, you do not need to use a line for the backup brick as you do in a 12-inch wall. After laying the backup tier up to the height of the second header course, lay the second header course in the face tier.

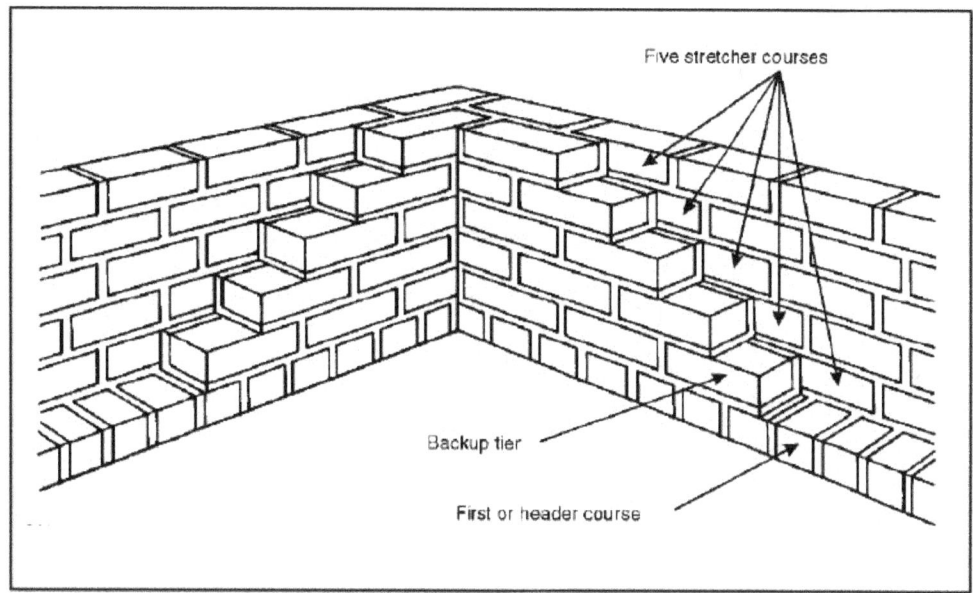

Figure 9-26 Laying backup brick for the corner lead of an 8 inch common bond brick wall

9-51. When the wall for the entire building is laid up to a height that includes the second header course, continue laying the corner leads up six more courses. Then construct the wall between the leads as described above. Repeat the entire procedure until the wall is laid to the required height.

LAYING A 12-INCH COMMON-BOND BRICK WALL

9-52. Figure 9-27, page 9-32, shows how to lay the first three courses of a 12-inch common bond-brick wall. Note that the construction is similar to that of an 8-inch wall, except that it includes a second tier of backup brick (see view 3 of figure 9-27).

- Step 1. Lay two overlapping header courses first (see view 1 of figure 9-27), and build the corner leads.
- Step 2. Lay the two tiers of backing brick, using a line for the inside tier.
- Step 3. Lay the second course as shown in view 2 of figure 9-27, and the third course as shown in view 3 of figure 9-27.

PROTECTING WORK INSIDE WALLS

9-53. Each night cover the tops of all work completed inside the brick walls to protect them from weather damage. Use the boards or tarpaulins secured by loose bricks.

Chapter 9

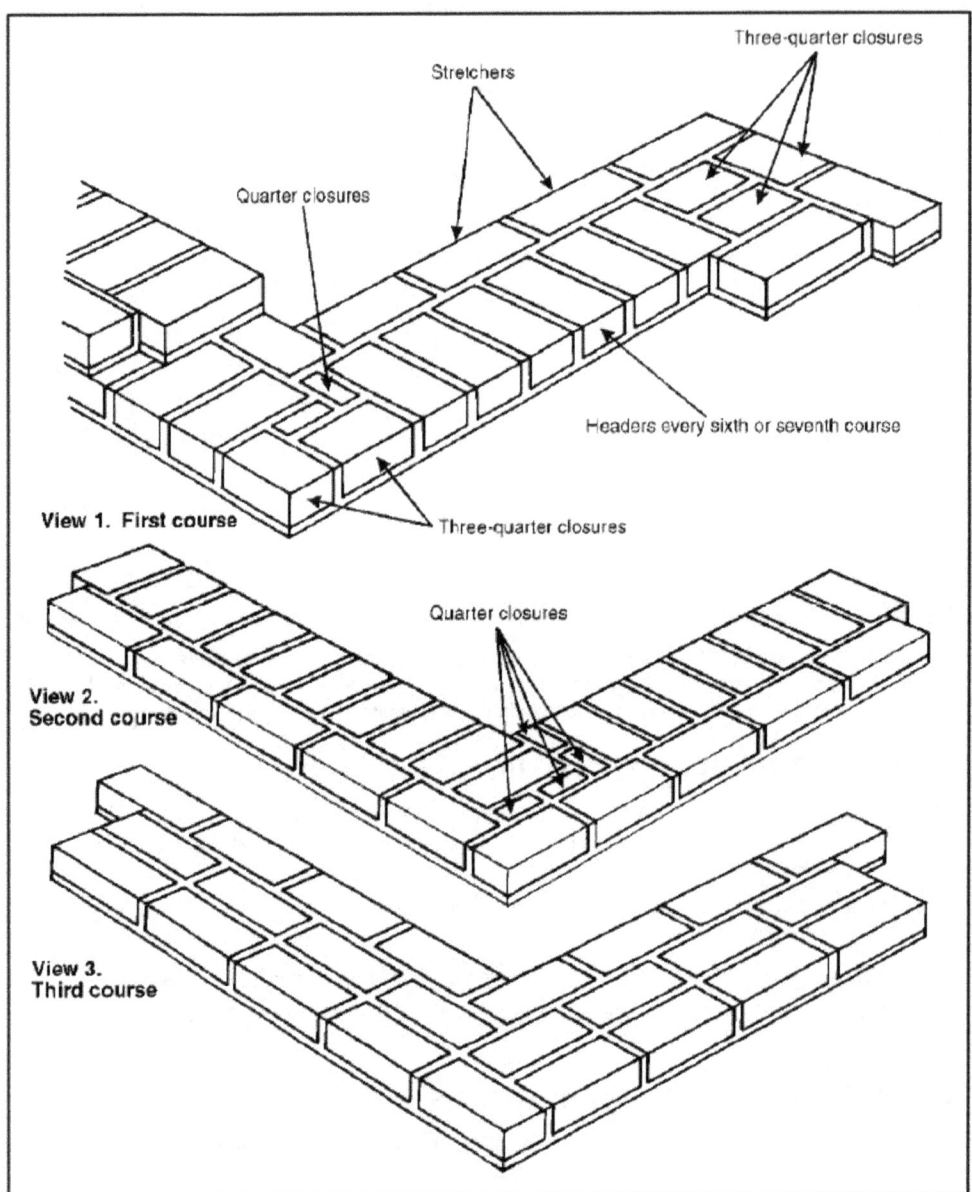

Figure 9-27. Laying a 12-inch common-bond brick wall

USING A TRIG

9-54. When building a long wall, erect a third lead between the corner leads. Then stretch a line from the left-hand lead to the middle lead to the right-hand lead. Now use a trig to keep the line from sagging or being windblown toward or away from the wall face. A trig (see figure 9-28) is a second short piece of line that loops around the main line and fastens to the top edge of a previously laid brick in the middle lead. A piece of broken brick rests on top of the trig to hold it in position.

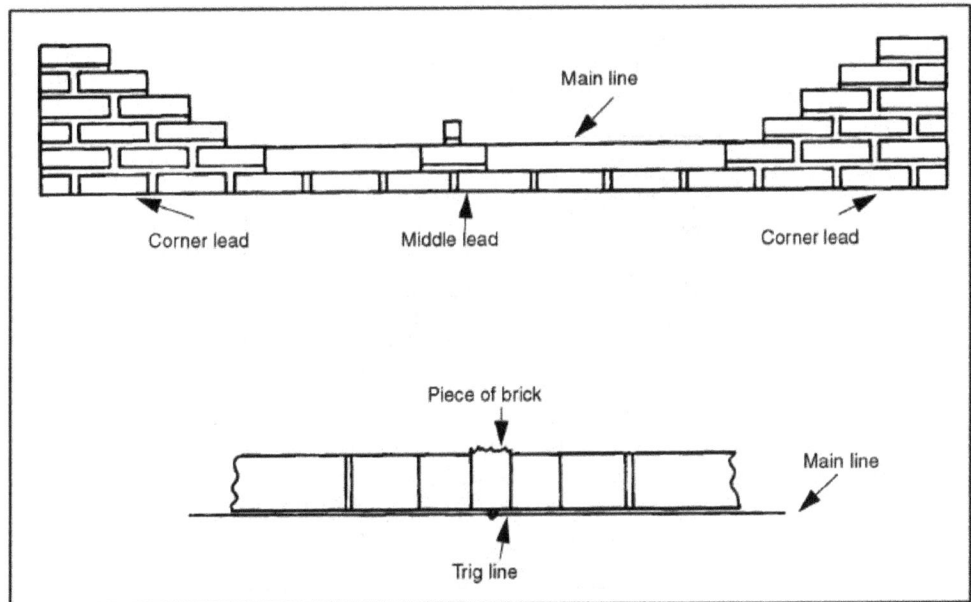

Figure 9-28. Use a trig to support the line when building a long wall

CONSTRUCTING WINDOW AND DOOR OPENINGS

9-55. You must plan ahead when laying any wall containing windows, making sure to leave openings of the correct size as the bricklaying proceeds.

WINDOW OPENINGS

9-56. Procedures to use when constructing window openings are as follows:

- Planning. First, find out the specified distance from the foundation to the bottom of the window sill. The height of the wall to the top of a full course must equal that distance. Now, calculate how many courses will bring the wall up to that height. For example, if the sill is 4 feet 4 1/4 inches above the foundation using 1/2-inch mortar joints, you must lay 19 courses before you reach the bottom of the sill. Calculate it this way: each brick plus one mortar joint equals 2 1/4 plus 1/2 equals 2 3/4 in per course, and 4 feet 4 1/4 inches divided by 2 3/4 inches equals 19 courses.
- Marking. Lay the corner leads and the wall in between them so that the top of the last course is not more than 1/4 inch above the top of the window frame. Use a pencil to mark the top of each course on the window frame itself. If the mark for the top of the last course does not come out to the proper level, change the joint thickness you plan to use until it does.
- Laying the brick. Lay the corner leads with joints of the calculated thickness. When the corner leads are built, install a line as described earlier and stretch it across the bottom of the window openings. Lay the brick for the wall between the leads up to sill height using the calculated joint thickness. If the window openings are planned properly, you can lay the face tier brick with a minimum of cutting.
- Laying the rowlock sill course. When the wall reaches sill height, lay the rowlock sill course as shown in figure 9-29, page 9-34. Pitch the course downward away from the window. The

rowlock normally takes up a vertical space equal to two courses of brick. Finish the exterior joint surfaces carefully to make them watertight.
- Placing the frame. As soon as the mortar sets, place the window frame on the rowlock sill course, bracing it firmly until the masonry reaches about ingathered of the way up the frame. (But do not remove the braces for several days so that the wall above the window frame sets properly). Now lay the rest of the wall around the frame until the top of the last course is not more than 1/4 inch above the window frame.

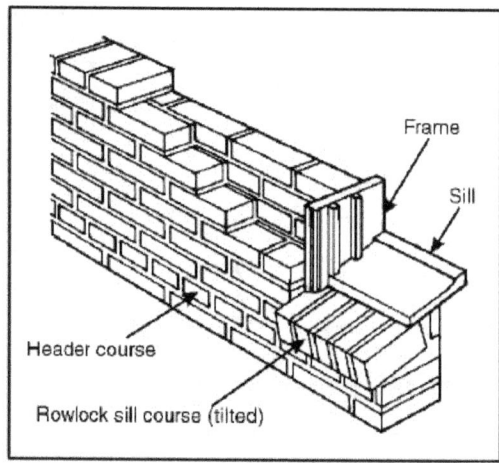

Figure 9-29. Constructing a window opening

Door Opening

9-57. Use the same procedure to construct a door opening (see figure 9-30) as for a window opening. To anchor the door frame to the masonry using screws or nails, cut pieces of wood to the size of a half closure and lay them in mortar the same as brick. Place the wood blocks at several points along the top and sides of the door opening.

Brick and Tile Masonry

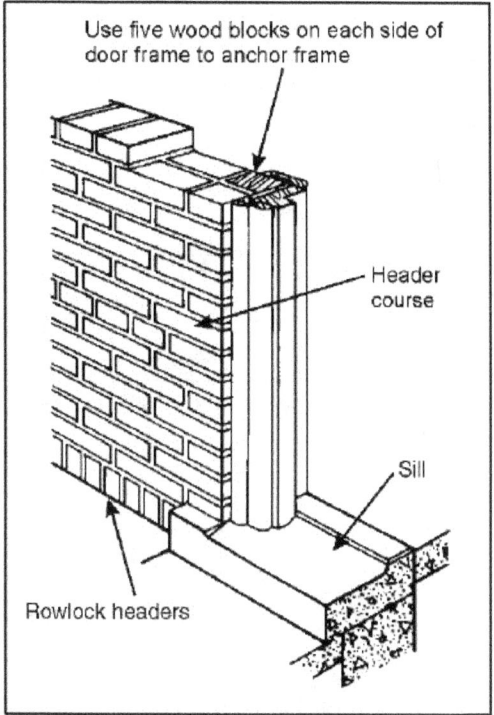

Figure 9-30. Constructing a door opening

LINTELS

9-58. The lintel above a window or door carries the weight of the wall above it. It rests on top of the last brick course that is almost level with the top of the window or door frame, and its sides bed firmly in mortar. Close any space between the window or door frame and the lintel with blocking, and weather strip it with bituminous materials. Then continue the wall above the window or door when the lintel is in place.

CONSTRUCTION

9-59. Lintels are made from steel, precast reinforced-concrete beams, or wood. Do not use wood lintels if possible. In reinforced brick masonry, properly installed steel reinforcing bars support the brick above wall openings.

INSTALLATION

9-60. The placement and relative positioning of lintels are determined by both the wall thickness and the type of window or door specified. This information is usually on the building drawings. If the lintel size is not specified, table 9-5, page 9-36, gives size and quantities of double-angle steel and wood lintels to use for various opening widths in both 8- and 12-inch walls. Figures 9-31 and 9-32, pages 9-36 and 9-37, show how to place different kinds of lintels in different wall thickness. Figure 9-31 shows how to install a doubleness steel lintel in an 8-inch wall. The angle is 1/4-inch thick, which allows the two angle legs that project up into the brick to fit exactly into the 1/2-inch joint between the face and backing ties.

Chapter 9

Table 9-5. Lintel sizes for 8-inch and 12-inch walls

Wall Thickness, in inches	Span				
	3 feet		4 Feet, Steel Angles*	5 Feet, Steel Angles*	6 Feet, Steel Angles*
	Steel Angles	Wood			
8	2-3 x 3 x ¼	2 x 8 2-2x4	2-3 x 3 x ¼	2-3 x 3 x ¼	2-3 ½ x 3 ½ x ¼
12	3-3 x 3 x ¼	2 2x 12 2-2 x 6	3-3 x 3 1/4	3-3 ½ x 3 ½ x ¼	3-3 ½ x 3 ½ x ¼
				7 Feet, Steel Angles*	8 Feet, Steel Angles*
				2-3 ½ x 3 ½ x ¼ 3-4 x 4 x ¼	2-3 ½ x 3 ½ x ¼ 3-4 x 4 x 4 ¼
* Wood lintels should not be used for spans over 3 feet since they will burn out in case of fire and allow the bricks to fall.					

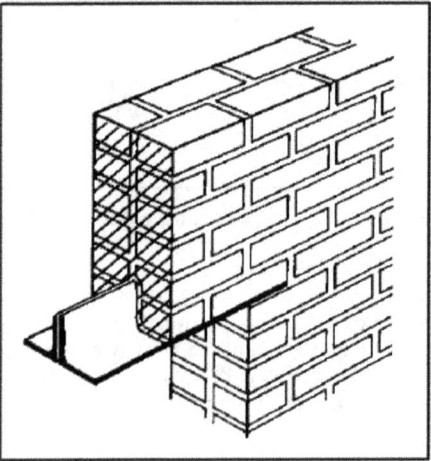

Figure 9-31. Installing a double-angle steel lintel in an 8 inch wall

Brick and Tile Masonry

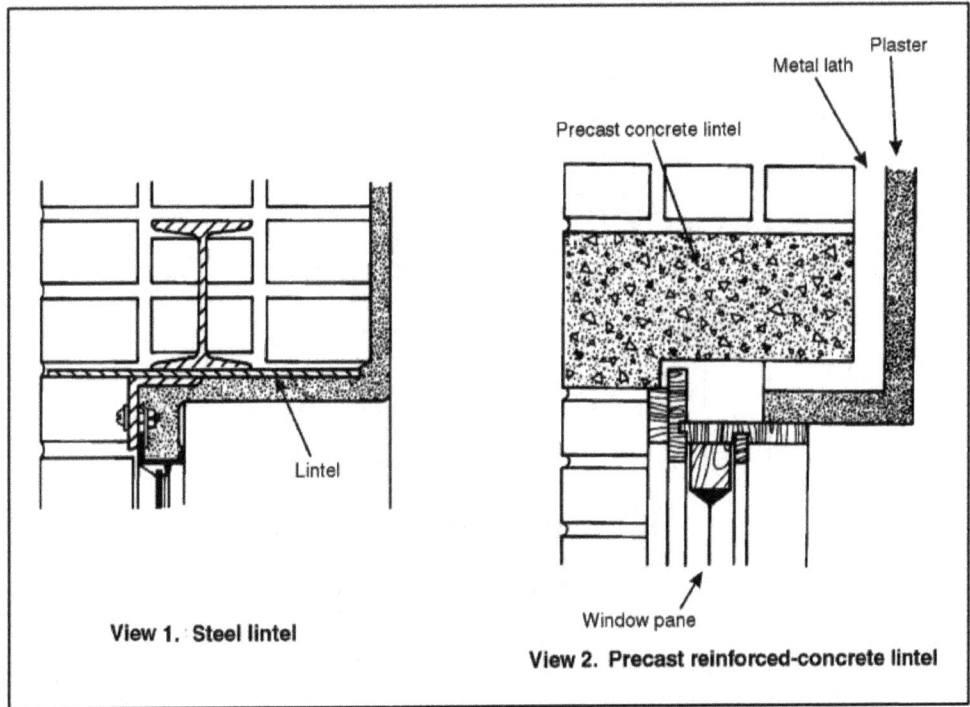

Figure 9-32. Installing lintels in a 12 inch wall

CORBELING

9-61. Corbeling consists of brick courses projecting beyond the wall face to increase its thickness or form a self-supporting shelf or ledge (see figure 9-33, page 9-38). The portion of a chimney exposed to weather is frequently corbeled to increase its thickness for better weather resistance. Corbeling usually requires various-sized bats (a broken brick with one end whole, the other end broken off). Use headers as much as possible, but the first projecting course can be a stretcher course if necessary. No course should extend more than 2 inches beyond the course underneath it, and the total corbel projection should not be greater than the wall thickness.

Chapter 9

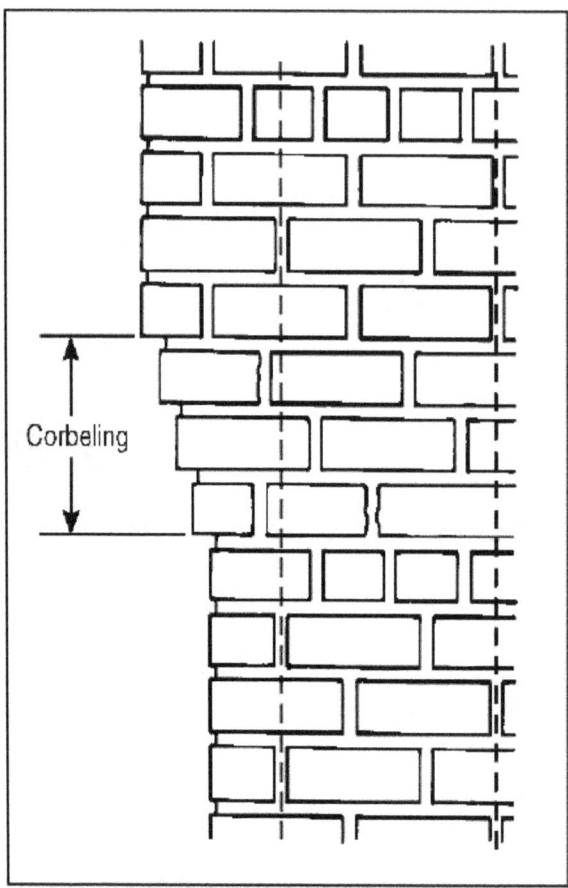

Figure 9-33. Constructing a corbelled brick wall

9-62. Corbel construction requires good workmanship for maximum strength. Make all mortar joints carefully and fill them completely with mortar. When the corbel must withstand large loads, consult a qualified engineer.

ARCHES

9-63. A well-constructed brick arch can support a heavy load, mainly due to its curved shape. Figure 9-34 shows two common arch shapes.

Brick and Tile Masonry

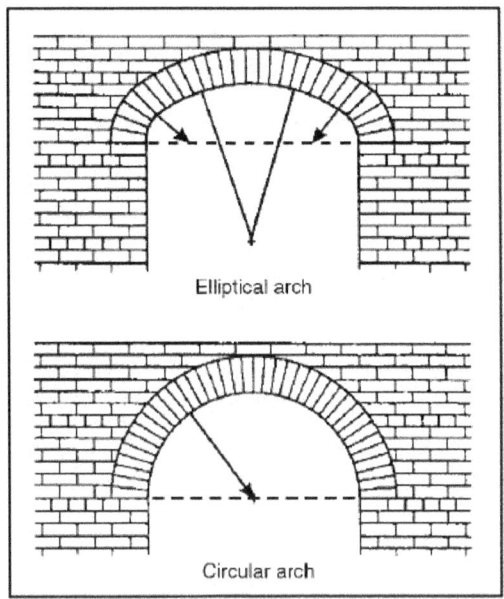

Figure 9-34. Common shapes

9-64. Brick arches require full mortar joints. Note that the joint width is narrower at the bottom of the arch than at its top, but it should not narrow to less than 1/4 inch at any point. As laying progresses, make sure that the arch does not bulge out of position.

BUILDING A TEMPLATE

9-65. Construct a brick arch over a temporary wood support called a template (see figure 9-35) that remains in place until the mortar sets.

9-66. You can obtain the template dimensions from the construction drawings. For arches spanning up to 6 feet, use 3/4-inch plywood to make the template. Cut two pieces to the proper curvature and nail them to 2 by 2 spaces that provide a surface wide enough to support the brick. Use wedges to hold the template in position until the mortar hardens enough to make the arch self-supporting. Then drive out the wedges.

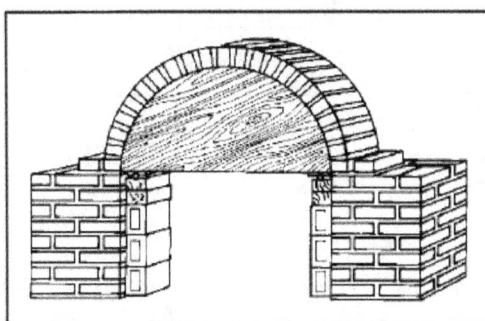

Figure 9-35. Using a template to construct an arch

Chapter 9

LAYING OUT THE ARCH

9-67. Lay out the arch carefully to avoid cutting the brick. Use an odd number of bricks so that the key or middle brick falls into place at the exact arch center or crown. The key or middle brick is the last one laid. To determine how many bricks an arch requires, lay the template on its side on level ground and set a trial number of bricks around the curve. Adjust the number of bricks and the joint spacing (not less than 1/4 inch) until the key brick is at the exact center of the curve. Mark the positions of the bricks on the template and use them as a guide when laying the brick.

WATERTIGHT WALLS

9-68. Water does not usually penetrate brick walls through the mortar or brick, but through cracks between the brick and the mortar. Cracks are more likely to occur in head joints than in bed joints.

PREVENTING CRACKS

9-69. Sometimes a poor bond between the brick and the mortar causes cracks to form and sometimes mortar shrinkage is responsible. Do not change the position of a brick after the mortar begins to set, because this destroys the bond between the brick and mortar and a crack will result. Wet high-suction bricks and other bricks as necessary during hot weather so that they do not absorb too much moisture from the mortar and cause it to shrink. You can reduce both the size and number of cracks between the mortar and the brick by tooling the exterior faces of all mortar joints to a concave finish. To obtain watertightness, completely fill all head and bed joints with mortar.

PARGING

9-70. Figure 9-36 shows a good way to produce a watertight wall called parging or back plastering. Parging means to plaster the back of the brick in the face tier with not less than 3/8 inch of rich cement mortar before laying the backing bricks. Because you cannot plaster over mortar protruding from the joints, first cut all joints flush with the back of the face tier.

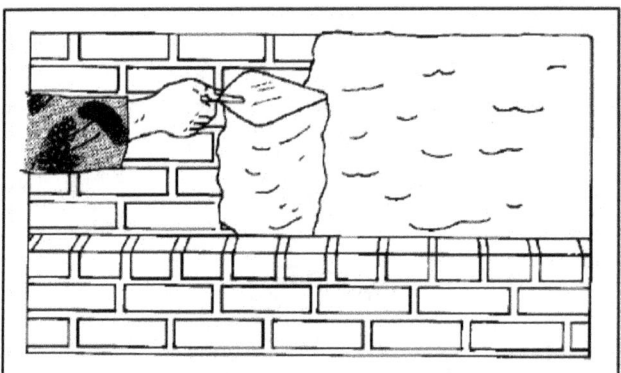

Figure 9-36. Parging the back of the face tier for watertightness

WATERPROOF WITH MEMBRANES

9-71. If a wall is subject to much water pressure, use a membrane to waterproof it. A properly installed membrane adjusts to any shrinkage or settlement without cracking.

CONSTRUCTING DRAINS

9-72. If the wall is subject to much groundwater, or the surrounding soil does not drain well, construct tile drains or French drains around the wall base (see figure 9-37). If drainage tile is not available, an 8-inch layer of coarse loose rock or stone will do the job. This is called a French drain.

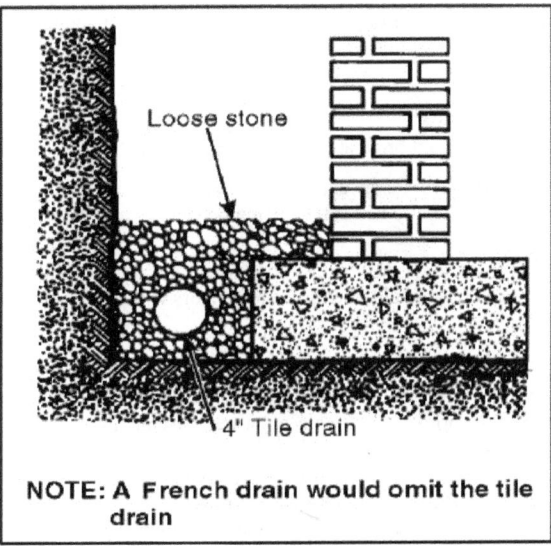

Figure 9-37. Draining a wall around its foundation

REPAIRING CRACKS

9-73. Before applying waterproof or portland-cement paints, repair all cracks. To repair mortar joint cracks, first chip out the mortar around the full width of the crack to a depth of about 2 inches. Carefully scrub the hole with clean water. While the hole surfaces are still wet, apply a coating of cement mortar made with enough water to form a thick liquid. Before the coating sets, fill the hole with prehydrated mortar, which is recommended for tuck-pointing in paragraph 9-91.

Note. Repair cracks in bricks the same way as cracks in concrete (see paragraphs 5-98 and 5-108).

USING WATERPROOF COATING

9-74. Walls can be waterproofed in three ways:

- Bituminous mastic. To make a below-grade foundation wall watertight, apply two coats of bituminous mastic to the exterior brick surface. You can apply asphalt or coal tar pitch using a mop.
- Waterproof paints. You can improve the watertightness of above-grade brick walls by applying a transparent, waterproof paint, such as a water solution of sodium silicate. Varnish, certain white and color waterproof paints, and high-quality oil-base paints are also effective. Apply them according to the directions.
- Portland-cement paint. This paint generally gives excellent results if you apply it when the wall is at least 30 days old. Use type 2, class A portland-cement paint and follow the manufacturer's instructions for mixing and applying it. Remove all efflorescence from the surface (see

paragraph 9-96) and dampen the surface with a water spray before applying the paint. Use whitewash or calcimine-type brushes or a spray gun, but spraying reduces the paint's rain resistance.

FIRE-RESISTANT BRICK

9-75. To line furnaces, incinerators, and so forth, use fire-resistant brick to protect the supporting structure or outer shell from intense heat. The outer shell probably consists of common brick or steel, neither of which has good heat resistance.

TYPES

9-76. The two types of fire-resistant brick are fire bricks and silica bricks.

- Fire bricks are made from a special clay, called fire clay, or that withstand higher temperatures and are heavier and usually larger than common brick. Their standard size is 9 by 4 1/2 by 2 1/2 inches.
- Silica bricks resist acid gases; however, do not use silica brick if it will be alternately heated and cooled. Therefore, you should line most incinerators with fire brick rather than with silica brick.

LAYING FIRE BRICK

9-77. Fire brick requires thin mortar joints, especially if the brick is subject to such high temperatures as those in incinerators. Store the bricks in a dry place until you use them.

MORTAR

9-78. Use a mortar made from fire clay and water mixed to the consistency of thick cream. Obtain fire clay by grinding used fire brick.

Laying procedure

9-79. To lay fire bricks, use the following procedures:

Step 1. Dip the brick in the mortar, covering all surfaces except the top bed.

Step 2. Lay the brick and tap it firmly into place with a bricklayer's hammer.

Step 3. Make the mortar joints as thin as possible, and fit the bricks together tightly. Remember that any heat that migrates through the cracks between the fire bricks will damage the outside shell of the incinerator or furnace.

Step 4. Stagger the head joints the same way you do in ordinary brick construction. The bricks in one course should lap those in the course underneath by one-half brick.

Laying Silica Brick

9-80. Lay silica bricks without mortar, fitting them so close together that they fuse at the joints at high temperatures. Stagger the head joints as you do in ordinary brick construction.

TYPES OF WALLS

9-81. The basic types of walls are: hollow and partition.

HOLLOW

9-82. Hollow walls consist of an inner and an outer wythe separated by an air space. The two most important types of hollow walls are the cavity wall and the rowlock wall. Partition walls divide the interior space in a one-story building. They may be load-bearing or nonload-bearing walls.

Cavity

9-83. A *cavity* wall is a watertight wall that can be plastered without furring or lathing. It looks the same on the exterior as a solid wall without header courses (see figure 9-38). Instead of headers, metal ties are installed every sixth course on a 24-inch center that holds the two tiers together. To prevent water penetration to the inner tier, angle the ties downward from the inner to outer tier. A 2-inch cavity or air space between the two brick wythes drains any water that penetrates the outer tier. The air space also provides good heat and sound insulation. The bottom of the cavity is above ground level. It is drained by weep holes in the vertical joints in the first course of the exterior tier. Make the weep holes simply by leaving the mortar out of the joint. Space them about a 24-inch interval.

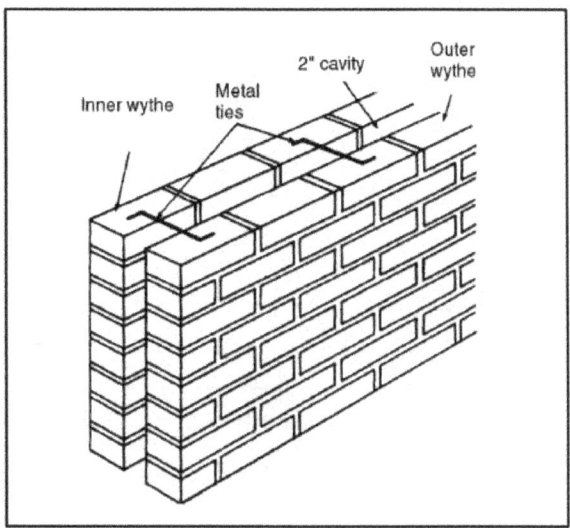

Figure 9-38. Construction details of a cavity wall

Rowlock

9-84. A *rowlock* is a header laid on its face or edge. A rowlock also has a 2-inch cavity between the wythes as shown in figure 9-39 page 9-44. In this type of rowlock wall, the face tier is loose like a common-bond wall having a full header course every seventh course. However, the bricks in the inner or backing tier are laid on edge. A header course ties the outer and backing ties together. For an all-rowlock wall, lay the brick on edge in both the inner and outer tiers. Install a header course every fourth course (that is, three rowlock courses to every header course). The rowlock wall is not as watertight as the cavity wall, because water will follow along any crack in the header course and pass through to the interior surface.

Chapter 9

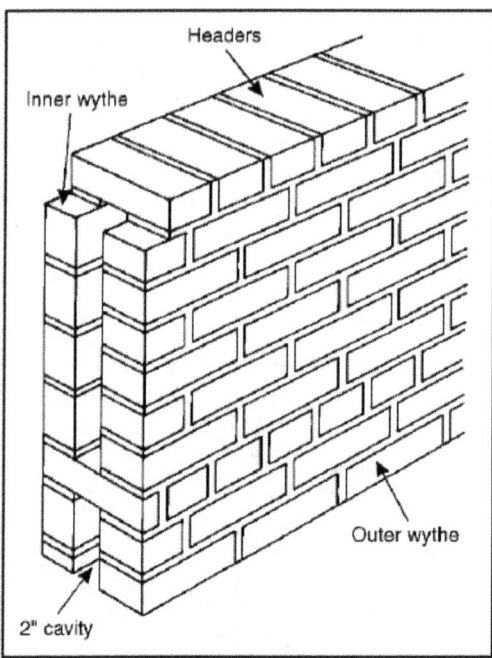

Figure 9-39. Construction details of a rowlock wall

PARTITION

9-85. A partition wall that carries very little load requires only one wythe producing a wall 4 inches thick. You can lay a wall of this thickness without headers.

LAYING HOLLOW AND PARTITION WALLS

9-86. Lay the brick for hollow and partition walls as described starting in paragraph 9-28 for making bed joints (f), head joints (h), cross joints (j), and closures (K and l). Use a line the same as for a common bond wall. Erect the corner leads first, and then build the wall between them.

MANHOLES

9-87. Sewer systems require manholes (see figure 9-40) for cleaning and inspection. The manhole size largely depends on the sewer size. Manholes are either circular or oval to reduce the stresses from both water and soil pressures. A 4-foot diameter manhole is satisfactory for small, straight-line sewers. Construction details of a typical manhole are shown in view 2 of figure 9-40. Although both the bottom and walls of a manhole are sometimes made from brick, the bottom is normally made from concrete because it is easier to cast in the required shapes. However, you can construct the walls more economically from brick, because it requires no form work.

Brick and Tile Masonry

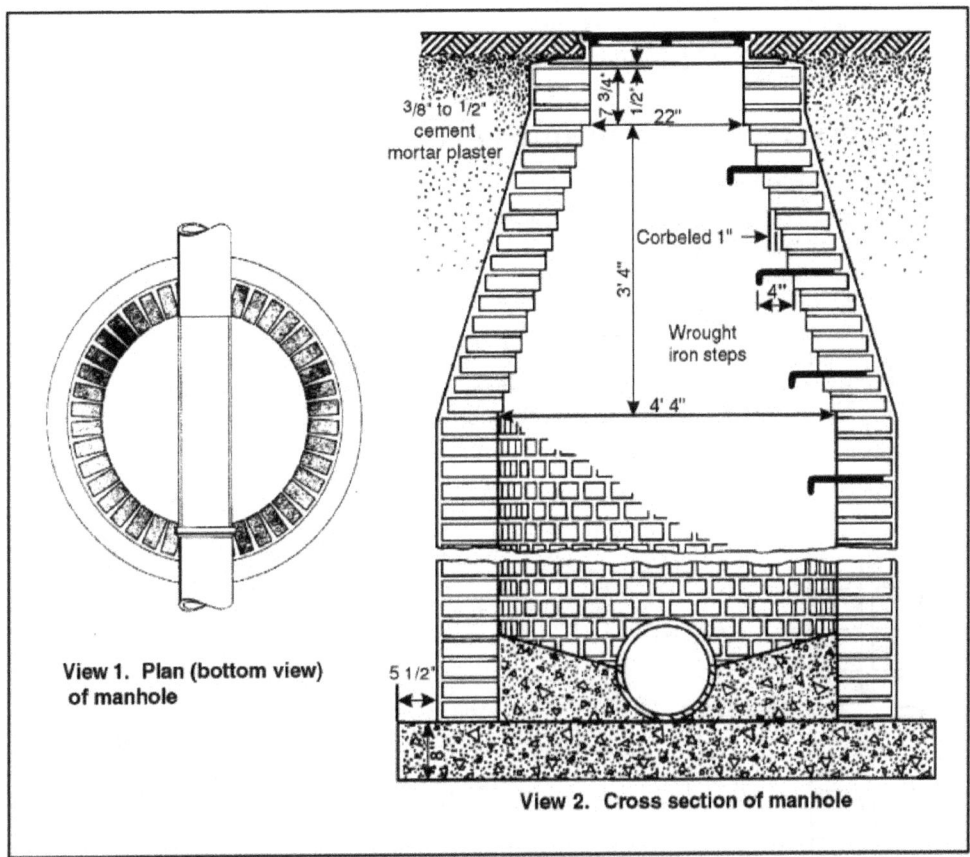

Figure 9-40. Construction details of a sewer manhole

WALL THICKNESS

9-88. The wall thickness of a manhole depends on its depth and diameter. You can use an 8-inch wall for manholes up to 8 feet in diameter and less than 15 feet deep. A qualified engineer should design any manholes over 15-feet deep.

MANHOLE CONSTRUCTION

9-89. Use only headers for an 8-inch wall, but no line. Use a mason's level to make sure that all bricks in a particular course are level. Span the manhole with a straightedge or place a mason's level on a straight surfaced 2 by 4 across the manhole, to make sure the brick rises to the same level all around. Because the wall appearance is not important, some irregularities in both brick position and mortar joint thickness are permissible. All joints should be either full or closure joints.

- Laying the first course. Place a 1-inch mortar bed on the foundation. Lay the first course on the mortar bed, followed by the succeeding header courses.
- Corbeling. To reduce the manhole diameter to fit the frame and cover, corbel the brick inward as shown in view 2 of figure 9-40. No brick should project more than 2 inches beyond the brick underneath it. Space the wrought-iron steps about every 15 inch vertically, and embed them in a

cross mortar joint. When complete, plaster the wall on the outside at least 3/8-inch thick with the same mortar used in laying the brick.
- Placing. Spread a 1-inch mortar bed on top of the last course, and place the base of the manhole frame in the bed.

SUPPORTING BEAMS ON A BRICK WALL

9-90. The following beams are used on brick walls.
- Wood Beams. Figure 9-41 shows how to support a wood beam on a brick wall. Note the wall tie. Keep mortar away from the beam as much as possible, because wood can dry rot when completely encased in mortar. Protect the beam with a beam box (see figure 9-41). Cut the end of the beam at an angle so that, in case of fire, it will fall without damaging the wall above the beam. For an 8- or 12-inch wall, let the beam bear on the full width of the inside tier.
- Steel Beams. When a brick wall must support a steel beam, insert a steel bearing plate set in mortar under the beam. A properly designed bearing plate prevents the beam from crushing the brick. The size of the bearing plate depends on the size of the beam and the load it carries.

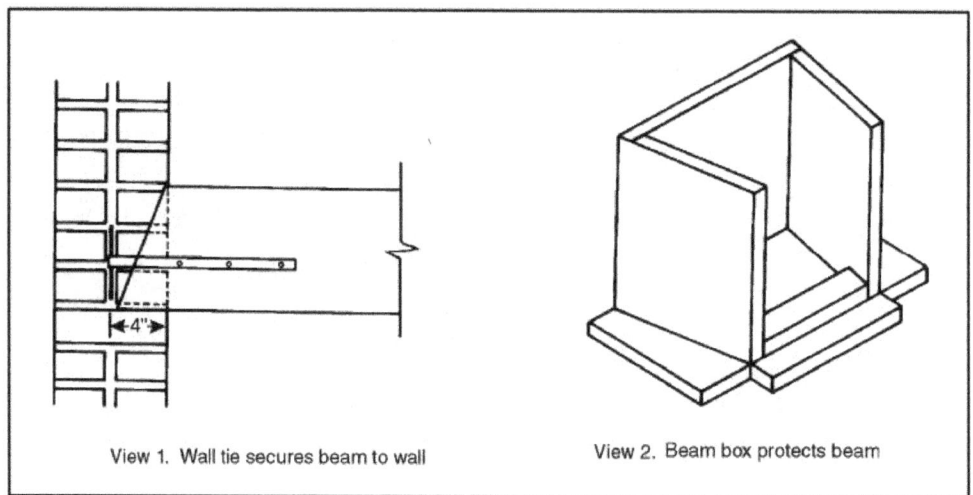

Figure 9-41. Supporting a wood beam on a brick wall

MAINTAINING AND REPAIRING BRICK WALLS

9-91. A well-constructed brick masonry wall requires little maintenance or repair. It can be more expensive to repair old masonry properly than to completely remove and replace just the disintegrated portion. Good mortar, proper joint finishing, and adequate flashing add little to the initial cost, but reduce maintenance cost throughout the life of the masonry.

TUCK-POINTING

9-92. Tuck-pointing during routine maintenance means to cut out all loose and disintegrated mortar to a depth of at least 1/2 inch and replace it with new mortar. Use a chisel having a cutting edge about 1/2 inch wide. To stop leakage, cut out all the mortar in the affected area, and replace it with new mortar.

PREPPING THE MORTAR JOINT

9-93. After cutting out the defective mortar, remove all dust and loose material with a brush or a water jet. If you use a water jet, no further joint wetting is required. If not, moisten the joint surfaces.

PREPARING MORTAR FOR TUCK-POINTING

9-94. Use portland-cement-lime prehydrated Type S mortar or prehydrated prepared mortar made from type II masonry cement. Prehydrating mortar greatly reduces the amount of shrinkage. Mix the dry ingredients with just enough water to produce a damp mass of a consistency that retains its form when you compress it into a ball with your hands. Allow the mortar to stand for at least 1 hour, but not more than 2 hours. Then mix the mortar with enough water to produce a stiff, but workable consistency.

FILLING THE JOINT

9-95. Filling a joint with mortar is called *repointing* and is done with a pointing trowel. Before filling the joint, allow the moisture used in preparing the joint to absorb. Then pack the prepared prehydrated mortar into the joint tightly in thin layers about 1/4 inch thick, and finish them to a smooth concave surface using a pointing tool. Push the mortar into the joint in one direction only from the starting point, using a forward motion to reduce the risk of forming air pockets.

CLEANING NEW BRICK AND REMOVING STAINS

9-96. A skilled bricklayer can build a masonry wall that is almost free from mortar stains. However, most new brick walls still need some cleaning.

- **Step 1.** Remove large mortar particles adhering to brick with a putty knife or chisel. Remove mortar stains with an acid solution of one part commercial muriatic acid to nine parts water. Acid must be poured into the water, not the water into the acid. Before applying the acid, thoroughly soak the masonry surface with water to prevent the stain from being drawn into the brick pores.
- **Step 2.** Apply the acid solution with a long-handled stiff-fiber brush. Take all precautions to prevent the acid from getting on your hands, arms, or clothing, and wear goggles to protect your eyes. Protect door and window frames. Scrub an area of 15 to 20 square feet with the acid solution, and then wash it down immediately with clear water. Make sure that you remove all acid before it can attack the mortar joints.
- **Step 3.** Removing any efflorescence. Efflorescence is a white deposit that forms on the surface of brick walls. It consists of soluble salts leached from the brick by penetrating water that dissolves the salt in the brick. When the water evaporates, the salt remain. Because efflorescence requires the presence of both water and salts, proper brick selection and a dry wall will keep it to a minimum. However, if simply scrubbing the wall with water and a stiff brush does not remove the efflorescence, you can remove it with the acid solution described above for cleaning new masonry.

CLEANING OLD BRICK

9-97. The principal ways to clean old brick masonry are sandblasting, steam cleaning with water jets, or using cleaning compounds. The type of brick and the nature of the stain will determine which method you use. Many cleaning compounds that do not affect the brick will damage the mortar. Rough-textured brick is more difficult to clean than smooth-textured brick.

9-98. Sometimes you cannot clean rough-textured brick without removing part of the brick itself, which changes the appearance of the wall.

Chapter 9

SANDBLASTING

9-99. Sandblasting consists of using compressed air to blow hard sand through a nozzle against a dirty surface, thereby removing enough of the surface to eliminate the stain. Place a canvas screen around the scaffold to salvage most of the sand. The disadvantage of sandblasting is that it leaves a rough-textured surface that collects soot and dust. Moreover, sandblasting usually cuts so deeply into the mortar joints that you may have to repoint them. After sandblasting, apply a transparent waterproofing paint to the surface to help prevent future soiling by soot and dust. Never sandblast glazed surfaces.

STEAM CLEANING WITH WATER JETS

9-100. Steam cleaning means to project a finely divided spray of steam and water at high velocity against a dirty surface. This removes grime effectively without changing the surface texture, which gives steam cleaning an advantage over sandblasting.

- Equipment. Use a portable, truck-mounted boiler to produce the steam at a pressure ranging from 140 to 150 psi. Each cleaning nozzle requires about a 12-horse-power boiler. The velocity of the steam and water spray as it strikes the surface is more important than the volume of the spray.
- Procedure. Use one garden hose to carry water to the cleaning nozzle and another to supply rinse water. Experiment to determine the best angle and distance from the wall to hold the nozzle. Adjust the steam and water valves until you obtain the most effective spray. Pass the nozzle back and forth over no more than a 3-square-feet area at one time. Rinse it immediately with clean water before moving to the next area.
- Additives. To aid cleaning action, add sodium carbonate, sodium bicarbonate, or trisodium phosphate to the water entering the nozzle. Reduce a lot of the salt or efflorescence remaining on the surface by washing it down with water before and after steam cleaning.
- Hand tools. Use steel scrapers or wire brushes to remove any hardened deposits that remain after steam cleaning. Be careful not to cut into the surface. After removing the deposits, wash down the surface with water and steam and clean it again.

CLEANING COMPOUNDS

9-101. You can use one of several cleaning compounds, depending on the nature of the stain. Most cleaning compounds contain salts that will cause efflorescence if the cleaning solution penetrates the surface. You can prevent this by thoroughly wetting the surface before applying the solution. You can remove whitewash, calcimine, or paint coatings with a solution of one part acid to five parts water. Use fiber brushes to scrub the surface with the solution while it is still foaming. After removing the coating, wash down the wall with clean water until you remove the acid completely.

PAINT REMOVERS

9-102. Apply paint remover with a brush to remove oil paint, enamels, varnishes, shellacs, or glue sizing. Leave the remover on until the coating is soft enough to scrape off with a putty knife. The following are effective paint removers:

- Commercial. When using commercial paint removers, the manufacturer's instructions should be followed.
- Chemical. Use a solution of 2 pounds of trisodium phosphate in 1 gallon of hot water. Another solution is 1 1/2 pounds of caustic soda in 1 gallon of hot water.
- Blasting and torching. Sandblasting or burning off with a blowtorch will also remove paint.

REMOVING MISCELLANEOUS STAINS

9-103. The following are procedures for removing different types of stains:

- Iron stains. Mix seven parts lime-free glycerin into a solution of one part sodium citrate in six parts lukewarm water. Add whiting or kieselguhr to make a thick paste and apply it to the stain with a trowel. Scrape off the paste when it dries. Repeat the procedure until the stain disappears; then, wash down the surface with water.
- Tobacco stains. Dissolve 2 pounds of trisodium phosphate in 5 quarts of water. Next, in an enameled pan, mix 12 ounces of chloride of lime in enough water to make a smooth thick paste. Then mix the trisodium phosphate with the lime paste in a 2-gallon stoneware jar. When the lime settles, draw off the clear liquid and dilute it with an equal amount of water. Make a stiff paste by mixing the clear liquid with powdered talc, and apply it to the stain with a trowel followed by washing the surface.
- Smoke stains. Apply a smooth, stiff paste made from trichlorethylene and powdered talc. Cover the container when you are finished to prevent evaporation. If a slight stain still remains after several applications, wash down the surface and then follow the procedure described above for removing tobacco stains. Use the paste only in a well-ventilated space because its fumes are harmful.
- Copper and bronze stains. Mix one part ammonium chloride (salammoniac) in dry form to four parts powdered talc. Add ammonia water and stir the solution to obtain a thick paste. Apply the paste to the stain with a trowel, and allow it to dry. Several applications may be necessary. Then wash down the surface with clear water.
- Oil stains. Make a solution of 1 gallon trisodium phosphate to 1 gallon of water, adding enough whiting to form a paste. Trowel the paste over the stain in a layer 1 1/2 inch thick, and allow it to dry for 24 hours. Remove the paste and wash down the surface with clean water.

FLASHING

9-104. Flashing is an impermeable membrane placed in brick masonry at certain locations to exclude water or to collect any moisture that penetrates the masonry and direct it to the wall exterior. Flashing can be made from copper, lead, aluminum, or bituminous roofing paper. Copper is best, although it stains the masonry as it weathers. Use lead-coated copper if such staining is unacceptable. Bituminous roofing papers are cheapest, but not as durable. They will probably require periodic replacement in permanent construction, and their replacement cost is greater than the initial cost of installing high-quality flashing. Corrugated copper flashing sheets produce a good bond with the mortar. They also make interlocking watertight joints at points of overlap.

PLACEMENT

9-105. Install flashing at both the head and sill of window openings and at the intersection between a wall and roof. The flashing edges should turn upward as shown in figure 9-41, page 9-46, to prevent drainage into the wall. Always install flashing in mortar joints. You can provide drainage for the wall above the flashing either by placing 1/4-inch cotton-rope drainage wicks at 18-inch spacings in the mortar joint just above the flashing or placing dowels in the proper mortar joint as you lay the brick and then remove them to make drainage holes. Where chimneys pass through the roof, the flashing should extend completely through the chimney wall and turn upwards 1 inch against the flue lining.

INSTALLATION

9-106. Using the following steps to install flashing:

Step 1. Spread a 1/2-inch mortar bed on top of the brick, and then push the flashing sheet down firmly into the mortar. Spread a 1/2-inch mortar bed on the flashing, and then force the brick or sill onto the top of the flashing.

Step 2. Figure 9-42, page 9-50, shows the proper flashing installations at both the head and the sill of a window. Note that the flashing fits under the face tier of brick at the steel lintel, then bends behind the face tier and over the top of the lintel.

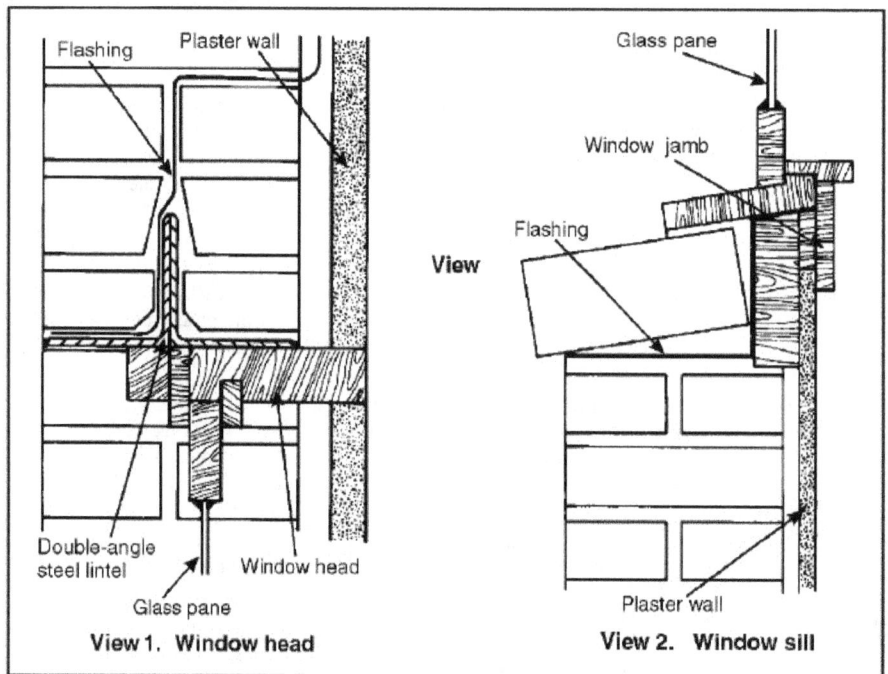

Figure 9-42. Installing flashing at window opening

Step 3. Figure 9-43 shows how to install flashing between the roof and the wall to prevent leakage at the intersection. Fit and caulk the upper end of the flashing into the groove of the raggle block as shown in figure 9-43.

FREEZE PROTECTION DURING CONSTRUCTION

9-107. Masonry walls built during cold weather may leak, because either the mortar froze before it set or the materials and walls were not adequately protected against freezing temperatures. During cold weather, prevent future wall leakage by—

- Storing materials properly.
- Heating mortar ingredients.
- Heating masonry units.
- Taking special precautions during placement.
- Protecting completed work.

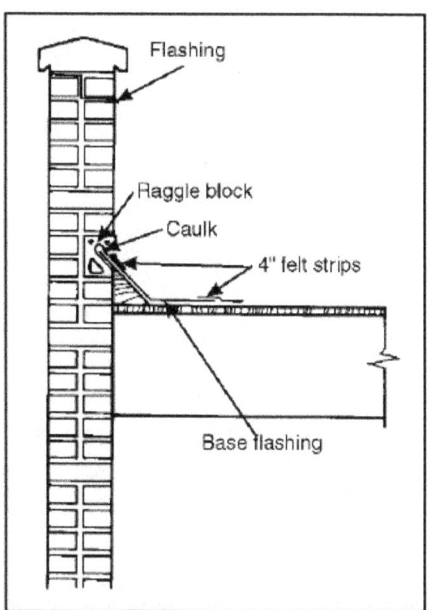

Figure 9-43. Flashing installation at intersection of roof and walls

STORING MATERIALS

9-108. Careless materials storage can cause time delays and/or contribute to poor workmanship, because you must remove all ice and snow and thaw masonry units before construction can proceed. Instead, completely cover all masonry units and mortar materials with tarpaulins or building paper. Store them on plank platforms either thick enough or raised high enough to prevent moisture absorption from the ground.

HEATING MORTAR INGREDIENTS

9-109. Heat both water and sand to a temperature not exceeding 160°F. Make sure that the temperature of the mortar—when you use it—is at least 70°F but not more than 120°F. Use steel mortar boxes on small jobs, and raise them about 1 foot above the ground so that you can supply heat to keep the mortar warm after mixing. Never add salt water to mortar to lower its freezing point.

CONSIDERING TEMPERATURE VARIATIONS

9-110. If the outside air temperature is below 40°F, the brick temperature when you lay it should be above 40°F on both sides of the masonry for at least 48 hours for Type M or S mortar, or for at least 72 hours for Type N mortar. If you use high-early-strength cement, reduce these time periods to 24 and 48 hours, respectively. Note that the use of high-early-strength cement in a mortar does not alter the setting rate much, but it does increase the rate of strength gain, thereby providing greater resistance to further freeze damage.

HEATING MASONRY UNITS

9-111. To prevent the warm mortar from cooling suddenly as it contacts the cold bricks, preheat all masonry units to about 40°F whenever the outside temperature is below 18°F. This requires careful planning and timing. When heat is required, provide inside brick storage so that you can supply heat at minimum expense.

Chapter 9

LAYING PRECAUTIONS

9-112. In below-freezing weather, sprinkle any high-suction brick with warm water just before you lay it. Never lay masonry units on snow- or ice-covered mortar beds, because little or no bond will exist between the mortar and units when the base thaws. Keep the tops of unfinished walls carefully covered whenever work stops. If the covering comes off and ice or snow collects on the wall top, remove it with live steam before continuing.

PROTECTING COMPLETED WORK

9-113. How you protect masonry from freezing varies with weather conditions and the individual job. Job layout, desired rate of construction, and the prevailing weather conditions all determine the amount of protection and the type of heat necessary to maintain above-freezing temperatures within the wall until the mortar sets properly.

MATERIAL QUANTITIES REQUIRED

9-114. See table 9-6 for the quantities of brick and mortar required for various masonry wall thickness.

SECTION IV - REINFORCED BRICK MASONRY

APPLICATIONS AND MATERIALS

9-115. When added strength is needed, brick walls, columns, beams, and foundations are reinforced the same as in concrete construction.

APPLICATIONS

9-116. Because brick masonry in tension has low strength compared with its compressive strength, it is reinforced with steel when subject to tensile stresses. Like concrete construction, the reinforcing steel is placed in either the horizontal or vertical mortar joints of beams, columns, walls, and footings. Reinforced brick structures can resist earthquakes that would severely damage nonreinforced brick structures. The design of such structures by qualified engineers is similar to that of reinforced concrete structures.

MATERIALS

9-117. Materials used for reinforced brick are—

- Brick. The same brick is used for reinforced brick masonry as for ordinary brick masonry. However, it should have a compressive strength of at least 2,500 pounds per square inch.
- Steel. The reinforcing steel is the same as that in reinforced concrete and is fabricated and stored the same way. Do not use high-grade steel except in emergencies, because masonry construction requires many sharp bends.
- Mortar. Use Type N mortar for its high strength.
- Wire. Use 18-gage, soft-annealed iron wire to tie the reinforcing steel.

Table 9-6. Quantities of materials required for brick walls

Wall Area in, Square Feet	Wall Thickness, in Inches							
	4		8		12		16	
	Number of Bricks	Mortar, in Cubic Feet	Number of Bricks	Mortar, in Cubic Feet	Number of Bricks	Mortar, in Cubic Feet	Number of Bricks	Mortar, in Cubic Feet
1	6.17	.08	12.33	.2	18.49	.32	24.65	.44
10	61.7	.8	123.3	2	184.9	3.2	246.5	4.4
100	617	8	1,233	20	1,849	32	2,465	44
200	1,234	16	2,466	40	3,698	64	4,930	88
300	1,851	24	3,699	60	5,547	96	7,395	132
400	2,468	32	4,932	80	7,396	128	9,860	176
500	3,085	40	6,165	100	9,245	160	12,325	220
600	3,702	48	7,398	120	11,094	192	14,790	264
700	4,319	56	8,631	140	12,943	224	17,253	308
800	4,936	64	9,864	160	14,792	256	19,720	352
900	5,553	72	10,970	180	16,641	288	22,185	396
1,000	6,170	80	12,330	200	18,490	320	24,650	440

Note. Quantities are based on a 1/2-inch-thick mortar joint; a 3/8-inch-thick joint uses 80 percent of these quantities; a 5/8-inch-thick joint uses 120 percent of these quantities.

CONSTRUCTION METHODS

9-118. Reinforcing steel can be placed in both horizontal and vertical mortar joints.

LAYING REINFORCED BRICK

9-119. Lay the brick the same way as ordinary brick masonry, with mortar joints 1/8 inch thicker than the diameter of the reinforcing bar. This provides 1/16 inch of mortar between the brick surface and the bar. Thus, large steel bars require mortar joints thicker than 1/2 inch.

PLACING THE STEEL

9-120. Embed all reinforcing steel firmly in mortar.

- Horizontal bars. Lay horizontal bars in the mortar bed, and then push them down into position. Spread more mortar on top of the bars, smooth it out until you produce a second bed joint of the proper thickness. Lay the next course in this mortar bed, following the same procedure as laying brick without reinforcing steel.
- Stirrups. Most stirrups are Z-shaped as shown in figure 9-44, page 9-54, to fit the mortar joints. Insert the lower leg under—and contacting—the horizontal bars, which requires a thicker joint at that point.
- Vertical bars. Hold vertical bars in the vertical mortar joints with wood templates having holes drilled at the proper bar spacing, or by wiring them to horizontal bars. Then lay the brick up around the vertical bars.
- Spacing. The minimum center-to-center spacing between parallel bars is 1-1/2 times the bar diameter.

Chapter 9

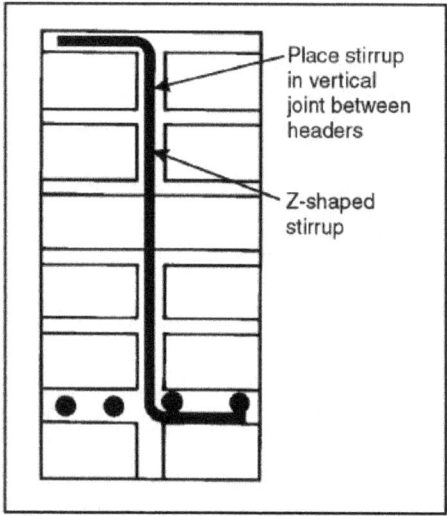

Figure 9-44. Reinforced brick-masonry beam construction

USING FORMWORK

9-121. Reinforced brick-masonry walls, columns, and footings need no formwork. However, reinforced brick beams and lintels—which act as beams—require formwork for the same reason that reinforced concrete beams do.

- The form consists only of support for the underside of the beam, no side formwork. The form for the beam underside is the same and is supported the same as that for concrete beams.
- When a beam joins a wall or another beam, cut the form short by 1/4 inch, and fill the gap with mortar. This allows the lumber to swell and makes form removal easy.
- Wait at least 10 days before removing the form from the beam underside.

BEAMS AND LINTEL CONSTRUCTION

9-122. Lintels are placed above the windows and doors to carry the weight of the wall above them. Lintels can be made of steel, precast reinforced concrete beams, or wood.

BEAM DIMENSIONS

9-123. Beam width and depth depend upon brick dimensions, mortar joint thickness, and the load that the beam will support. However, beam width usually equals the wall thickness, that is 4, 8, 12, or 16 inches. Beam depth should not exceed approximately three times its width.

BEAM CONSTRUCTION PROCEDURES

9-124. The following procedures will be used in beam construction:

Step 1. Lay the first course on the form using full head joints, but no bed joint (see figure 9-44).

Step 2. Spread a mortar bed about 1/8 inch thicker than the diameter of the horizontal reinforcing bars on the first course, and embed the bars.

Step 3. Slip the legs of any stirrups under the horizontal bars as shown in figure 9-44. Be sure to center the stirrup in the vertical mortar joint.

Step 4. After properly positioning the stirrups and the horizontal bars, spread more mortar on the bed joints if necessary, and smooth its surface. Then lay the remaining courses in the normal way.

Step 5. Lay all bricks for one course before proceeding to the next course to ensure a continuous bond between the mortar and steel bars. Often, three or four bricklayers must work on one beam to spread the bed-joint mortar for the entire course, place the reinforcing steel, and lay the brick before the mortar sets.

LINTEL CONSTRUCTION

9-125. The steel bars should be 3/8 inch in diameter, or less if you must maintain a 1/2-inch mortar joint. Place the bars in the first and fourth mortar joints above the opening (see figure 9-45). They should extend 15 inches into the brick wall on each side of the opening. Table 9-7 gives the number and diameter of bars required for different width wall openings. See paragraph 9-55 for how to place the wall above a window or door opening.

FOUNDATION FOOTINGS

9-126. Footings are the enlargements at the lower end of a foundation wall, pier, or column and are required to distribute the load equally.

9-127. Large footings usually require reinforcing steel because they develop tensile stresses. As in all brick foundations, lay the first course in a mortar bed about 1 inch thick, spread on the subgrade.

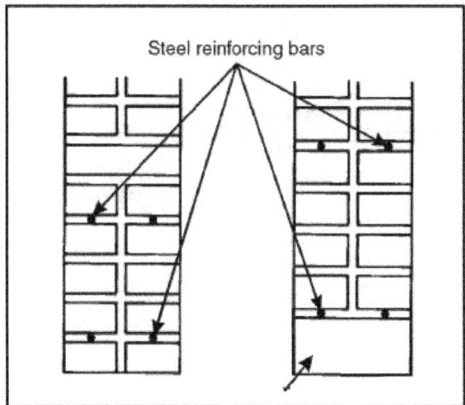

Figure 9-45. Reinforced brick-masonry lintel construction

Table 9-7. Quantities of bars required for lintels

Width of Wall Opening, in Feet	Bar Quantity	Bar Diameter, in Inches
6	2	1/4
9	3	1/4
12	3	3/8

WALL FOOTING

9-128. Figure 9-46, page 9-56, shows a typical wall footing with steel dowels extending above it to tie the footing and wall together. The number 3 bars running parallel to the wall prevent perpendicular cracks from forming.

Chapter 9

COLUMN FOOTING

9-129. Reinforced brick-column footings are usually square or rectangular as shown in figure 9-47. The dowels not only anchor the column to the footing, but transfer stress from one to the other. Note that both layers of horizontal steel are in the first mortar joint, which is accepted practice for small bars. When using large bars, place one layer in the second mortar joint, and reduce the bar spacing in this joint.

COLUMNS AND WALLS

9-130. Load-carrying capacity increases when brick columns and walls have steel reinforcement.

9-131. At least 1 1/2 inches of mortar or brick should cover the reinforcing bars. Install 3/8-inch diameter steel hoops or ties at every course (see figure 9-48, page 9-58) to hold bars in place. Use circular hoops whenever possible. Lap-weld their ends, or bend them around the reinforcing bars as shown in figure 9-48.

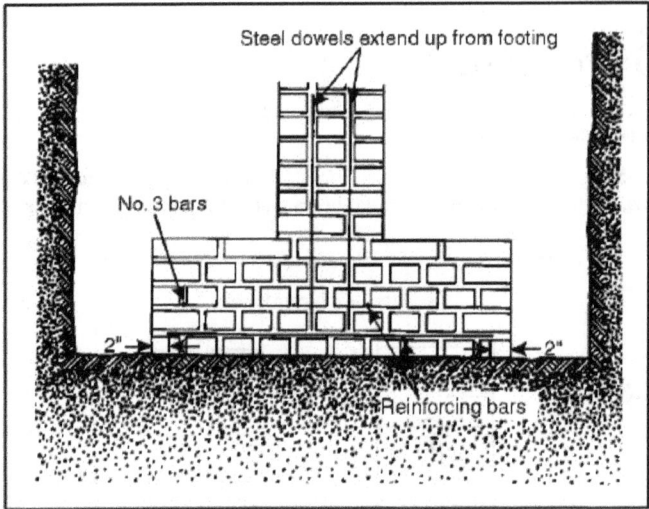

Figure 9-46. Reinforced brick masonry-wall footing construction

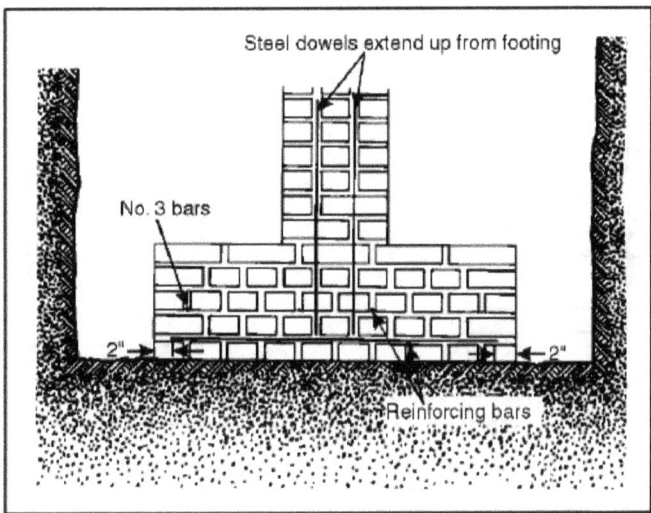

Figure 9-47. Reinforced brick masonry-column footing construction

- Holding steel in place. When the column footing is complete, tie the reinforcing steel to the dowels projecting from the footing. Slip the necessary number of hoops over the dowels and fasten them temporarily some distance above the course being laid, but within your reach. You do not need to wire the hoops to the dowels. Hold the tops of the dowels in position either with a wood template or by tying them securely to a hoop near the top of the column.
- Laying the brick. Lay the brick as described in paragraph 9-23. Place the hoops in a full mortar bed, and smooth it out before laying the next course. You can use brick bats in the column core or wherever it is inconvenient or impossible to use full-size bricks. After laying each course, fill the core and any space around the reinforcing bars with mortar. Then push any necessary bats into the mortar until they are completely embedded. Now spread the next mortar bed and repeat the procedure.

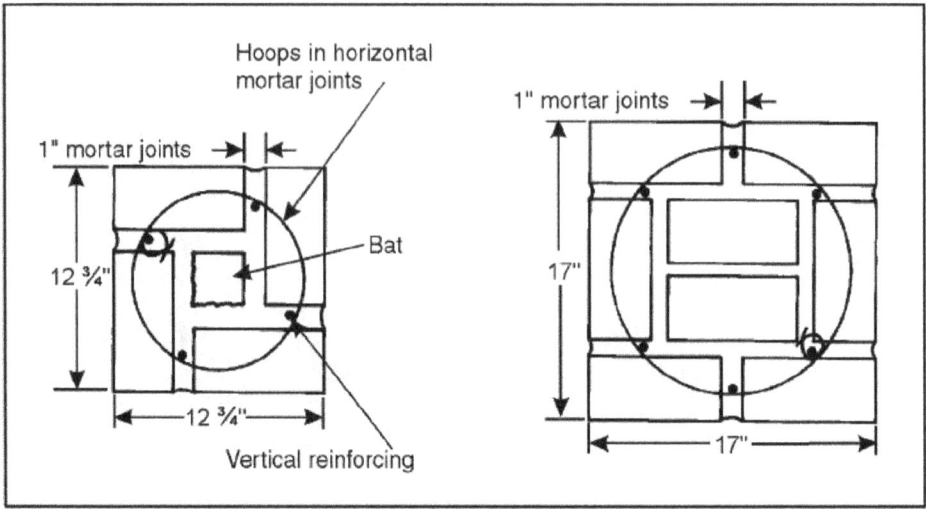

Figure 9-48. Reinforced brick masonry-column construction

9-132. Reinforce brick masonry walls with both horizontal and vertical bars. Place the bars as described in paragraph 9-119. Then wire the vertical stirrups to the dowels projecting out of the wall footing.

- Constructing corner leads. Place the reinforcing bars in corner leads as shown in figure 9-49. Use the same size bars as in the rest of the wall, and let them extend 15 inches. The horizontal bars in the remaining wall should overlap the corner bars by the same 15 inches. As for beams, you must lay all brick in one course between the corner leads before laying any other brick, because you must embed the entire bar in mortar at one time.
- Laying the remaining wall. As you lay the remaining brick, fill all spaces around the reinforcing bars with mortar.

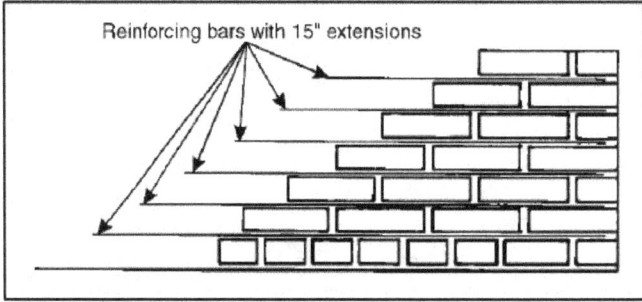

Figure 9-49. Corner lead for reinforced brick masonry wall

SECTION V - STRUCTURAL CLAY-TILE MASONRY

STRUCTURAL CLAY TILE

9-133. Structural clay tile is either a hollow or cored burned-clay masonry unit having cores that are parallel either in the vertical or horizontal direction. The manufacturing process and the type of clay used are the same as it is for brick.

9-134. Hollow masonry units made from burned clay or shale are usually called simply clay tile. Figure 9-50, page 9-60, shows several common types and sizes. These stretcher units are made by forcing a plastic clay through special dyes, then cutting the tiles to size and burning them the same way as brick. The amount of burning depends upon the tile grade.

9-135. The hollow spaces in the tile are called cells, the external wall is called a shell, and the partitions between cells are called webs. The shell should be at least 3/4 inch thick and the web 1/2 inch thick.

9-136. Side-construction tile has horizontal cells, whereas end-construction tile has vertical cells. Neither is better than the other, and both are available in the types described below.

9-137. The two basic categories of structural clay tile—load-bearing and nonload-bearing—differ in their characteristics. Load-bearing structural clay tile further subdivides into three categories: load-bearing wall tile, structural facing tile, and ceramic glazed structural facing tile. Nonload-bearing structural clay tile further subdivides into three categories: nonload-bearing partition and furring tile, fireproofing tile, and screen tile.

LOAD BEARING

9-138. Load-bearing tile has three types, divided by use: wall, facing, and glazed facing. Load-bearing wall tile includes—

- Wall tile for constructing exposed or faced load-bearing walls. This tile carries the entire load, including the facing of stucco, plaster, stone, or other material.
- Back-up tile for backing up combination walls of brick or other masonry in which both the facing and the backing support the wall load. Headers bond the facing or outer tier to the backing tile. The inside face is scored so that you can plaster it without lath.

9-139. The ASTM covers two grades of wall and back-up tile based on weather resistance. Grade LB is suitable for general construction that is not exposed to weathering, or exposed to weathering but protected by at least 3 inches of facing. Grade LBX can be used in masonry exposed to weathering with no facing material.

Chapter 9

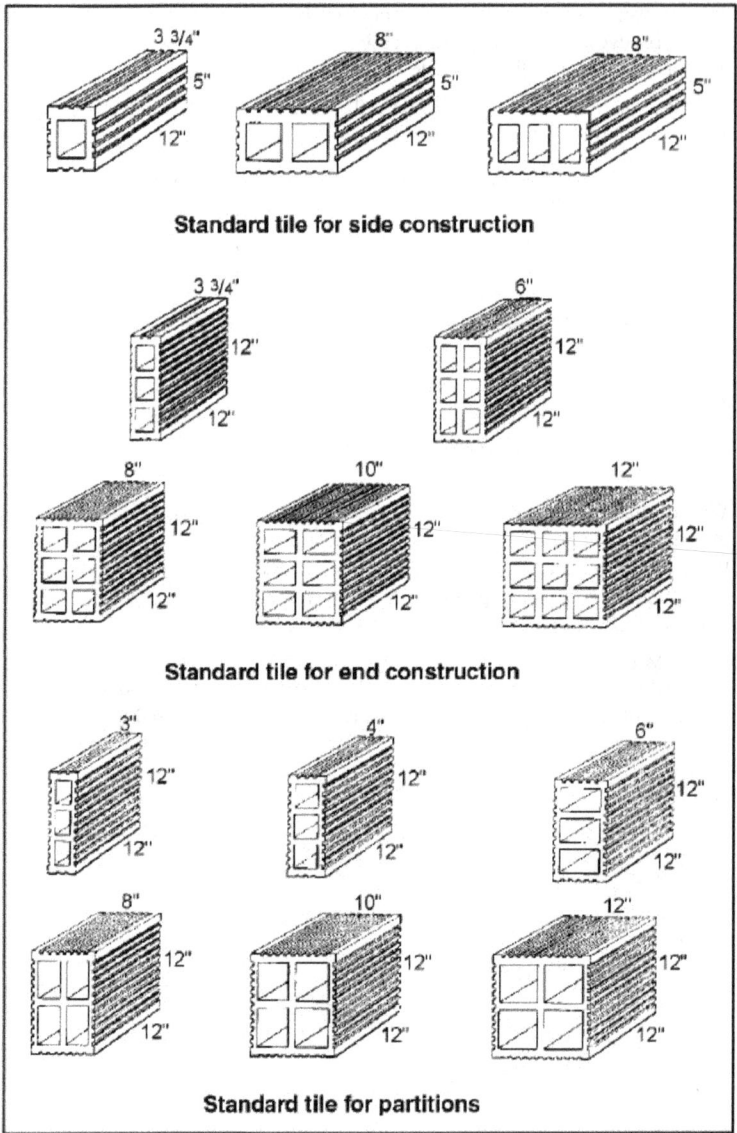

Figure 9-50. Types of structural clay tile

- Structural facing tile is divided into two classes, based on the thickness of the face shell: standard and special duty. The ASTM grades each of these classes by factors affecting appearance.
  - Type FTX is suitable for both exterior and interior walls and partitions. It has an excellent appearance and is easy to clean.
  - Type FTS, although inferior in quality to Type FTX, is suitable for both exterior and interior walls where some surface finish defects are not objectionable.
- Ceramic glazed structural facing tile has an exposed surface of either a ceramic or salt glaze, or a clay coating. Use this tile where you need a stainproof, easily cleaned surface. It is available in

many colors and produces a durable wall having a pleasing appearance. Ceramic glazed facing tile is divided into two types and two grades:
- Type I is suitable for general use where only one finished face will be exposed.
- Type II is suitable for use where the two opposite finished faces will be exposed.
- Grade S (select) is suitable for use with comparatively narrow mortar joints.
- Grade SS (select sized or ground edge) is suitable for uses requiring very small variations in face dimensions.

Nonload Bearing

9-140. Nonload-bearing tile include three types:
- Nonload-bearing partition and furring tile.
  - Partition tile is suitable for constructing nonload-bearing interior partitions or for backing nonload-bearing combination walls.
  - Furring tile is suitable for lining the wall interiors to provide both a plaster base and an air space between plaster and the wall.
- Fireproofing tile protects structural members including steel girders, columns, and beams.
- Screen tile is available in a large variety of patterns, sizes, and shapes, and in a limited number of colors. The surfaces may be smooth, scored, combed, or roughened. Screen tile is divided into two types and three grades:
  - Type STX has an excellent appearance and minimum size variation.
  - Type STA varies more in size.
  - Grade SE has high resistance to weathering, freezing, and thawing.
  - Grade ME has moderate resistance to weathering.
  - Grade NE is suitable for interior use only.

Special Units

9-141. Besides the standard units shown in figure 9-50, you can order special clay tile units to use at windows, door openings, and at corners. Consult a manufacturer's catalog for any special units you need.

Quantities of Materials Required

9-142. Tables 9-8 and 9-9, page 9-62, give the number of structural clay tiles—both side construction and end construction—and the amount of mortar required for walls of different thickness covering varying areas.

Table 9-8. Quantities of materials required for side construction of hollow clay-tile walls

Wall Area, in Square Feet	4-inch Thick Qall and 4 x 5 x 12-inch Tile		8-inch Thick Qall and 8 x 5 x 12-inch Tile	
	Number of Tiles	Mortar, in Cubic Feet	Number of Tiles	Mortar, in Cubic Feet
1	2.1	.045	2.1	.09
10	21	.45	21	.9
100	210	4.5	210	9.0
200	420	9.0	420	18
300	630	13.5	630	27
400	840	18.0	840	36
500	1,050	22.5	1,050	45
600	1,260	27.0	1,260	54
700	1,470	31.5	1,470	63
800	1,680	36.0	1,680	72
900	1,890	40.5	1,890	81
1,000	2,100	45.0	2,100	90

Note. Quantities are based on 1/2-inch thick mortar joints.

Table 9-9. Quantities of materials required for end construction of hollow clay-tile walls

Wall Area, in Square Feet	4-inch Thick Wall and 4 x 12 x 12-inch Tile		6-inch Thick Wall and 6 x 12 x 12-inch Tile		8-inch Thick Wall and 8 x 12 x 12-inch Tile		10-inch Thick Wall and 10 x 12 x 12-inch Tile	
	Number of Tiles	Mortar, in Cubic Feet	Number of Tiles	Mortar, in Cubic Feet	Number of Tiles	Mortar, in Cubic Feet	Number of Tiles	Mortar, in Cubic Feet
1	.93	.025	.93	.036	.93	.049	.93	.06
10	9.3	.25	9.3	.36	9.3	.49	9.3	.6
100	93	2.5	93	3.6	93	4.9	93	6
200	186	5.0	186	7.2	186	9.8	186	12
300	279	7.5	279	10.8	279	14.7	279	18
400	372	10.0	372	14.4	372	19.6	372	24
500	465	12.5	465	18.0	465	24.5	465	30
600	558	15.0	558	21.6	558	29.4	558	36
700	651	17.5	651	25.2	651	34.3	651	42
800	744	20.0	744	28.8	774	39.2	774	48
900	837	22.5	837	32.4	837	44.1	837	54
1,000	930	25.0	930	36.0	930	49.0	930	60

Note. Quantities are based on 1/2-inch thick mortar joints.

PHYSICAL CHARACTERISTICS OF STRUCTURAL CLAY TILE

STRENGTH

9-143. The compressive strength of an individual clay tile depends on its ingredients and the method of manufacture, plus the thickness of its shell and webs. You can predict that tile masonry will have a minimum compressive strength of 300 pounds per square inch based on the cross section. Tile masonry has low tensile strength—in most cases less than 10 percent of its compressive strength. Other physical properties are—

- Abrasion resistance. Like brick, the abrasion resistance of clay tile depends mainly on its compressive strength. The stronger the tile, the greater its resistance to wear, but abrasion resistance decreases as the amount of absorbed water increases.
- Weather resistance. Structural clay facing tile has excellent resistance to weathering. Freezing and thawing produces almost no deterioration. Tile absorbing up to 16 percent of its weight of water satisfactorily resists freezing and thawing effects. When masonry is exposed to weather, use only portland-cement-lime mortar or mortar prepared from masonry cement.
- Heat- and sound-insulating properties. Because of the dead air space in its cells, clay tile has better heat-insulating qualities than solid unit walls. Its sound penetration-resistance compares favorably with that of solid masonry walls, but is somewhat less.
- Fire resistance. Structural clay-tile walls have much less fire resistance than solid masonry walls. However, you can improve it by plastering the wall surface. Partition walls 6 inches thick will resist a fire for one hour, if the fire's temperature does not exceed 1,700°F within that hour.
- Weight. Structural clay-tile weighs about 125 pounds per cubic foot. However, because hollow cell size varies, actual tile weight depends on the manufacturer and the type. A 6-inch tile wall weighs approximately 30 pounds per square foot, whereas a 12-inch tile wall weighs approximately 45 pounds per square feet.

CONSTRUCTION DETAILS AND METHODS

9-144. Refer to the paragraphs listed to obtain specific information applicable to structural clay-tile construction:

- Tools and equipment (paragraphs 7-2 and 7-3).
- Finishing joints (paragraph 9-36).
- Bricklayer's duties (paragraph 9-37).
- Bricktender's duties (paragraph 9-38).
- Watertight walls (paragraph 9-68).
- Maintaining and repairing brick walls (paragraph 9-91)
- Cleaning new brick and removing efflorescence (paragraph 9-96).
- Cleaning old brick (paragraph 9-97).
- Freeze protection (paragraph 9-107).

APPLICATIONS

9-145. The three practical uses for structural tile are discussed below.

- Exterior walls. You can use structural clay tile for either load bearing or nonload-bearing exterior walls. It is suitable for both below-grade and above-grade construction.
- Partition walls. Nonload-bearing partition walls ranging from 4 to 12 inches thick are often built of structural clay tile. They are easy to construct, lightweight, and have good heat- and sound-insulating properties.
- Backing for brick walls. Structural clay tile can be used as a backing unit for a brick wall.

MORTAR JOINTS

9-146. The general procedure for making mortar joints for structural clay tile is the same as for brick. Mortar joints for end-construction units are described as follows:

- Bed joint. To make a bed joint, spread 1 inch of mortar on the bed tile shells, but not on the webs (see view 1 of figure 9-51). Spread the mortar about 3 feet ahead of laying the tile. Because the head joints in clay-tile masonry are staggered, the position of a tile in one course does not match the tile underneath it. Therefore, the webs do not make contact, and any mortar you spread on them is useless.
- Head joint. Form the head joint by spreading plenty of mortar along each tile edge, as shown in view 2 of figure 9-51. Because a clay tile unit is heavy, use both hands to push it into the mortar bed until it is properly positioned. The mortar joint should be about 1/2 inch thick, depending upon the type of construction. Use enough mortar that it squeezes out of the joints, and then cut the excess off with a trowel. The head joint need not be solid like a head joint in brick masonry, unless it is subject to weather.
- Closure joints. Use the procedure described in paragraph 9-33 for making closure joints in brick masonry.

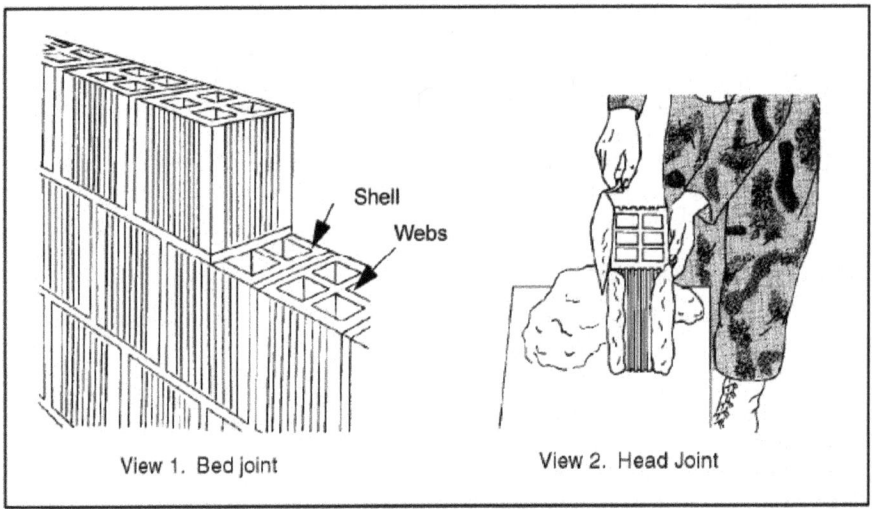

Figure 9-51. Laying end-construction clay tile

9-147. Mortar joints for side-construction units are described as follows:

- Bed joint. Spread the mortar about 1 inch thick, approximately 3 feet ahead of laying the tile. You need not make a furrow as you must for brick bed joints.
- Head joint. Use one of the two following methods:
  - Method A. Spread as much mortar on both edges of the tile as will adhere (see view 1 of figure 9-52). Then push the tile into the mortar bed against the tile already laid until it is properly positioned. Cut off the excess mortar.
  - Method B. Spr39ead as much mortar as will adhere on the interior shell of the bed tile and on the exterior shell of the unit you are placing (see view 2 of figure 9-52). Then push the tile into place and cut off the excess mortar.
- Mortar joint thickness. Make the mortar joints about 1/2 inch thick, depending upon the type of construction.

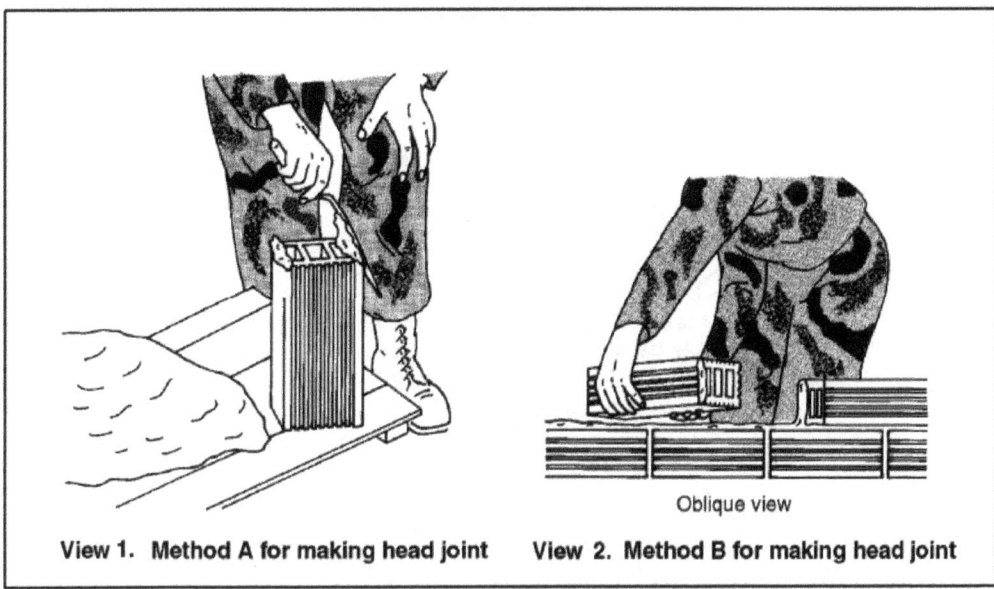

View 1. Method A for making head joint   View 2. Method B for making head joint

Figure 9-52. Laying side-construction clay

LAYING AN 8-INCH BRICK WALL WITH A 4-INCH HOLLOW-TILE BACKING

9-148. When laying an 8-inch brick wall with a 4-inch structural hollow-tile backing, the brick wall has six stretcher courses between the header courses. The side-construction backing tile is 4 by 5 by 12 inches in size. The 5-inch tile height equals the height of two brick courses plus a 1/2-inch mortar joint.

9-149. Lay the tile bed joint so that the top of the tile is level with every second brick course. Therefore, the thickness of the tile bed joint depends upon the thickness of the brick bed joint.

Laying Out the Wall

9-150. Lay out the first brick course temporarily without mortar, as described in paragraph 9-44. This determines the number of bricks in one course.

Laying the Corner Leads

9-151. As shown in view 1 of figure 9-53, page 9-66, the first brick course in the corner lead is the same as the first course of the corner lead for a solid 8-inch brick wall, except that you lay one more brick next to brick (p) as shown in step 5 of figure 9-22, page 9-27. Lay all the bricks for the corner lead before laying any tile. Then lay the first tile course as shown in view 2 of figure 9-53. Complete the corner lead as shown in view 3, figure 9-53.

LAYING AN 8-INCH STRUCTURAL CLAY-TILE WALL

9-152. Use 8- by 5- by 12-inch side-construction tile in a half-lap bond to construct an 8-inch structural clay-tile wall. You can insert a 2- by 5- by 8-inch soap at the corners as shown in figure 9-54, page 9-67. A soap is a thin end-construction tile.

Chapter 9

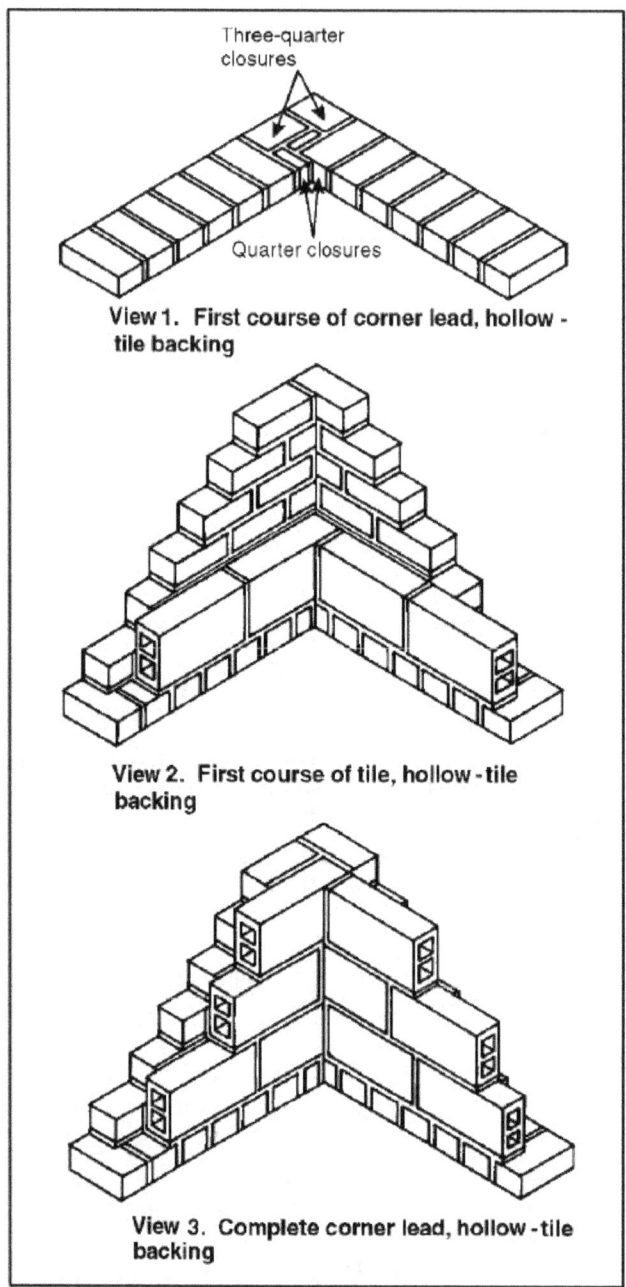

Figure 9-53. Laying a corner lead with hollow-tile backing

Laying Out The Wall

9-153. Follow the procedure described in paragraph 9-43.

LAYING THE CORNER LEADS

9-154. Lay tiles a and b first (see figure 9-54), and check their level as you lay them. To avoid exposing open cells on the wall face, use either end-construction tile for corner tiles b, g, and h, or a soap as shown in figure 9-54. Lay tile b so that it projects 6 inch from the inside corner, as shown, to start the half-lap bond. Now lay tiles c and d, and check their level as you lay them. Next, lay tiles e and f, and check their level. Lay the remainder of the corner tile, and check the level of each as you lay it. After erecting the leads, lay the wall between them using a line.

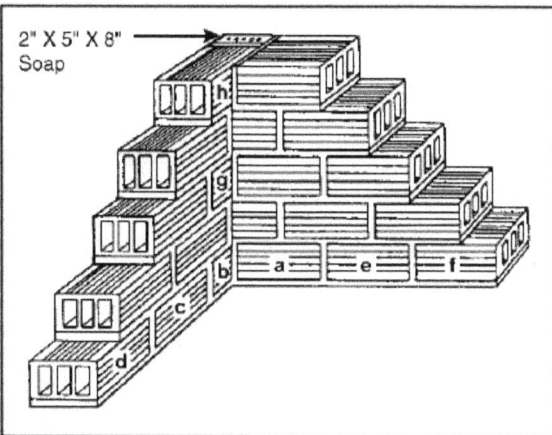

Figure 9-54. Constructing the corner lead of an 8 inch structural clay tile wall

This page intentionally left blank.

Appendix A
Conversion Table

Metric To English			English to Metric		
Multiply	*By*	*To Obtain*	*Multiply*	*By*	*To Obtain*
Length					
Centimeters	0.0394	Inches	Inches	2.54	Centimeters
Meters	3.28	Feet	Feet	0305	Meters
Meters	1.094	Yards	Yards	0.914	Meters
Kilometer	0.621	Miles (stat.)	Miles (stat.)	1.609	Kilometers
Kilometers	0.540	Miles (Naut.)	Miles (naut.)	1.853	Kilometers
Area					
Square	0.1550	Square	Square	6.45	Square
Square	10.76	Square feet	Square feet	0.0929	Square
Square	1.196	Square yards	Square yards	0.836	Square
Volume					
Cubic	0.610	Cubic inches	Cubic inches	16.39	Cubic
Cubic	35.3	Cubic feet	Cubic feet	0.0283	Cubic
Cubic	1.308	Cubic yards	Cubic yards	0.765	Cubic
Milliliters	0.0338	US liq ounces	US liq ounces	29.6	Milliliters
Liters	1.057	US liq quarts	US liq quarts	0.946	Liters
Liters	0.264	US liq	US liq	3.79	Liters
Weight					
Grams	0.0353	Ounces	Ounces	28.4	Grams
Kilograms	2.20	Pounds	Pounds	0.454	Kilograms
Metric tons	1.102	Short tons	Short tons	0.907	Metric tons
Metric tons	0.984	Long tons	Long tons	1.016	Metric tons

This page intentionally left blank.

Appendix B
Method of Making Slump Test

This test method covers the procedures to use both in the laboratory and in the field to determine the consistency of the concrete. Although not a precise method, it gives sufficiently accurate results. The slump test does not apply if the concrete contains aggregates much larger than 2 inches in size.

APPARATUS

B-1. The following apparatus are available in the mobile laboratory: a metal mold and a tamping rod.

METAL MOLD

B-2. The metal mold is a piece of 16-gage galvanized sheet metal shaped like a 12-inch high truncated (cut off) cone having an 8-inch diameter base and a 4-inch diameter top (see figure B-1). Both the base and top are open, parallel to each other, and perpendicular to the axis of the cone.

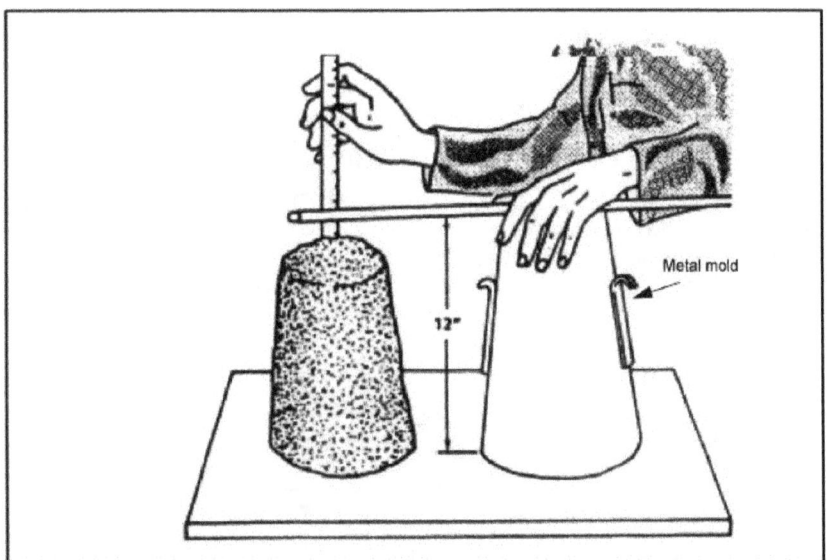

Figure B-1. Measuring slump

TAMPING ROD

B-3. The tamping rod is 5/8 inch in diameter, 24 inches long, and has a 1-inch taper at one end terminating in a rounded tip bullet nose. It have a radius of about 1/4 inch.

TEST SPECIMENS

B-4. The concrete samples that make up the test specimens must be representative of an entire batch. Note for future reference the work location of the concrete batch sampled. Take all concrete samples at the

Appendix B

mixer, or if you are using ready-mixed concrete, during discharge from the transportation vehicle. Start taking these samples by repeatedly passing a scoop or pail through the concrete stream at the beginning of discharge, and continuing until the entire batch discharges. Take a sample of paving concrete immediately after depositing the batch on the subsurface. Collect at least five samples from different portions of the pile and thoroughly mix them to form the test specimen. Transport the specimens to the testing site to counteract segregation. Mix the concrete using a shovel until it is uniform in appearance.

PROCEDURE

B-5. Dampen the mold and place it on a flat, moist, nonabsorbent, and firm surface. Fill the mold immediately with three equal layers of a concrete specimen described in paragraph B-5. As you fill the mold, rotate each scoopful of the concrete around the top edge of the mold as the concrete slides from it. This ensures a symmetrical concrete distribution within the mold. Tamp each layer 25 strokes with the tamping rod, distributing the strokes uniformly over the cross section of the mold, and penetrating the underlying layer. Tamp the bottom layer throughout its depth. After tamping the top layer, strike off the surface with a trowel so that the concrete fills the mold exactly. Without any delay, carefully lift the mold straight up from the concrete and place it beside the specimen.

SLUMP MEASUREMENT

B-6. Place the tamping rod across the top of the mold. Measure the distance between the bottom of the rod and the displaced original center of the specimen's top surface. If a shearing off of concrete from one side or portion of the mass occurs, disregard the test and make a new test on another portion of the sample.

Note. If two consecutive tests on a sample of concrete show a falling away or shearing off of a portion of the concrete from the mass of the specimen, the concrete probably lacks the necessary plasticity and cohesiveness for the slump test to be applicable.

SUPPLEMENTARY TEST PROCEDURE

B-7. After completing the slump measurement, tap the side of the specimen gently with the tamping rod. How the concrete mix behaves under this treatment is a valuable indication of its cohesiveness, workability, and placement. A well-proportioned, workable mix will gradually slump (fall or flatten out) but still retain its original consistency, whereas a poor mix will crumble, segregate, and fall apart.

Appendix C
Field Test for Moisture Deformation on Sand

C-1. Sands used as the fine aggregate in concrete may contribute a significant amount of moisture to the concrete mix. This moisture should be accounted for by decreasing the amount of water added to the dry materials in order to maintain the W/C ratio that is a part of the concrete design. The following procedure can be used as a field test for estimating the amount of moisture in sand. The procedure allows for some variation in estimating; therefore, the percentage of moisture determined is somewhat judgmental.

C-2. The samples used for this test should be taken from a depth of 6 to 8 inches below the surface of the piled sand. This negates the effect of evaporation at the surface of the pile. Squeeze a sample of sand in the hand. Then, open the hand and observe the sample. The amount of FSM can be estimated from the following criteria:

- Damp sand (0% to 2% FSM). The sand will fall apart. The damper the sand the more it tends to cling together (see figure C-1).
- Wet sand (3% to 4% FSM). The sand will clings together without excess water being forced out (see figure C-2, page C-2).
- Very wet sand (5% TO 8% FSM). The sand ball will glisten or sparkle with water. The sand will have moisture on it and may even drip (see figure C-3, page C-2).

C-3. The percentage of FSM determined by this method approximates the amount of water by weight on the sand. Use these estimates to adjust the mix design as indicated in paragraph 3-16 and 3-17.

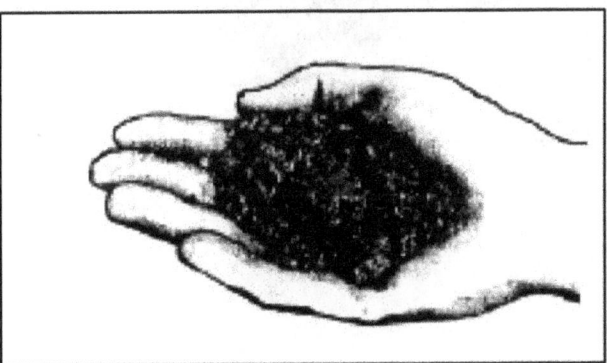

Figure C-1. Damp sand

Appendix C

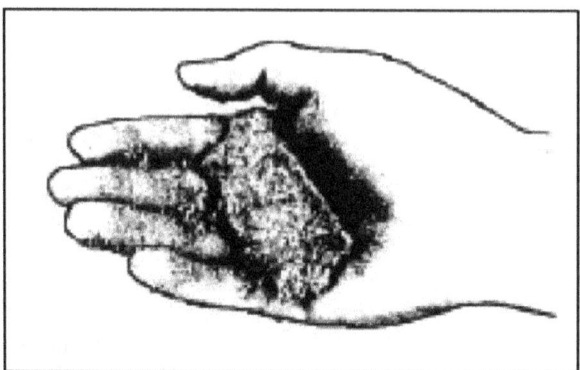

Figure C-2. Wet sand

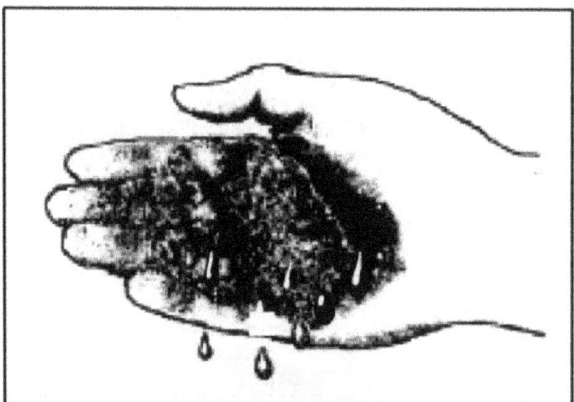

Figure C-3. Very wet sand

Glossary

ABBREVIATIONS AND ACRONYMS

AASHTO	American Association of State Highway and Transportation Officials
ACI	American Concrete Institute
ASTM	American Society for Testing and Materials
ATTN	attention
BB	bean bolster
BF	bulking factor
CA	coarse aggregate(s)
CESL	corrected equivalent static load
CMC	continuous high chair
CL	critical load
cu ft	cubic foot
cu yd	cubic yard (s)
D	Depth
DA	Department of Army
DC	District of Columbia
DL	dead load
DRUW	dry-rotted unit weight
EM	enlisted men
ESL	equivalent static load
F	fahrenheit
$f'c$	specified compressive strength of concrete, psi
$f't$	flexural tensile stress
FA	fine aggregate(s)
FM	field manual
FSM	free surface moisture
ft	Foot
gal	gallon(s)
gm	gram(s)
HC	high chair
HP	horsepower
HQ	headquarters
hr	hour(s)
inch	inch, inches

Glossary

K		a constant, usually between 8-10
kg		kilogram(s)
lb		pound(s)
lb/cu ft		pounds per cubic foot
lb/cu yd		pounds per cubic yard
LF		linear foot
lin		Linear
liq		Liquid
LL		live load
m		Meter
max		Maximum
MCP		maximum concrete pressure
min		Minimum
mm		Millimeter
MPa		Megapascal
MO		Missouri
MOS		military occupational specialty
MSA		maximum-size aggregate
MSS		maximum stud spacing
MWS		maximum wale spacing
NCO		noncommissioned officer
Nos		Numbers
OC		on center
pH		denotes hydrogen-ion activity
ppm		parts per million
psi		pounds per square inch
PV		partial volume
R		flexural strength, in psi
S4S		surfaced on four sizes
SB		slab bolster
SG		specific gravity
SL		static load
sq		Square
SSD		saturated, surface dry
TL		total load
TM		technical manual
TOE		table of organization and equipment
TRADOC		United States Army Training and Doctrine Command
ULJ		uniform load on the joist
ULS		uniform load on a stud

Glossary

ULSstr	uniform load on the stringer
ULW	uniform load on wale
US	United States
USAES	United States Army Engineer School
vol	Volume
w	Width
W/C	water to cement
wt	Weight
yd	yard(s)

SYMBOLS

Δ	delta; change in x
°	Degrees
:	to (ratio)
ϕ	Diameter
θ	theta, an angular measurement, degrees
$\leq$	less than, or equal to
$\geq$	greater than, or equal to
$\perp$	perpendicular to
$\parallel$	parallel to
'	Feet
"	Inches
@	approximately or at
%	Percent
$\pm$	plus or minus
d_b	diameter of bar

This page intentionally left blank.

References

SOURCE USED

These are the sources quoted or paraphrased in this publication.

ARMY PUBLICATIONS

FM 3-34.13. *Carpentry.* 3 October 1995.

FM 5-103. *Survivability.* 10 June 1985.

FM 5-170. *Engineer Reconnaissance.* 5 May 1998. (superseded by FM 3-34.170)

FM 5-430-00-1. *Planning and Design of Roads, Airfields and Heliports in the Theater of Operations - Road Design, Volume I.* 26 August 1994.

FM 5-430-00-2. *Planning and Design of Roads, Airfields and Heliports in the Theater of Operations - Airfield and Heliport Design, Volume II.* 29 September 1994.

TM 5-337. *Paving and Surfacing Operations.* 21 February 1966. (superseded by FM 5-436)

TM 5-349. *Arctic Construction.* 19 February 1962.

TM 5-624. *Maintenance and Repairs of Surface Area.* 27 October 1995.

TM 5-3895-221-15. *Operator, Organizational, Direct Support, General Support, and Depot Maintenance Manual: Mixer, Concrete, Trailer Mounted: Gasoline Driven, 16 cubic feet (Chain Belt Model HBG).* 30 August 1962. (rescinded)

TM 5-3895-372-10. *Operation Instructions for Concrete-Mobile Mixer Body, M919, Model 8CM-24/F.* 5 October 1979.

TM 5-331D. *Utilization of Engineer Construction Equipment, Volume D-1 Asphalt and Concrete Equipment.* 21 April 1969. (superseded by FM 5-436)

NONMILITARY PUBLICATION

ACI 211.1-81. *Guide for using Admixture in concrete.* September 1981. (superseded by ACI 211.1-91)

ACI 306.R-78. *Cold Weather Concreting.* 1978. (superseded by ACI 306R-10)

ACI 318-83. *Building Codes Requirements for Reinforced Concrete.* 1983. (superseded by ACI 318-83[86])

ASTM A615. *Deformed and Plain Billet Steel Bars for Concrete Reinforcement.* 1996. (superseded by ASTM A615/A615M-09B)

ASTM A616. *Rail Steel Deformed and Plain Bars for Concrete Reinforcement.* 1996. (superseded by ASTM A996/A996M-09B)

ASTM A617. *Axle-Steel Deformed and Plain Bar for Concrete Reinforcement.* 1997. (superseded by ASTM A996/A996M-09B)

ASTM C29. *Bulk Density "Unit Weight" and Voids in Aggregate.* 1997. (superseded by ASTM C29/C29M-09)

ASTM C33. *Concrete Aggregates.* 1997. (superseded by ASTM C33/C33M-11a)

ASTM C40. *Organic Impurities in Fine Aggregates for Concrete.* 1992. (superseded by ASTM C40/C40M-11)

ASTM C87. *Effect of Organic Impurities in Fine Aggregates on Strength of Mortar.* 1983. (superseded by ASTM C87/C87M-10)

ASTM C117. *Material Finer Than 75 (number 200) Sieve in Mineral Aggregate by Washing.* 1995.

ASTM C123. *Lightweight Pieces in Aggregate.* 1996. (superseded by ASTM C123/C123M-11)

References

ASTM C150. *Portland Cement.* 1997. (superseded by ASTM C150/C150M-11)

ASTM C260. *Air-Entraining Admixture for Concrete.* 1995. (superseded by ASTM C260/C260M-10a)

ASTM C816. *Sulfur in Graphite by Combustion-Iodometric Titration Method.* 1985. (superseded by ASTM C816/C816M-85(2010)e1

DOCUMENTS NEEDED
These documents must be available to the intended users of this publication.

Department of Army (DA) Form 2028. *Recommended changes to Publications and Blank Forms.* February 1974.

READING RECOMMENDED
This reading contain relevant supplemental information.

TM 3-34.01. *Engineer Data.* 14 September 1987.

Index

A

Abrasion Resistance, 2-6
Absolute-volume method, 3-10
Admixture
  coloring agents, 2-15
Admixtures
  accelerators, 2-16, 5-40
  Accelerators, 7-7
  Air-Entraining Materials, 2-14
  Dampproofing and permeability-reducing agents, 2-16
  Definition, 2-13
  Gas-forming agents, 2-16
  Grouting agents, 2-16
  Pozzolans, 2-16
  Purpose, 2-13
  Retarding admixtures, 2-16
  Water-reducing agents, 2-16
  Workability agents, 2-16
Aggregate
  Abrasion resistance, 2-14
  Absorption and surface moisture, 2-6
  Bulk Unit Weight, 2-10
  Bulking, 2-11
  characteristics, 2-5
  Chemical stability, 2-6
  Coarse aggregate, 2-7
  Definition, 2-5
  Fine aggregate, 2-7
  Fineness Modulus, 2-7
  Freeze and Thaw Resistance, 2-6
  Gap-Graded Aggregate, 2-10
  Grading, 2-7
  Handling and Storing, 2-12
  Heating, 5-39
  Impurities, 2-12
  Measuring, 5-5
  Particle Shape, 2-7
  selection, 2-8
  Specific Gravity, 2-10
  Surface Texture, 2-7
Air-entrained concrete
  air-entraining agents, 2-14
  Air-entraining agents, 2-14
  considerations, 3-4
  Description, 2-13
  Factors affecting air content, 2-14, 3-4
  Properties, 2-14
  Recommended Air Content, 2-15
  Tests for Air Content, 2-15
Air-entraining cement
  Properties, 3-7
Air-Entraining Portland Cement
  Description, 2-2
  Properties, 2-2
Anchor Bolts, 4-48, 8-19

B

Batching plants, 5-3
Battens, 4-4
Blocks, 8-1
Bonds, bricklaying
  American, 9-8
  Common, 9-8
  Dutch, 9-7
  English, 9-7, 9-8
  English cross, 9-7, 9-8
  Flemish, 9-7, 9-8
  Mortar, 9-7
  Pattern, 9-7
  Running, 9-7
  Stack, 9-8
  Structural, 9-7
  Types of Bonds, 9-7
Brick
  Classification, 9-1
  Cutting, 9-1
  Sizes, 9-1
  Surfaces, 9-2
  Types, 9-1
  Weight, 9-1
Brick masonry
  Abrasion resistance, 9-5
  Bonds, 9-7
  Bricklayer's Tools And Equipment, 7-1
  Buckets, 5-18
  buggies, 5-14
  Bulk Unit Weight, 2-10
  Bulking, 2-11
  Bull header, 9-6
  Bull stretcher, 9-6
  Characteristics, 9-1
  Definition, 9-1
  expansion and contraction, 9-5
  Fire Resistance, 9-4
  Heat-insulating properties, 9-5
  Joints, 9-1, 9-4
  Sound insulating properties, 9-6
  Strength, 9-3
  Terms, 9-6
  Ties, 9-7, 9-8
  Weather resistance, 9-4
Bulk unit weight, 2-6

C

Cement
  description, 2-1
  Measuring, 5-5
  Packaging, 2-3
  Portland, 2-1
  Storage, 2-3
Cleaning
  Acid, 5-30
  Block Walls, 8-25
  Concrete, 5-29
  Sandblasting, 5-29
  with Mortar, 5-29
Clinkers, 2-1
Cold Weather Concreting
  considerations, 5-37
  Curing, 5-41
  Effects, 5-38
  Techniques, 5-39
Columns
  Forms, 4-4, 4-30, 4-35
Concrete
  Absolute-Volume Method, 3-10
  Aggregate, 2-5, 3-14
  blocks, 8-1
  Brooming, 5-28
  buckets, 5-3, 5-18
  cleaning, 5-29
  cold-weather operations, 5-37
  Consolidating, 5-21
  Creep, 6-3
  Curing, 5-30
  Finishing, 2-15, 5-25
  Floating, 5-26
  freezing, 5-41
  grout, 2-16
  Handling and Transporting, 5-13
  Hot Weather operations, 5-34
  hydration, 5-30
  masonry units, 8-1
  Materials, 5-2
  measuring mix materials, 3-5
  Measuring Mix Materials, 5-5

Index

Mixing, 5-6, 5-7
Mortar, 5-29
Patching, 5-44
Pavement, 5-28
Placement, 5-19
Placing, 5-20
Placing Concrete
 Underwater, 5-24
proportioning, 3-1
Pumps, 5-18
Remixing, 5-12
Repair, 5-42
Screeding, 5-25
segregation, 5-13
Segregation, 5-20
shrinkage, 5-35, 5-44, 5-45
Slump, B-1
spading, 5-21
strength, 5-11, 5-19, 5-21, 5-30
Trial-Batch Method, 3-6
Troweling, 5-27
Variation in Mixtures, 3-13
vibration, 2-14
Vibration, 5-21
yield, 3-14
Construction site preparation
 Approach Roads, 5-2
 batching plants, 5-3
 Clearing and Draining, 5-2
 Excavation, 5-3
 Reconnaissance, 5-1
 Safety Facilities, 5-3
 Stockpiling, 5-2
Course, 9-6
curing
 cold weather, 5-37
Curing
 compounds, 5-34
 Factors, 5-30
 Hydration, 5-30
 Methods, 5-31
 strength, 5-30

D
DEFINITION, 2-13

E
Excavation
 considerations, 5-1
 Considerations, 5-3
 Hand, 5-4
 Machine, 5-3

F
Fineness Modulus, 2-7
Finishing
 Brooming, 5-28
 cleaning, 5-29
 floating, 5-26

Pavement, 5-28
Premature, 2-15
purpose, 5-25
Repairing, 5-42
rubbed finish, 5-28
Screeding, 5-25
troweling, 5-27
Troweling, 5-27
Floating, 5-26
Floor forms, 4-18
Footings
 Column, 4-35
 concrete masonry, 8-7
 Forms, 4-35
 rubble stone masonry, 8-30
 Wall, 4-34
Form work
 Bracing, 4-15
 Characteristics, 4-1
 Column, 4-4, 4-30
 column and footing, 4-35
 Construction, 4-41
 Design, 4-5
 floor, 4-18
 Footing and Pier, 4-33
 Form Failure, 4-41
 Foundation, 4-32
 Importance, 4-1
 Joists, 4-18
 Management Aspects, 4-5
 Materials, 4-1
 Oiling, 4-40
 Overhead Slab, 4-18
 Panel Wall, 4-35
 pavement, 4-2
 Pavement, 4-40
 Removal, 5-41
 Safety, 4-41
 slab, overhead, 4-18
 Stair, 4-38
 Steel, 4-2, 4-40
 Stringer, 4-18
 Stripping, 4-41, 5-41
 Studs, 4-3
 Tie rods, 4-4
 Tie wires, 4-4
 TIME ELEMENT, 5-5
 use, 4-1
 Wales, 4-3
 Wall Footing, 4-34
 Wetting, 4-40
 Yokes, 4-4
Form Work
 Battens, 4-4
Formwork
 Column and Footing, 4-35
Foundation Forms, 4-32

G
grading, 2-8
Grading, 2-7, 2-8
Grouts, 2-16

H
Handling and transporting concrete
 delivery methods, 5-14
 principles, 5-13
Header, 9-6
Hot-weather concreting
 Cooling materials, 5-35
 Effects, 5-34
 Precautions, 5-36
 Problems, 5-34
hydration, 1-1, 2-3, 5-30
hydraulic, 1-1

J
Joints
 Bed, 8-30
 Construction, 4-46
 Control, 4-45, 8-20
 expansion or contraction joints, 4-42
 Head, 8-31
 Tooling, 8-19

M
Mason
 Duties, 8-25
 Helper, 8-25
 Tools And Equipment, 7-1
Masonry, concrete
 Closure Block, 8-17
 Curing, 8-5
 Footings, 8-7, 8-30
 joints, 7-2, 7-5, 7-7
 Lintels, 8-25
 lintels and sills, 8-25
 Making Blocks, 8-4
 Materials Tower, 7-11
 Patching, 8-25
 Planning, 8-5
 Sills, 8-25
 sizes and shapes, 8-2, 8-3
 terminology, 7-1
 Terminology, 9-6
 Tooling, 8-19
 Tools and Equipment, 7-1
 units, 8-1
 Walls, 8-7
Masonry,concrete
 Cleaning, 8-25
Mixing water
 IMPURITIES, 2-3
 measuring, 5-5
 measuringr, 5-6

Index

PURPOSE, 2-3
Mixing, concrete
  16-Cubic Foot Mixer, 5-9
  batch plants, 5-3
  Central Mix Plants, 5-8
  Hand, 5-6
  Machine, 5-7
  Measuring Mix Materials, 5-5
  Mixing plants, 5-7
  operations, 5-5
  Principles, 5-5
  remixing, 5-12
  Site Mix Plants, 5-8
  time, 5-6, 5-7
  WATER, 2-3
Mortar
  accelerators, 7-7
  Antifreeze Materials, 7-7
  board, 7-3
  Bond, 7-5
  box, 7-3
  brick, 7-5
  bricks, 9-3
  cement, 2-2
  Cements, 2-1
  concrete block, 8-18
  concrete blocks, 7-5
  Mixing, 7-6
  Plasticity, 7-5
  Repairing, 7-7
  Retempering, 7-7
  Storing, 7-6
  Strength and Durability, 7-5
  Tuck-pointing, 7-7
  Types, 7-5
  water, 9-4
  Water Quality, 7-7
  Water Retentivity, 7-5

N

Natural cement
  characteristics, 2-1
  defininition, 2-1
  types, 2-1
  uses, 2-1
newlink box, 7-3

P

parging, 8-10
Patching
  new concrete, 5-44
  Old Concrete, 5-45
Placing concrete
  Compaction, 5-20
  consolidating, 5-21
  fresh concrete on hardened concrete, 5-20
  Hand Method, 5-22
  Methods, 5-14
  on Slopes, 5-21
  placing, 5-19, 5-20
  Preliminary Preparation, 5-19
  rate, 5-20
  Segregation, 5-20
  Tremie Method, 5-24
  Underwater, 5-24
  V bration, 5-21
Portland cement
  ASTM Types, 2-1
  description, 2-1
  Other ASTM cements, 2-2
  Packaging and Shipping, 2-3
  Storage, 2-3
Precast Concrete
  Advantages, 6-17
  Characteristics, 6-16
  Definition, 6-16
  Design, 6-18
  Erecting, 6-20
  Prefabricating, 6-18
  prestressed, 6-17
  Products, 6-17
  Transportation, 6-19
Proportioning concrete mixtures
  Absolute-Volume Method, 3-10
  Trial-Batch Method, 3-6
  Variation in Mixtures, 3-13
Pumps, concrete, 5-18
PURPOSE, 2-3

R

r clay til, 7-5
Reinforced-Concrete
  Basis of Design, 6-4
  Bending Strength, 6-2
  Bond Strength, 6-2
  Creep, 6-3
  Definitions, 6-1
  Design, 6-3
  Homogeneous Beams, 6-3
  Neutral Axis, 6-3
  Principles, 6-1
  Shear Strength, 6-1
  Slab and Wall, 6-7
  Specifications, 6-4
  Tensile Strength, 6-1
Reinforcing steel
  bars, 6-7
  Beams, 6-14
  Bolsters, 6-9
  Chairs, 6-9
  Columns, 6-14
  Floor Slabs, 6-14
  Footings, 6-16
  Grades, 6-8
  On-Site Prefabrication, 6-13
  Placement, 6-13
  Spacers, 6-13
  Splices, 6-11
  Stirrups, 6-9
  Support, 6-13
  Walls, 6-16
Remixing Concrete, 5-12
Repairing
  Block Walls, 8-25
  New Concrete, 5-42
  Old Concrete, 5-45
Retarding admixtures, 2-16
Rowlock, 9-6
Rubbed finish, 5-28
Rubble Stone Masonry
  Bonding, 8-31
  Footings, 8-30
  Joints, 8-30
  Laying, 8-30
  Materials, 8-30
  types, 8-29
  uses, 8-29
runways for placing concrete, 5-14

S

Safety
  Construction site, 5-3
  form work, 4-41
Sand test, C-1
scaffolds
  considerations, 7-7
  Foot, 7-8
Scaffolds
  Outrigger, 7-9
  Putlog, 7-9
  Steel, 7-10
  Trestle, 7-8
Screeding, 5-25
Sieves, 2-7
Sills, 8-25
Slump, 3-5
Slump test, 3-5
Slump Test, B-1
Soldier, 9-6
Specific Gravity, 2-10
Spreader, 4-3
Stair Forms, 4-38
Steel Forms, 4-40
Strength
  form work, 4-1
  mortar, 7-5
Stretcher, 9-6

Index

stucco, 2-3
Subgrade Preparation, 5-19

T

Tremie, 5-24
Trial-Batch Method, 3-6
Trowel, 7-2
Troweling, 5-27

V

Vibration, 5-21
voids, 2-9

W

Wales, 4-3
Walls
    Basement, 8-10
    Bearing, 8-24
    concrete masonry, 8-7
    Footings, 8-7
    Forms, 4-2
    Intersecting, 8-24
    panel, 4-35
    Weathertight Concrete Masonry, 8-11
Warehouse pack, 2-3
Water
    Heating, 5-39
    IMPURITIES, 2-3
    measuring, 5-5
    mixing, 2-1, 5-34, 5-35
    PURPOSE, 2-3
    Sea, 2-4
Water-cement ratio, 3-1, 5-5
Waterproofing
    parging, 8-10
Windows
    openings, 8-25
Wythe, 9-6

Y

Yokes, 4-4

TM 3-34.44 (FM 5-428)/MCRP 3-17.7D
23 July 2012

By order of the Secretary of the Army:

RAYMOND T. ODIERNO
General, United States Army
Chief of Staff

Official:

JOYCE E. MORROW
Administrative Assistant to the
Secretary of the Army
1029402

RICHARD P. MILLS
Lieutenant General, USMC
Deputy Commandant for
Combat Development and Integration

DISTRIBUTION:
Active Army, Army National Guard, and U.S. Army Reserve: Not to be distributed; electronic media only.

www.ingramcontent.com/pod-product-compliance
Lightning Source LLC
Chambersburg PA
CBHW080456110426
42742CB00017B/2901